AF905316

Springer Proceedings in Mathematics & Statistics

Volume 440

This book series features volumes composed of selected contributions from workshops and conferences in all areas of current research in mathematics and statistics, including data science, operations research and optimization. In addition to an overall evaluation of the interest, scientific quality, and timeliness of each proposal at the hands of the publisher, individual contributions are all refereed to the high quality standards of leading journals in the field. Thus, this series provides the research community with well-edited, authoritative reports on developments in the most exciting areas of mathematical and statistical research today.

Vladimir Rovenski · Paweł Walczak · Robert Wolak
Editors

Differential Geometric Structures and Applications

4th International Workshop on Differential Geometry, Haifa, Israel, May 10–13, 2023

Springer

Editors
Vladimir Rovenski
Department of Mathematics
University of Haifa
Haifa, Israel

Paweł Walczak
Katedra Geometrii
University of Lodz
Lodz, Poland

Robert Wolak
Instytut Matematyki
Uniwersytet Jagiellonski
Krakow, Poland

ISSN 2194-1009 ISSN 2194-1017 (electronic)
Springer Proceedings in Mathematics & Statistics
ISBN 978-3-031-50585-0 ISBN 978-3-031-50586-7 (eBook)
https://doi.org/10.1007/978-3-031-50586-7

Mathematics Subject Classification: 53-XX, 53C15, 53C20, 53C21, 53C25, 53C43, 53D15, 58-XX, 58J05, 58K05

© The Editor(s) (if applicable) and The Author(s), under exclusive license to Springer Nature Switzerland AG 2024

This work is subject to copyright. All rights are solely and exclusively licensed by the Publisher, whether the whole or part of the material is concerned, specifically the rights of translation, reprinting, reuse of illustrations, recitation, broadcasting, reproduction on microfilms or in any other physical way, and transmission or information storage and retrieval, electronic adaptation, computer software, or by similar or dissimilar methodology now known or hereafter developed.
The use of general descriptive names, registered names, trademarks, service marks, etc. in this publication does not imply, even in the absence of a specific statement, that such names are exempt from the relevant protective laws and regulations and therefore free for general use.
The publisher, the authors, and the editors are safe to assume that the advice and information in this book are believed to be true and accurate at the date of publication. Neither the publisher nor the authors or the editors give a warranty, expressed or implied, with respect to the material contained herein or for any errors or omissions that may have been made. The publisher remains neutral with regard to jurisdictional claims in published maps and institutional affiliations.

This Springer imprint is published by the registered company Springer Nature Switzerland AG
The registered company address is: Gewerbestrasse 11, 6330 Cham, Switzerland

Paper in this product is recyclable.

Preface

This volume contains a collection of articles written by the participants and their collaborators of the 4th International workshop *Differential Geometric Structures and Applications* held at the University of Haifa from May 10 to May 13, 2023, as well as of the previous workshops in this series held at University of Haifa.

The 1st International Workshop in the series, named *Reconstruction of Geometrical Objects Using Symbolic Computations* was held on September 2008.

The 2nd named *Geometry and Symbolic Computations* in May 2013. The papers related to this workshop were published in the same Springer Proceedings series: https://link.springer.com/book/10.1007/978-3-319-04675-4.

The 3rd International Workshop, called *Geometric Structures and Interdisciplinary Applications* in May 2018.

The detailed information about all four workshops can be found on V. Rovenski's homepage: http://math.haifa.ac.il/ROVENSKI/rovenski.html.

The 4th workshop was organized by one of the editors (V. Rovenski) and sponsored by the Center for Computational Mathematics and Scientific Computation and the Department of Mathematics at the University of Haifa.

In the workshop participated researchers from 11 countries, namely: Brazil, Czech Republic, India, Japan, Poland, Romania, Russia, Serbia, Spain, Ukraine, and, of course, Israel.

The discussion of geometry by the participants continued during the excursion day of the workshop, where the participants got acquainted with the sights of Jerusalem.

Two editors of this volume (V. Rovenski and P. Walczak) formed the Scientific Committee of the workshop, and Dr. Irina Albinsky was the workshop secretary.

The papers presented in this volume are closely related to the lectures given at the workshop which was designed to cover different aspects of modern differential geometry together with some applications. Some of them are survey articles often complemented with sections containing new results of the authors.

Five papers of the volume are on topics related to the geometry and topology of Riemannian manifolds. Another three investigate some aspects of the theory of special contact manifolds and their generalizations. The remaining papers reflect the current interest of the authors and are devoted, respectively, to harmonic measures, geometry of vector fields on Lie groups, Smith-Gysin sequence, twistor bundles and some applications of geometry.

We express our gratitude to all the participants, the contributors of the volume, the sponsors and everyone who helped us while organizing the conference and preparing the volume for publication. In particular, we would like to mention Dr. I. Albinsky who organized the culture programs, registration of participants and opening/closing ceremonies.

Haifa, Israel
Lodz, Poland
Krakow, Poland
September 2023

Vladimir Rovenski
Paweł Walczak
Robert Wolak

Contents

Some Topics in Sasakian Geometry, a Survey

Aleksy Tralle

Abstract In the seminal book *Sasakian Geometry* by C. Boyer and K. Galicki, the authors formulated a research program for studying topological properties and answering questions about the existence of Sasakian structures. We survey recent progress in this topic.

Keywords Smale-Barden manifold · Sasakian structure · Seifert bundle

1 Introduction and Preliminaries

Sasakian geometry has become an important and active subject, especially after the appearance of the fundamental treatise of C. Boyer and K. Galicki [5]. The chapter "K-contact and Sasakian structures" of this book contains an extended discussion on the topological problems in the theory of Sasakian, and, more generally, K-contact manifolds. This article is a survey based on recent results [3, 4, 16, 17, 26, 34–39]. The aim of this survey is to attract the attention of researchers to the theory of Seifert bundles over cyclic symplectic or Kähler orbifolds, which may be of importance for the development of Sasakian geometry. The foundations of this theory were laid by J. Kollár [30, 31]. However, in its original version, it imposed restrictions of smoothness on the base cyclic orbifolds. Papers [16, 34, 35, 39] contain further development of Kollár's theory, which includes non-smooth orbifolds. This enabled the authors to solve several open problems posed in [5]. Also, we include into this survey several results on rational homotopy properties of Sasakian manifolds [3, 4, 12, 13, 26, 38] and on the classification of Smale-Barden manifolds with Sasakian structures [37]. These results answer questions posed in [5]. It should be mentioned that results on rational homotopy properties of Sasakian manifolds use A. Tievsky's

To Vladimir Rovenski's 70-th birthday.

A. Tralle (✉)
Faculty of Mathematics and Computer Science, University of Warmia and Mazury, Słoneczna 54, 10-710 Olsztyn, Poland
e-mail: tralle@matman.uwm.edu.pl

© The Author(s), under exclusive license to Springer Nature Switzerland AG 2024
V. Rovenski et al. (eds.), *Differential Geometric Structures and Applications*, Springer Proceedings in Mathematics & Statistics 440,
https://doi.org/10.1007/978-3-031-50586-7_1

model (in the sense of rational homotopy theory) [46]. This construction is contained in the Ph.D. thesis of Tievsky and may be not well known.

In this survey, we draw the reader's attention to the use of this model, which may be of value in establishing new topological properties of Sasakian manifolds. It should be noted that although we focus on the particular articles mentioned above, they are inspired and motivated by the results of C. Boyer, K. Galicki and their co-authors [5–10]. It would also be interesting to think over the more general approach to "Sasaki-like" structures developed by D.S. Patra and V. Rovenski in [41] in the framework of this article.

Let us mention that we freely use the basics of symplectic topology [33], the theory of compact complex surfaces [2], complex and symplectic surgery [25] and some facts from algebraic geometry [28]. All the required material is contained in these sources, but we do not discuss it.

The structure of this survey is as follows. In Sect. 1 we define Sasakian and K-contact manifolds and formulate Rukimbira's theorem [43], which enables us to consider these manifolds as total spaces of Seifert bundles over singular cyclic orbifolds. Also, we discuss some basic facts of rational homotopy theory used in Sect. 7 stressing Tievsky's algebraic models of Sasakian manifolds. In Sect. 2 we discuss methods of construction of singular cyclic orbifolds and Seifert bundles. This is a core section. Section 3 describes applications of the theory developed in Sect. 2 to several existence problems posed in [5]. Section 7 is devoted to rational homotopy properties of Sasakian manifolds. In particular, it is shown that triple Massey products do not obstruct Sasakian structures in the simply connected case, but higher order Massey products do. This leads to a solution of the problem posed in [5] about the existence of simply-connected K-contact manifolds with no Sasakian structures. Finally, let us note that each of the questions from [5] is discussed separately in Sects. 3 and 7.

Basically, we omit the proofs of the described results providing the appropriate references. However, in order to show the method of Seifert fibrations and the use of Tievsky's model as well as symplectic surgery and Donaldson's hyperplane theorem, we show parts of the proofs of Theorems 12, 17 and 19.

1.1 Sasakian and K-Contact Manifolds

Let M be a $(2n+1)$-dimensional manifold. An *almost contact metric structure* on M consists of a quadruplet $(\eta\,,\xi\,,\phi\,,g)$, where η is a differential 1-form, ξ is a nowhere vanishing vector field (the *Reeb* vector field), ϕ is a C^∞ section of $\mathrm{End}(TM)$ and g is a Riemannian metric on M, satisfying the following conditions:

$$\eta(\xi)=1,\quad \phi^2=-\mathrm{id}+\xi\otimes\eta,\quad g(\phi X\,,\phi Y)=g(X\,,Y)-\eta(X)\eta(Y)\,,\qquad (1)$$

for all vector fields X, Y on M. Thus, the kernel of η defines a codimension one distribution $\mathcal{D}=\ker(\eta)$, and there is the orthogonal decomposition of the tangent bundle TM of M

$$TM = \mathcal{D} \oplus \mathcal{L},$$

where $\mathcal{L}$ is the trivial line subbundle of TM generated by ξ. Note that conditions in (1) imply that

$$\phi(\xi) = 0, \quad \eta \circ \phi = 0.$$

For an almost contact metric structure (η, ξ, ϕ, g) on M, the fundamental 2-form F on M is defined by

$$F(X, Y) = g(\phi X, Y),$$

where X and Y are vector fields on M. Hence,

$$F(\phi X, \phi Y) = F(X, Y),$$

that is F is compatible with ϕ, and $\eta \wedge F^n \neq 0$ everywhere.

An almost contact metric structure (η, ξ, ϕ, g) on M is said to be a *contact metric* if

$$g(\phi X, Y) = d\eta(X, Y).$$

In this case, η is a *contact form*, meaning

$$\eta \wedge (d\eta)^n \neq 0$$

at every point of M. If (η, ξ, ϕ, g) is a contact metric structure such that ξ is a Killing vector field for g, then (η, ξ, ϕ, g) is called a K-*contact* structure. A manifold with a K-contact structure is called a K-*contact manifold.*

There is the notion of integrability of an almost contact metric structure. More precisely, an almost contact metric structure (η, ξ, ϕ, g) is called *normal* if the Nijenhuis tensor N_ϕ associated to the tensor field ϕ, defined by

$$N_\phi(X, Y) := \phi^2[X, Y] + [\phi X, \phi Y] - \phi[\phi X, Y] - \phi[X, \phi Y],$$

satisfies the equation

$$N_\phi = -d\eta \otimes \xi.$$

This last equation is equivalent to the condition that the almost complex structure J on $M \times \mathbb{R}$ given by

$$J\Big(X, f\frac{\partial}{\partial t}\Big) = \Big(\phi X - f\xi, \eta(X)\frac{\partial}{\partial t}\Big)$$

is integrable, where f is a smooth function on $M \times \mathbb{R}$ and t is the standard coordinate on $\mathbb{R}$ (see [44]). In other words, ϕ defines a complex structure on the kernel $\ker(\eta)$ compatible with $d\eta$.

A *Sasakian structure* is a normal contact metric structure, in other words, an almost contact metric structure (η, ξ, ϕ, g) such that

$$N_\phi = -d\eta \otimes \xi, \quad d\eta = F.$$

If (η, ξ, ϕ, g) is a Sasakian structure on M, then (M, η, ξ, ϕ, g) is called a *Sasakian manifold.*

Riemannian manifolds with a Sasakian structure can also be characterized in terms of the Riemannian cone over the manifold. A Riemannian manifold (M, g) admits a compatible Sasakian structure if and only if $M \times \mathbb{R}^+$ equipped with the cone metric $h = t^2 g + dt \otimes dt$ is Kähler, see [5]. Furthermore, in this case the Reeb vector field is Killing, and the covariant derivative of ϕ with respect to the Levi-Civita connection of g is given by

$$(\nabla_X \phi)(Y) = g(\xi, Y)X - g(X, Y)\xi,$$

where X and Y are vector fields on M.

In the study of Sasakian manifolds, the *basic cohomology* plays an important role. Let (M, η, ξ, ϕ, g) be a Sasakian manifold of dimension $2n + 1$. A differential form α on M is called *basic* if

$$\iota_\xi \alpha = 0 = \iota_\xi d\alpha,$$

where ι_ξ denotes the contraction of differential forms by ξ. We denote by $\Omega^k_B(M)$ the space of all basic k-forms on M. Clearly, the de Rham exterior differential d of M takes basic forms to basic forms. Denote by d_B the restriction of the de Rham differential d to $\Omega^*_B(M)$. The cohomology of the differential complex $(\Omega^*_B(M), d_B)$ is called the *basic cohomology*. These basic cohomology groups are denoted by $H^*_B(M)$. When M is compact, the dimensions of the de Rham cohomology groups $H^*(M)$ and the basic cohomology groups $H^*_B(M)$ are related as follows:

Theorem 1 ([5, Theorem 7.4.14]) *Let (M, η, ξ, ϕ, g) be a compact Sasakian manifold of dimension $2n + 1$. Then, the Betti number $b_r(M)$ and the basic Betti number $b^B_r(M)$ are related by*

$$b^B_r(M) = b^B_{r-2}(M) + b_r(M)$$

for $0 \leq r \leq n$, In particular, if r is odd and $r \leq n$, then $b_r(M) = b^B_r(M)$.

A Sasakian structure on M is called *quasi-regular* if there is a positive integer δ satisfying the condition that each point of M has a foliated coordinate chart (U, t) with respect to ξ (the coordinate t is in the direction of ξ) such that each leaf for ξ passes through U at most δ times. If $\delta = 1$, then the Sasakian structure is called *regular* (see [5, p. 188]).

If X is a compact Kähler manifold, whose Kähler form ω defines an integral cohomology class, then the total space of the circle bundle

$$S^1 \hookrightarrow M \to X \tag{2}$$

with Chern class $[\omega] \in H^2(M, \mathbb{Z})$ is a regular Sasakian manifold with a contact form η that satisfies the equation $d\eta = \pi^*(\omega)$, where π is the projection in (2). This description is known and can be derived from [29]. In the same way, one defines the *regularity* and *quasi-regularity* of K-contact manifolds.

Let (X, ω) be a symplectic manifold such that the cohomology class $[\omega]$ is integral. Consider the principal S^1-bundle $\pi : M \to X$ given by the cohomology class $[\omega] \in H^2(X, \mathbb{Z})$. Such fiber bundles are called *Boothby-Wang fibrations* [11]. We have the following result (which is also known, and can be derived, for example, from [32]).

Theorem 2 *Any Boothby-Wang fibration admits a K-contact structure on the total space.*

It is important to note the following result of Rukimbira.

Proposition 1 ([43]) *If a compact manifold M admits a K-contact structure, it admits a quasi-regular K-contact structure. If a compact manifold M admits a Sasakian structure, it admits a quasi-regular Sasakian structure.*

It follows that analyzing topological properties of K-contact/Sasakian manifolds, we can always assume that we are given *quasi-regular* structures. It follows that our approach *is based on constructing Seifert bundles over Kähler or symplectic cyclic orbifolds (or manifolds, for regular structures)*. In some cases, it is useful to have an intermediate class of *semi-regular* Sasakian structures, see [36]. A cyclic orbifold is *smooth*, if all its points are smooth (see [36, Proposition 2 and Definition 3]). Seifert bundles over smooth orbifold yield semi-regular Sasakian structure. We do not present all the details here, because we are interested mainly in the general case.

1.2 Formal Manifolds and Massey Products

In this section, some definitions and results about minimal models and Massey products are reviewed [18, 22, 47]. These will be used in Sect. 7.

We work with *differential graded commutative algebras*, or DGAs, over the field $\mathbb{R}$ of real numbers. The degree of an element a of a DGA is denoted by $|a|$. A DGA $(\mathcal{A}, d)$ is *minimal* if:

1. $\mathcal{A}$ is free as an algebra, that is, $\mathcal{A}$ is the free algebra $\bigwedge V$ over a graded vector space $V = \bigoplus_i V^i$,
2. there is a collection of generators $\{a_\tau\}_{\tau \in I}$ indexed by some well ordered set I, such that $|a_\mu| \leq |a_\tau|$ if $\mu < \tau$ and each da_τ is expressed in terms of preceding a_μ, $\mu < \tau$. This implies that da_τ has no a linear part.

In our context, the main example of DGA is the de Rham complex $(\Omega^*(M), d)$ of a differentiable manifold M, where d is the exterior differential.

Given a differential graded commutative algebra $(\mathcal{A}, d)$, we denote its cohomology by $H^*(\mathcal{A})$. The cohomology of a differential graded algebra $H^*(\mathcal{A})$ is naturally a DGA with the product structure inherited from that on $\mathcal{A}$ and with the differential being identically zero. According to [18, p. 249], an *elementary extension* of a differential graded commutative algebra $(\mathcal{A}, d)$ is any DGA of the form

$$(\mathcal{B} = \mathcal{A} \otimes \bigwedge V, d_{\mathcal{B}}),$$

where $d_{\mathcal{B}}|_{\mathcal{A}} = d$, $d_{\mathcal{B}}(V) \subset \mathcal{A}$ and $\bigwedge V$ is the free algebra over a finite-dimensional vector space V, whose elements have all the same degree. Morphisms between DGAs are required to preserve the degree and to commute with the differential. We shall say that $(\bigwedge V, d)$ is a *minimal model* of the differential graded commutative algebra $(\mathcal{A}, d)$ if $(\bigwedge V, d)$ is minimal and there exists a morphism of differential graded algebras

$$\rho : (\bigwedge V, d) \to (\mathcal{A}, d)$$

inducing an isomorphism $\rho^* : H^*(\bigwedge V) \xrightarrow{\sim} H^*(\mathcal{A})$ of cohomologies.

In [27], Halperin proved that any connected differential graded algebra $(\mathcal{A}, d)$ has a minimal model unique up to isomorphism. A *minimal model* of a connected differentiable manifold M is a minimal model $(\bigwedge V, d)$ for the de Rham complex $(\Omega^*(M), d)$ of differential forms on M. The importance of the minimal model comes from the fact that it is a *homotopy invariant* [22, 47].

Recall that a minimal algebra $(\bigwedge V, d)$ is called *formal* if there exists a morphism $\psi : (\bigwedge V, d) \to (H^*(\bigwedge V), 0)$ of differential algebras inducing the identity map on cohomology. Also a differentiable manifold M is called *formal* if its minimal model is formal. Many examples of formal manifolds are known: spheres, projective spaces, compact Lie groups, symmetric spaces, flag manifolds, and all compact Kähler manifolds [21, 47].

Massey products are known to be obstructions to formality [18]. Let $(\mathcal{A}, d)$ be a DGA (in particular, it can be the de Rham complex of differential forms on a differentiable manifold). Suppose that there are cohomology classes $[a_i] \in H^{p_i}(\mathcal{A})$, $p_i > 0$, $1 \le i \le 3$, such that $a_1 \cdot a_2$ and $a_2 \cdot a_3$ are exact. Write $a_1 \cdot a_2 = da_{1,2}$ and $a_2 \cdot a_3 = da_{2,3}$. The *(triple) Massey product* of the classes $[a_i]$ is defined to be

$$\begin{aligned} \langle [a_1], [a_2], [a_3] \rangle &= [a_1 \cdot a_{2,3} + (-1)^{p_1+1} a_{1,2} \cdot a_3] \\ &\in \frac{H^{p_1+p_2+p_3-1}(\mathcal{A})}{[a_1] \cdot H^{p_2+p_3-1}(\mathcal{A}) + [a_3] \cdot H^{p_1+p_2-1}(\mathcal{A})}. \end{aligned}$$

Now we move on to the definition of higher Massey products (see [47]). Given

$$[a_i] \in H^*(\mathcal{A}), \quad 1 \le i \le t, \quad t \ge 3,$$

the Massey product $\langle [a_1], [a_2], \ldots, [a_t] \rangle$, is defined if there are elements $a_{i,j}$ on $\mathcal{A}$, with $1 \le i \le j \le t$ and $(i, j) \ne (1, t)$, such that

$$\begin{aligned} a_{i,i} &= a_i\,, \\ d\,a_{i,j} &= \sum_{k=i}^{j-1} (-1)^{|a_{i,k}|} a_{i,k} \cdot a_{k+1,j}\,. \end{aligned} \tag{3}$$

Then the *Massey product* $\langle [a_1], [a_2], \ldots, [a_t] \rangle$ is the set of cohomology classes

$$\langle [a_1], [a_2], \ldots, [a_t] \rangle$$

$$= \left\{ \left[\sum_{k=1}^{t-1} (-1)^{|a_{1,k}|} a_{1,k} \cdot a_{k+1,t} \right] \mid a_{i,j} \text{ as in (3)} \right\} \subset H^{|a_1|+\ldots+|a_t|-(t-2)}(\mathcal{A})\,.$$

We say that the Massey product is *zero* if $0 \in \langle [a_1], [a_2], \ldots, [a_t] \rangle$.

It should be mentioned that for $\langle a_1, a_2, \ldots, a_t \rangle$ to be defined, it is necessary that all the lower order Massey products $\langle a_1, \ldots, a_i \rangle$ and $\langle a_{i+1}, \ldots, a_t \rangle$ with $2 < i < t-2$ are defined and trivial.

Massey products are related to formality by the following well-known result.

Theorem 3 ([18, 47]) *A DGA, which has a non-zero Massey product, is not formal.*

1.3 Tievsky's Model

Let (M, η, ξ, ϕ, g) be a compact Sasakian manifold. Tievsky in [46], motivated by the approach in [18] and using the basic cohomology of M, found a model for M, that is, a DGA with the same minimal model as the manifold M.

Theorem 4 ([46]) *Let (M, η, ξ, ϕ, g) be a compact Sasakian manifold. Then a model for M is given by the DGA $(H^*_B(M) \otimes \bigwedge(x),\, D)$, where $|x| = 1$, $D(H^*_B(M)) = 0$ and $Dx = [d\eta]_B$. Therefore, $(H^*_B(M) \otimes \bigwedge(x),\, D)$ is an elementary extension of the DGA $(H^*_B(M),\, 0)$ (the differential is zero).*

2 Quasi-regular Sasakian Manifolds as Seifert Bundles Over Cyclic Orbifolds

Thus, *regular* K-contact structures are Boothby-Wang fibrations. Let us discuss the *quasi-regularity condition* for K-contact and Sasakian structures. First, we refer to [5] for the general theory of (cyclic) orbifolds, and we freely use this theory without further explanations concentrating on *constructions* of Kähler and symplectic *singular* orbifolds. Quasi-regular Sasakian and K-contact structures arise as *Seifert bundles* over such orbifolds. Full details can be found in [34].

2.1 Symplectic and Kähler Singular Orbifolds and Seifert Bundles

Definition 1 ([34]) A cyclic singular symplectic (Kähler) orbifold X is a symplectic orbifold, whose isotropy set is of dimension 0, that is, a finite set P of points called a singular set.

For a cyclic singular symplectic 4-manifold, a singular point is an isolated isotropy point $x \in P \subset X$. A local model around x is of the form $\mathbb{C}^2/\mathbb{Z}_d$, where $\xi = \exp(2\pi i/d)$ acts as

$$\xi \cdot (z_1, z_2) = (\xi^{e_1} z_1, \xi^{e_2} z_2),$$

and $\gcd(e_1, d) = \gcd(e_2, d) = 1$. We will write $d = d(x)$.

Definition 2 ([34]) A sing-symplectic surface is a symplectic 2-orbifold $D \subset X$ such that if $x \in D$ is a singular point, then D is fixed by the isotropy subgroup. Two sing-symplectic surfaces D_1 and D_2 intersect nicely if at at every intersection point $x \in D_1 \cap D_2$ there is an adapted Darboux chart (z_1, z_2) centered at x such that $D_1 = \{(z_1, 0)\}$ and $D_2 = \{(0, z_2)\}$ in a model $\mathbb{C}^2/\mathbb{Z}_d$, where $\mathbb{Z}_d$ is a subgroup of $U(1) \times U(1)$.

The construction of such orbifolds is based on the following result (note that, for simplicity, we restrict ourselves to dimension 4).

Theorem 5 ([34]) *Let X be a cyclic singular 4-manifold with the set of singular points P. Let D_i be embedded sing-symplectic surfaces intersecting nicely, and take coefficients $m_i > 1$ such that $\gcd(m_i, m_j) = 1$ if D_i and D_j have a non-empty intersection. Then there exists an orbifold X with isotropy surfaces D_i of multiplicities m_i, and singular points $x \in P$ of multiplicity $m = d(x)\prod_{i \in I_x} m_i$, $I_x = \{i \mid x \in D_i\}$.*

Note that if one replaces the word "symplectic" by "Kähler", one obtains cyclic singular Kähler orbifold. If one takes $P = \emptyset$ one obtains smooth cyclic orbifolds as in [31]. Finally, it is important to understand that Theorem 5, indeed, shows how to construct Kähler or symplectic orbifolds, which are non-smooth: one must construct the data

$$X, P, D_i \subset X, m_i, m = d(x)\prod_{i \in I_x} m_i, I_x = \{i \mid x \in D_i\}, \tag{4}$$

satisfying the restrictions given by Theorem 5.

Definition 3 Let X be a cyclic n-orbifold. A Seifert bundle over X is an oriented $(n+1)$–manifold endowed with a smooth S^1-action and a continuous map $\pi : M \to X$ such that for an orbifold chart $(U, \tilde{U}, \mathbb{Z}_m, \varphi)$ there is a commutative diagram

$$\begin{array}{ccc} (S^1 \times \tilde{U})/\mathbb{Z}_m & \xrightarrow{\cong} & \pi^{-1}(U) \\ \downarrow & & \downarrow{\scriptstyle \pi} \\ \tilde{U}/\mathbb{Z}_m & \xrightarrow{\cong} & U \end{array}$$

where the action of $\mathbb{Z}_m$ on S^1 is by multiplication by $\exp(2\pi i/m)$, and the top diffeomorphism is S^1-equivariant.

The description of Sasakian manifolds is contained in the following result.

Theorem 6 ([5, Theorems 7.5.1 and 7.5.2]) *Let M be a manifold endowed with a quasi-regular Sasakian structure (η, ξ, ϕ, g). Then the space of leaves of the foliation $\mathcal{F}_\xi$ determined by the Reeb vector field has a natural structure of a cyclic Kähler orbifold, and the projection $M \to X$ is a Seifert bundle. Conversely, if (X, ω) is a cyclic Kähler orbifold and M is the total space of the Seifert bundle determined by the class $[\omega]$ of the Kähler form, then M admits a quasi-regular Sasakian structure.*

In the same way, one can describe quasi-regular K-contact manifolds replacing cyclic Kähler orbifolds by cyclic symplectic orbifolds (see [34] for the details).

2.2 *Constructions of Cyclic Kähler Orbifolds and Seifert Bundles*

An important basic tool of constructing cyclic Kähler orbifolds in this work is by blowing-down complex surfaces along chains of smooth rational curves of negative self-intersection < -1 (i.e. no (-1)-curves). We freely use the theory of complex surfaces referring to [2, 25, 28].

Proposition 2 ([2]) *Consider the action of the cyclic group $\mathbb{Z}_m$ on $\mathbb{C}^2$ given by $(z_1, z_2) \mapsto (\eta z_1, \eta^r z_2)$, where $\eta = e^{2\pi i/m}$, $0 < r < m$ and $\gcd(r, m) = 1$. Then write a continuous fraction*

$$\frac{m}{r} = [b_1, \ldots, b_l] = b_1 - \cfrac{1}{b_2 - \frac{1}{b_3 - \ldots}}\,.$$

The resolution of $\mathbb{C}^2/\mathbb{Z}_m$ has an exceptional divisor formed by a chain of smooth rational curves of self-intersection numbers $-b_1, -b_2, \ldots, -b_l$.

Proposition 3 ([16, Lemma 3.1]) *Conversely, let X be a smooth complex surface containing a chain of smooth rational curves $E_1, \ldots, E_l$ of self-intersections $-b_1, -b_2, \ldots, -b_l$, with all $b_i \geq 2$, intersecting transversally (so that $E_i \cap E_{i+1}$ are nodes, $i = 1, \ldots, l-1$). Let $\pi : X \to \bar{X}$ be the contraction of $E = E_1 \cup \ldots \cup E_l$. Then $\bar{X}$ has a cyclic singularity at $p = \pi(E)$, with an action given by Proposition 2. Moreover, if D is a curve intersecting transversally a tail of the chain (that is, either E_1 or E_l at a non-nodal point), then the push down curve $\bar{D} = \pi(D)$ is a sing-complex curve in $\bar{X}$ (in the sense of Definition 2, with appropriate modifications in the complex case).*

Let us mention the following. For the convenience of references, we will alternately follow the terminology and notation of [5, 30, 31, 34]. If X is a cyclic orbifold with singular set P and a family of surfaces D_i with multiplicities m_i, we will say that we are given a divisor $\cup D_i$ with multiplicities $m_i > 1$. The formal sum

$$\Delta = \sum_i \left(1 - \frac{1}{m_i}\right) D_i \tag{5}$$

will be called the *branch divisor*.

Given a cyclic Kähler orbifold X, with singular points P, and branch divisor (5), we need an extra piece of information in order to determine a Seifert bundle $M \to X$. For each point $x \in P$ with multiplicity $m = dm_1m_2$, we have an adapted chart $U \subset \mathbb{C}^2/\mathbb{Z}_m$ with action

$$\exp(2\pi i/m)(z_1, z_2) = (e^{2\pi i j_1/m} z_1, e^{2\pi i j_2/m} z_2), \tag{6}$$

so that the local model of the Seifert fibration is as in Definition 3. We call $j_x = (m, j_1, j_2)$ the *local invariants* at $x \in P$. Note that $j_1 = m_1 e_1$ and $j_2 = m_2 e_2$. Assume that $D_1 = \{(z_1, 0)\}$ is one of the isotropy surfaces with multiplicity m_1. The local invariant of D_1 is by definition, $j_{D_1} = (m_1, j_2)$, where j_2 is considered modulo m_1. This also yields a *compatibility condition* for the local invariants of singular points and isotropy surfaces.

In order to construct Seifert bundles, we need to assign local invariants to each of the singular points $x \in P$ and each of the isotropy surfaces $D_i \subset X$, in a compatible way. For this we need to choose for each D_i some j_i with $\gcd(j_i, m_i) = 1$. Then we can assign local invariants at the singular points using the following result.

Proposition 4 ([34, Proposition 25]) *Suppose that X is a cyclic 4-orbifold such that each singular point $x \in P$ lies in a single isotropy surface D_i, if any. Take integers j_i with $\gcd(m_i, j_i) = 1$ for each D_i. Then there exist local invariants for X.*

Once compatible local invariants are fixed, a Seifert bundle is determined by its orbifold Chern class. Given a Seifert bundle $\pi : M \to X$, the order of a stabilizer (in S^1) of any point p in the fiber over $x \in X$ is denoted by $m = m(x)$, as in (6).

Definition 4 For a Seifert bundle $M \to X$ define the first Chern class as follows. Let $l = \operatorname{lcm}(m(x) \mid x \in X)$. Denote by M/l the quotient of M by $\mathbb{Z}_l \subset S^1$. Then $M/l \to X$ is a circle bundle with the first Chern class $c_1(M/l) \in H^2(X, \mathbb{Z})$. Define

$$c_1(M) = \frac{1}{l} c_1(M/l) \in H^2(X, \mathbb{Q}).$$

The first orbifold Chern class of the Seifert bundle can be calculated using the formula given by the following result, that we state only for complex orbifolds.

Proposition 5 ([34], Proposition 35) *Let X be a cyclic 4-orbifold with a complex structure and $D_i \subset X$ complex curves of X, which intersect transversally. Let $m_i > 1$ such that $\gcd(m_i, m_j) = 1$ if D_i and D_j intersect. Suppose that there are given local* *invariants* (m_i, j_i) *for each* D_i *and* j_p *for every singular point* p*, which are compatible. Choose any* $0 < b_i < m_i$ *such that* $j_i b_i \equiv 1 (\mathrm{mod}\ m_i)$*. Let* B *be a complex line bundle over* X*. Then there exists a Seifert bundle* $M \to X$ *with the given local invariants and the first orbifold Chern class*

$$c_1(M) = c_1(B) + \sum\nolimits_i \frac{b_i}{m_i}[D_i].$$

The set of all such Seifert bundles forms a principal homogeneous space under $H^2(X, \mathbb{Z})$*, where the action corresponds to changing* B*. Moreover, if* X *is a Kähler cyclic orbifold and* $c_1(M) = [\omega]$ *for the orbifold Kähler form, then* M *is Sasakian.*

Here, $[D_i]$ are understood as cohomology classes in $H^2(X, \mathbb{Q})$.

Proposition 6 ([34, Lemma 34]) *Assume that we are given a Seifert bundle* $M \to X$ *over a cyclic orbifold with the set of isotropy points* P *and branch divisor* $\sum(1 - \frac{1}{m_i})D_i$*. Let* $\mu = \mathrm{lcm}(m_i)$*. Then* $c_1(M/\mu) = \mu\, c_1(M)$ *is integral in* $H^2(X - P, \mathbb{Z})$*.*

Also, we need a method of calculating the cohomology of the total spaces of Seifert fibrations, in particular, we need to have a condition, which ensures $H_1(M, \mathbb{Z}) = 0$. Such method is contained in the following theorem. Recall that an element a in a free abelian group A is called *primitive* if it cannot be represented as $a = kb$ with non-trivial $b \in A$, $k \in \mathbb{N}$, $k > 1$.

Theorem 7 ([34]) *Suppose that* $\pi : M \to X$ *is a quasi-regular Seifert bundle over a cyclic orbifold* X *with isotropy surfaces* D_i *and set of singular points* P*. Let* $\mu = \mathrm{lcm}(m_i)$*. Then* $H_1(M, \mathbb{Z}) = 0$ *if and only if*

1. $H_1(X, \mathbb{Z}) = 0$,
2. $H^2(X, \mathbb{Z}) \to \oplus H^2(D_i, \mathbb{Z}_{m_i})$ *is onto,*
3. $c_1(M/\mu) \in H^2(X - P, \mathbb{Z})$ *is primitive.*

Moreover, $H_2(M, \mathbb{Z}) = \mathbb{Z}^k \oplus (\oplus_i \mathbb{Z}_{m_i}^{2g_i})$*,* $g_i =$ *genus of* D_i*,* $k + 1 = b_2(X)$*.*

Thus, if we want to test the assumptions of Theorem 7, we, in particular, calculate $H^2(X - P, \mathbb{Z})$ and check if $c_1(M/\mu)$ is primitive in $H^2(X - P, \mathbb{Z})$.

2.3 Fundamental Groups of Total Spaces of Seifert Bundles

Many questions in topology of Sasakian manifolds are much more difficult to answer under the condition of 1-connectedness. Since the basic tool of the studying Sasakian manifolds is via Seifert fibration $M \to X$, one needs methods of calculating the fundamental group of M. In particular, we need to check that $\pi_1(M) = 1$. We will systematically use the following.

Definition 5 The orbifold fundamental group $\pi_1^{\text{orb}}(X)$ is defined as

$$\pi_1^{\text{orb}}(X) = \pi_1(X - (\Delta \cup P))/\langle \gamma_i^{m_i} = 1 \rangle,$$

where $\langle \gamma_i^{m_i} = 1 \rangle$ denotes the following relation on $\pi_1(X - (\Delta \cup P))$: for any small loop γ_i around a surface D_i in the branch divisor, one has $\gamma_i^{m_i} = 1$.

We will use without further notice the following exact sequence:

$$\cdots \to \pi_1(S^1) = \mathbb{Z} \to \pi_1(M) \to \pi_1^{\text{orb}}(X) \to 1,$$

see [5, Theorem 4.3.18]. It is easy to see that if $H_1(M, \mathbb{Z}) = 0$ and $\pi_1^{\text{orb}}(X)$ is abelian, then $\pi_1(M)$ must be trivial. This holds since if $H_1(M, \mathbb{Z}) = 0$, then $\pi_1(M)$ has no abelian quotients. As $\pi_1^{\text{orb}}(X)$ is assumed abelian, we find that $\pi_1(M)$ is a quotient of $\mathbb{Z}$, hence again abelian. This implies that $\pi_1(M) = 0$.

The following result will be used in the calculations of the orbifold fundamental groups.

Proposition 7 ([40]) *If Z is a smooth simply-connected projective surface with smooth complex curves C_i intersecting transversally and satisfying $C_i^2 > 0$, then $\pi_1(Z - (C_1 \cup \ldots \cup C_r))$ is abelian.*

2.4 *The Sum Up of the Construction*

Thus, to construct a Seifert bundle $M \to X$ determining a quasi-regular Sasakian structure on M one needs to choose the following:

- orbifold data (4),
- compatible local data (m_i, j_i) and j_p for each singular point $p \in P$, as in Proposition 4,
- any integers such that $0 < b_i < m_i$ and $j_i b_i \equiv 1 \pmod{m_i}$.

These data will determine a Seifert bundle with $c_1(M)$ given by the formula in Proposition 5. Choosing these data in an appropriate way, one constructs quasi-regular Sasakian manifolds. The authors of [16, 34–36, 39] solve some problems posed in [5] choosing these data in a way to get the required conditions on the second Betti number $b_2(M)$ and on the triviality of $H_1(M, \mathbb{Z})$ using Theorem 7. By the method of Sect. 2.3 one derives $\pi_1(M) = 0$.

3 Existence Problems for Smale-Barden Manifolds

3.1 Smale-Barden Manifolds

A 5-dimensional simply connected manifold M is called a *Smale-Barden manifold.* These manifolds are classified by their second homology group over $\mathbb{Z}$ and the so-called *Barden invariant* [5]. In more detail, let M be a compact smooth oriented simply connected 5-manifold. Let us write $H_2(M,\mathbb{Z})$ as a direct sum of cyclic groups of prime power order

$$H_2(M,\mathbb{Z}) = \mathbb{Z}^k \oplus \big(\bigoplus_{p,i} \mathbb{Z}_{p^i}^{c(p^i)}\big),$$

where $k = b_2(M)$. Choose this decomposition in a way that the second Stiefel-Whitney class map $w_2 : H_2(M,\mathbb{Z}) \to \mathbb{Z}_2$ is zero on all but one summand $\mathbb{Z}_{2^j}$. The value of j is unique, it is denoted by $i(M)$ and is called the Barden invariant. The fundamental question arises, *which Smale-Barden manifolds admit Sasakian structures?* In the subsequent sections we present results of [16, 35–37, 39] in the context of this general problem. These articles solve several particular problems posed in [5]. Each of these questions is discussed separately.

3.2 Definite Sasakian Structures

Recall that the Reeb vector field ξ on a co-oriented contact manifold (M,η) determines a 1-dimensional foliation $\mathcal{F}_\xi$ called the *characteristic foliation.* For a manifold with a Sasakian structure (η,ξ,ϕ,g), one can define *basic Chern classes* $c_k(\mathcal{F}_\xi)$ of $\mathcal{F}_\xi$, which are elements of the basic cohomology $H_B^{2k}(\mathcal{F}_\xi)$ (see [5, Theorem/Definition 7.5.17]). We say that a Sasakian structure is positive (negative) if $c_1(\mathcal{F}_\xi)$ can be represented by a positive (negative) definite (1,1)-form. A Sasakian structure is called null, if $c_1(\mathcal{F}_\xi) = 0$. If none of them, it is called indefinite.

3.3 Rational Homology Spheres

Question 1 ([5, Open Problems 10.3.3 and 10.3.4]) Which simply connected rational homology 5-spheres admit negative Sasakian structures?

Contrary to this question, there are very few rational homology spheres, which admit positive Sasakian structures.

Theorem 8 ([5, Theorem 10.2.19], [30, Theorem 1.4]) *Suppose that a rational homology sphere M admits a positive Sasakian structure. Then M is spin and $H_2(M,\mathbb{Z})$ is one of the following:*

$$0,\ \mathbb{Z}_m^2,\ \mathbb{Z}_5^4,\ \mathbb{Z}_4^4,\ \mathbb{Z}_3^4,\ \mathbb{Z}_3^6,\ \mathbb{Z}_3^8,\ \mathbb{Z}_2^{2n},$$

where $n > 0$ and $m \geq 2$, where m not divisible by 30. *Conversely, all these cases do occur.*

Theorem 9 ([5, Theorem 10.3.14]) *Let M be a rational homology sphere. If it admits a Sasakian structure, then it is either positive, and the torsion in $H_2(M, \mathbb{Z})$ is restricted by Theorem 8, or it is negative. There exist infinitely many such manifolds, which admit negative Sasakian structures but no positive Sasakian structures. There exist infinitely many positive rational homology spheres, which also admit negative Sasakian structures.*

One can see that any torsion group, which is realizable by a simply connected Sasakian rational homology sphere M but which does not appear in the list given by Theorem 8 gives an example of an answer to Question 1.

The first result on negative Sasakian structures on rational homotopy spheres was obtained by R. Gomez [23].

Theorem 10 *Any simply connected rational homotopy sphere from the list in Theorem 8 except, possibly, $\mathbb{Z}_m^2, m < 5$ and $\mathbb{Z}_2^{2n}$, admits both negative and positive Sasakian structures.*

In the smaller class of semi-regular Sasakian structures, Question 1 is answered by the following result.

Theorem 11 ([37]) *Let $m_i \geq 2$ be pairwise coprime and $g_i = \frac{1}{2}(d_i - 1)(d_i - 2)$. Assume that* $\gcd(m_i, d_i) = 1$. *Let M be a Smale-Barden manifold with $H_2(M, \mathbb{Z}) = \oplus_{i=1}^r \mathbb{Z}_{m_i}^{2g_i}$ and spin, with the exceptions $\mathbb{Z}_m^2, \mathbb{Z}_2^{2n}, \mathbb{Z}_3^6$. Then M admits a negative semi-regular Sasakian structure. Conversely, if M is a simply-connected rational homology 5-sphere admitting a semi-regular Sasakian structure, then it must satisfy the above assumptions.*

Theorem 26 in [16] shows that Theorem 11 is not true in the quasi-regular case. The above discussion motivates the first problem we want to address, which certainly contributes to Question 1.

Question 2 Which simply connected rational homology spheres admit both negative and positive Sasakian structures?

We give a complete answer to the Question 2 in Theorem 12: all positive Sasakian simply connected rational homology spheres also admit negative Sasakian structures.

Also, C. Boyer and K. Galicki posed the following question.

Question 3 ([5, Open Problem 10.3.6]) Show that all $\#_k(S^2 \times S^3)$ admit negative Sasakian structures or determine precisely for which k this holds true.

In Theorem 13 we give a complete answer to this question: any $\#_k(S^2 \times S^3)$ admits negative Sasakian structures.

Let us characterize a definite Sasakian structure in a different way. Recall that a Sasakian structure is positive (negative), if $c_1(\mathcal{F}_\xi)$ can be represented by a positive (negative) definite $(1,1)$-form. If $\pi : M \to X$ is a Seifert bundle determined by a quasi-regular Sasakian structure, then by [5, Proposition 7.5.23], $c_1(\mathcal{F}_\xi) = \pi^* c_1^{\mathrm{orb}}(X)$. Since $c_1(\mathcal{F}_\xi)$ is represented by a basic form (see [5, Sect. 7.5.2]), it follows that $c_1(\mathcal{F}_\xi)$ is represented by a negative definite $(1,1)$-form if and only if $c_1^{\mathrm{orb}}(X) < 0$ (that is, represented by a negative definite orbifold $(1,1)$-form). Thus, we get the following (see also [7, Sect. 4]).

Proposition 8 *A 5-manifold M admits a quasi-regular negative Sasakian structure if and only if the base X of the corresponding Seifert bundle $M \to X$ has the property that the canonical class K_X^{orb} is ample.*

In the case of rational homology spheres Sasakian structures fall into two classes.

Proposition 9 ([5, Proposition 7.5.29]) *Let M be a $(2n+1)$-dimensional rational homology sphere. Any Sasakian structure on M is either positive or negative.*

Theorem 12 *Any simply connected rational homology sphere admitting a positive Sasakian structure admits also a negative Sasakian structure.*

In order to illustrate the methods we reproduce a part of the proof from [39] without any changes. By Theorem 10, it only remains to check that the homology groups $H_2(M,\mathbb{Z}) = \mathbb{Z}_2^{2n}$, $n > 0$, and $\mathbb{Z}_m^2$, $m < 5$, from Theorem 8 can be covered. Each of these cases is worked out separately in [39]. Here we show the proof only for $\mathbb{Z}_2^{2n}$ (the other case is similar in spirit).

We begin with some known facts on Hirzebruch surfaces [2, 25, 28]. Fix $n \geq 0$. By definition, the Hirzebruch surface $\mathbb{F}_n$ is the projectivization of the vector bundle

$$\mathcal{O}_{\mathbb{CP}^1}(n) \oplus \mathcal{O}_{\mathbb{CP}^1}.$$

For a holomorphic section $\sigma : \mathbb{CP}^1 \to \mathcal{O}_{\mathbb{CP}^1}(n)$, we denote by E_σ the image of $(\sigma, 1) : \mathbb{CP}^1 \to \mathcal{O}_{\mathbb{CP}^1}(n) \oplus \mathcal{O}_{\mathbb{CP}^1}$ in $\mathbb{F}_n$. The curve $E_0 \subset \mathbb{F}_n$ is called the zero section of $\mathbb{F}_n$. As $E_0 \equiv E_\sigma$ for all sections σ, we have that $E_0^2 = n$. Let C denote the fiber of the fibration $\mathbb{F}_n \to \mathbb{CP}^1$. Then

$$C^2 = 0, \quad E_0 \cdot C = 1.$$

It is known [45] that for $n > 0$ the surface $\mathbb{F}_n$ contains a unique irreducible curve E_∞ of negative self-intersection:

$$E_\infty \cdot E_\infty = -n.$$

This curve is called the section at infinity since it is given by the image of $(\sigma, 0) : \mathbb{CP}^1 \to \mathcal{O}_{\mathbb{CP}^1}(n) \oplus \mathcal{O}_{\mathbb{CP}^1}$. One can calculate

$$E_\infty \equiv E_0 - nC, \quad K_{\mathbb{F}_n} \equiv (n-2)C - 2E_0,$$

where $K_{\mathbb{F}_n}$ is the canonical divisor. Clearly

$$H^2(\mathbb{F}_n, \mathbb{Z}) = \mathbb{Z}\langle C, E_\infty \rangle.$$

Now we are ready to prove our first main result.

Proposition 10 *A simply connected rational homology sphere with $H_2(M, \mathbb{Z}) = \mathbb{Z}_2^{2n}$, $n \geq 1$, and spin admits a negative Sasakian structure.*

Proof Consider $\mathbb{F}_n$ with the zero section E_0, the section at infinity E_∞, and the fiber C of $\mathbb{F}_n \to \mathbb{CP}^1$. Let $\beta \geq 1$ and take a smooth divisor

$$D \equiv C + \beta E_0.$$

This exists because $C + \beta E_0$ is very ample (it can be represented by a Kähler form) and Bertini's theorem is applicable. By the genus formula $D^2 + K_{\mathbb{F}_n} \cdot D = 2g(D) - 2$, we get

$$g(D) = \frac{\beta^2 - \beta}{2} \cdot n. \tag{7}$$

One also calculate

$$D \cdot E_\infty = (C + \beta E_0) \cdot E_\infty = 1,$$

as $E_0 \cdot E_\infty = 0$. Consider also rational curves $D_i = E_{\sigma_i}$, $i = 1, \ldots, s$, which are sections, $D_i \equiv E_0$. These can be taken to intersect transversally (with each other and also with D).

Let X be the orbifold obtained by the blow-down of $\mathbb{F}_n$ along E_∞. By Proposition 3, X is a cyclic orbifold with a cyclic singularity p of degree $d(p) = n$. For simplicity we denote curves on $\mathbb{F}_n$ and on X by the same letters. Endow X with an orbifold structure with the branch divisor

$$\Delta = \left(1 - \frac{1}{m}\right) D + \sum\nolimits_{i=1}^{s} \left(1 - \frac{1}{m_i}\right) D_i\,, \tag{8}$$

where $m_1, \ldots, m_s$ are pairwise coprime, and the m_i are coprime to m and n (but possibly $\gcd(n, m) > 1$).

One can easily calculate that

$$H_2(X, \mathbb{Z}) = H^2(X - \{p\}, \mathbb{Z}) = \mathbb{Z}\langle C \rangle, \tag{9}$$

$$H_2(X - \{p\}, \mathbb{Z}) = H^2(X, \mathbb{Z}) = \mathbb{Z}\langle E_0 \rangle. \tag{10}$$

Note that C is a divisor passing through the singular point p. Since the (co)homology groups in (9) and (10) are dual, we can calculate the intersection numbers over $\mathbb{Q}$ and obtain

$$C^2 = \frac{1}{n}$$

in $H_2(X, \mathbb{Q})$. Indeed, we can write $E_0 \equiv a\, C$ for some $a \in \mathbb{Z}$, and substituting this to $n = E_0^2 = a\, E_0 \cdot C = a$, we get $E_0 \equiv n\, C$. Then $C^2 = \frac{1}{n^2} E_0^2 = \frac{1}{n}$. We also have

$$D \equiv C + \beta E_0 \equiv C + \beta n C = (1 + \beta n) C.$$

Considering X as an algebraic variety, we calculate the canonical divisor K_X from the adjunction formula (noting that C is sing-complex):

$$K_X \cdot C + C^2 = -\chi_{\mathrm{orb}}(C) = 2g(C) - 2 + \Big(1 - \frac{1}{n}\Big).$$

Since $g(C) = 0$ we get $K_X \cdot C + \frac{1}{n} = -2 + (1 - \frac{1}{n})$. Writing $K_X = b\, C$, $b \in \mathbb{Q}$, and substituting in the above, we get $b = -(n+2)$. So finally

$$K_X = -(n+2) C.$$

Since all D_i are homologous to E_0, we obtain the following formula for the orbifold canonical class:

$$\begin{aligned} K_X^{\mathrm{orb}} &= K_X + \Big(1 - \frac{1}{m}\Big) D + \sum\nolimits_{i=1}^{s} \Big(1 - \frac{1}{m_i}\Big) D_i \\ &= -(n+2) C + \Big(1 - \frac{1}{m}\Big)(1 + \beta n) C + \sum\nolimits_{i=1}^{s} \Big(1 - \frac{1}{m_i}\Big) n C \\ &= \Big(-(n+2) + \Big(1 - \frac{1}{m}\Big)(1 + \beta n) + n \sum\nolimits_{i=1}^{s} \Big(1 - \frac{1}{m_i}\Big)\Big) C. \end{aligned} \tag{11}$$

Now, consider a Seifert bundle

$$M \to X$$

determined by the orbifold data (8). We need to choose local invariants $0 < j_i < m_i$, $\gcd(j_i, m_i) = 1$, for each D_i, and $0 < j < m$, $\gcd(j, m) = 1$, for D. These will be fixed later. The local invariant j_p at the singular point is given by applying Proposition 4.

By Proposition 8, we know that the existence of a negative Sasakian structure on M is equivalent to the ampleness of the orbifold canonical bundle K_X^{orb}, and this is equivalent to the positivity of the coefficient in (11). Now we specify some coefficients in order to realize the desired cohomology groups. Put $m = 2$ and $\beta = 2$, so by (7), we have $g(D) = n$. By Theorem 7, this gives $H_2(M, \mathbb{Z}) = \mathbb{Z}_2^{2n}$. Calculating the coefficient in (11), we get

$$-\frac{3}{2} + n \sum\nolimits_{i=1}^{s} \Big(1 - \frac{1}{m_i}\Big).$$

Choosing m_i and s large, we can get positivity for any n. As we said, we take m, m_i all of them pairwise coprime, and also $\gcd(m_i, n) = 1$ for all i.

Note that we have done this calculation assuming that the assumptions of Theorem 7 are satisfied. The assumption (2) holds for D_i since $H^2(X, \mathbb{Z}) = \mathbb{Z}\langle E_0\rangle$, $D_i \equiv E_0$, and $\gcd(m_i, n) = 1$. It holds for D since $D \equiv (1 + n\beta)C$, so the map $H^2(X, \mathbb{Z}) \to H^2(D, \mathbb{Z}_m)$ is $[E_0] \mapsto 1 + n\beta$, and $\gcd(1 + n\beta, m) = 1$ for $\beta = m = 2$.

Let us now check assumption (3) of Theorem 7. By Proposition 5,

$$c_1(M) = c_1(B) + \frac{b}{m}[C + \beta E_0] + \sum\nolimits_i \frac{b_i}{m_i}[E_0],$$

for a line bundle B, so $c_1(B) = q[E_0]$, for some integer $q \in \mathbb{Z}$. As we take all $m, m_1, \ldots, m_s$ pairwise coprime, we have $\mu = m \cdot m_1 \cdots m_s$. We need to check that we can arrange for $\mu\, c_1(M)$ to be integral and primitive in $H^2(X - P, \mathbb{Z})$, which means that we need

$$c_1(M/\mu) = \mu\, c_1(M) = [C].$$

Note that since $c_1(M) > 0$, we have by Proposition 5 that M is Sasakian. We compute

$$\mu\, c_1(M) = b \cdot m_1 \cdots m_s[C + \beta E_0] + \sum\nolimits_i b_i \cdot m \cdot m_1 \cdots \hat{m}_i \cdots m_s[E_0] + \mu\, c_1(B).$$

In $H^2(X, \mathbb{Z})$ we have $[E_0] = n[C]$, so $c_1(M/\mu) = k[C]$ with

$$k = (1 + n\beta)bm_1 \cdots m_s + n\Big(\sum\nolimits_i b_i m m_1 \cdots \hat{m}_i \cdots m_s\Big) + \mu q.$$

Since we already fixed $m = \beta = 2$, we need to arrange b, b_i, m_i, q to get $k = 1$. First note that

$$\gcd((1 + n\beta)m_1 \cdots m_s, n m m_1 \cdots \hat{m}_i \cdots m_s) = 1,$$

by the choice of m, m_i. Then there are $\bar{b}, \bar{b}_i \in \mathbb{Z}$ such that

$$1 = (1 + n\beta)\bar{b}m_1 \cdots m_s + n\Big(\sum\nolimits_i \bar{b}_i m m_1 \cdots \hat{m}_i \cdots m_s\Big).$$

Take $\bar{b} = b + mq_0$, $\bar{b}_i = b_i + m_i q_i$, where $0 \le b < m$ and $0 \le b_i < m_i$. Then

$$1 = (1 + n\beta)bm_1 \cdots m_s + n\Big(\sum\nolimits_i b_i m m_1 \cdots \hat{m}_i \cdots m_s\Big) + \Big((1 + n\beta)q_0 + \sum n q_i\Big)\mu,$$

as required. Finally note that $\gcd(b, m) = 1$ and $\gcd(b_i, m_i) = 1$, so that we can take local invariants j, j_i with $bj \equiv 1 \pmod{m}$ and $b_i j_i \equiv 1 \pmod{m_i}$ and apply Proposition 4.

It remains to show that $\pi_1(M) = 1$. By Proposition 7 applied to $\pi_1(\mathbb{F}_n - S)$, $S = D \cup D_1 \cup \ldots \cup D_s$, we get that this group is abelian. Now take the curve E_∞, that intersects transversally S at one point. Take a loop δ around E_∞, then

$\pi_1(\mathbb{F}_n - (S \cup E_\infty))$ is generated by δ and $\pi_1(\mathbb{F}_n - S)$. Let $\gamma, \gamma_1, \dots, \gamma_s$ be loops around each of the divisors $D, D_1, \dots D_s$. Take a general fiber C of $\mathbb{F}_n \to \mathbb{CP}^1$, and intersecting with $S \cup E_\infty$, we get a homotopy $\delta = \gamma^\beta \prod_{i=1}^s \gamma_i$, since $\beta = C \cdot D$ and $1 = C \cdot D_i, i = 1, \dots, s$. Note also that δ, γ commute since the curves E_∞, D intersect transversally at one point. Also γ_i, γ_j commute since D_i, D_j intersect transversally (just take the link of the complement of the two curves around an intersection point, which is a 2-torus containing both loops; hence, they commute in this T^2). Analogously γ_i, γ commute. Finally, $\delta = \gamma^\beta \prod_{j=1}^s \gamma_j$, so it also commutes with γ_i in $\pi_1(\mathbb{F}_n - (S \cup E_\infty))$.

Now recall that $\pi_1^{\mathrm{orb}}(X)$ is the quotient of $\pi_1(X - S) = \pi_1(\mathbb{F}_n - (S \cup E_\infty))$ with the relations $\gamma^m = 1$, $\gamma_i^{m_i} = 1$. Then we have that $\pi_1^{\mathrm{orb}}(X)$ is also abelian. It follows that $\pi_1(M)$ is abelian. Since $H_1(M, \mathbb{Z}) = 0$ we get necessarily $\pi_1(M) = 1$. By Theorem 7, M is a rational homology sphere with $H_2(M, \mathbb{Z}) = \mathbb{Z}_2^{2n}$.

We omit the case $\mathbb{Z}_m^2$, $m < 5$ referring to [39]. □

Remark 1 In [42], a complete classification of Sasaki-Einstein rational homology spheres was obtained. This result is a positive type counterpart to the present work, which deals with the negative case. More results on the positive case can be found in [10].

4 Quasi-regular Negative Sasakian Structures on $\#_k(S^2 \times S^3)$

Theorem 13 *Any $\#_k(S^2 \times S^3)$ admits a quasi-regular negative Sasakian structure.*

This yields a complete answer to Question 3.

The methods of proof of this theorem are close to the proof of Proposition 10, they are based on finding suitable X, the pattern of curves and checking the ampleness of K_X^{orb} (see [39]).

5 Realization of Homological Data by Sasakian Manifolds in Dimension 5

It is an important open problem to describe the class of Smale-Barden manifolds, which admit Sasakian structures (see [5, Chap. 10]). Some necessary conditions for a Smale-Barden manifold to carry a Sasakian structure are known.

Definition 6 Let M be a Smale-Barden manifold. We say that M satisfies the *condition G-K* if the pair $(H_2(M, \mathbb{Z}), i(M))$ written in the form

$$H_2(M, \mathbb{Z}) = \mathbb{Z}^k \oplus \left(\bigoplus_{p,i} \mathbb{Z}_{p^i}^{c(p^i)}\right),$$

where $k = b_2(M)$, satisfies all of the following:

1. $i(M) \in \{0, \infty\}$,
2. for every prime p, $t(p) := \#\{i \mid c(p^i) > 0\} \leq k + 1$,
3. if $i(M) = \infty$, then $t(2) := \#\{i \mid c(2^i) > 0\} \leq k$.

Question 4 ([5, Question 10.2.1]) Suppose that a Smale-Barden manifold satisfies the condition G-K. Does it admit a Sasakian structure?

It is known from [5, Corollary 10.2.11] that the condition G-K is necessary for the existence of K-contact; hence, Sasakian structures. However, it is not known to what extent it is sufficient. Kóllar [30] have found subtle obstructions to the existence of Sasakian structures on Smale-Barden manifolds with $k = 0$. Also, the recent works [17] and [36] showed the existence of *homology Smale-Barden* and true Smale-Barden manifolds, which carry a K-contact but do not carry any *semi-regular* Sasakian structure. Therefore, finding sufficient conditions is an important problem.

As the condition G-K is a bound on the numbers $t(p)$ controlling the size of the torsion part at primes p, it is natural to start by focusing on the case of small values of $t(p)$. Let

$$\mathbf{t}(X) = \max\{t(p) \mid p \text{ is prime}\}.$$

The case of $\mathbf{t} = 0$ is that of the torsion-free Smale-Barden manifolds, where we have only regular Sasakian structures. The next case for analysis is $\mathbf{t} = 1$. In this direction, we prove the following characterization of Sasakian structures with $\mathbf{t} = 1$.

Theorem 14 *Let M be a Smale-Barden manifold, whose second integral homology has the form*

$$H_2(M, \mathbb{Z}) = \mathbb{Z}^k \oplus \left(\bigoplus_{i=1}^{r} \mathbb{Z}_{m_i}^{2g_i}\right),$$

and with $i(M) = 0, \infty$. Assume that $k \geq 1$, $m_i \geq 2, g_i \geq 1$, with m_i pairwise coprime, $1 \leq i \leq r$. Then M admits a semi-regular Sasakian structure.

Therefore, all Smale-Barden manifolds with $k \geq 1$ and $\mathbf{t} = 1$ admit Sasakian structures; moreover, they admit semi-regular Sasakian structures.

6 Null Sasakian Structures

Definition 7 A Sasakian structure is called *null* if $c_1(\mathcal{F}_\xi) = 0$.

It was shown in [5] that if a Smale-Barden manifold admits a null Sasakian structure, it must be homeomorphic to to the connected sum of at most 21 copies of $S^2 \times S^3$. Moreover, it is proved that any $M = \#_k(S^2 \times S^3)$ with $2 \leq k \leq 21$ admits a null Sasakian structure, except, possibly, $b_1(M) = 2$ and $b_2(M) = 17$. The following problem was posed in [5]

Question 5 (Open Problem 10.3.2 in [5]) Find examples of null Sasakian structures on $\#_2(S^2 \times S^3)$ and $\#_{17}(S^2 \times S^3)$, or show that none can exist.

The answer is given in [16].

Theorem 15 ([16], Theorem 1.5) *Any* $M = \#_k(S^2 \times S^3)$ *with* $2 \leq k \leq 21$ *admits a null Sasakian structure.*

The proof of this theorem uses the theory of singular Kähler orbifolds. In the context of null Sasakian structures, one needs to construct a cyclic K3-orbifold with the required restrictions on the second Betti number and the appropriate Seifert bundle.

7 Topology of Sasakian Manifolds

7.1 Boyer-Galicki Program

The following topological obstructions to the existence of Sasakian structures on a compact manifold M of dimension $2n + 1$, where known [5].

- vanishing of the odd Stiefel-Whitney classes,
- the inequality $1 \leq cup(M) \leq 2n$ on the cup-length,
- the evenness of the pth Betti number for p odd with $1 \leq p \leq n$,
- the estimate on the number of closed orbits of the Reeb vector field (there should be at least $n + 1$),

It is well-known that the theory of Sasakian manifolds is, in some sense, parallel to the theory of the Kähler manifolds. There is a deep theorem of Deligne, Griffiths, Morgan and Sullivan on the rational homotopy type of Kähler manifolds [18]. In the same spirit, rational homotopical properties of a manifold are related to the existence of suitable geometric structures on the manifold [21]. Therefore, it is important to build a version of such theory for compact Sasakian manifolds. In [5, Chap. 7], the authors pose the following problems.

1. There are obstructions to the existence of Sasakian structures expressed in terms of Massey products, which depend on basic cohomology classes of the related K-contact structure. Can one obtain a topological characterization of them?
2. Do there exist simply connected K-contact non-Sasakian manifolds (open Problem 7.4.1 in [5])?
3. What finitely presented groups can be realized as fundamental groups of compact Sasakian manifolds?

7.2 Rational Homotopy Properties of Sasakian Manifolds

Following the program formulated by Boyer and Galicki, the papers [3, 4] develop the theory of the rational homotopy type of such manifolds. The well-known theorem of Deligne, Griffiths, Morgan and Sullivan [18] provides formality as an obstruction to the existence of Kähler structure. Since triple Massey products obstruct formality, they also obstruct Kähler structures. Thus, the question about homotopic properties related to formality is natural in the Sasakian framework.

Non-formal and formal Sasakian manifolds.

It was shown in [4] that the triple Massey products do not obstruct Sasakian structures. The examples below show this.

Example 1 *(Non-formal non-simply connected case,* [4]*)* Recall from [15] that the real Heisenberg group H^{2n+1} admits a homogeneous regular Sasakian structure with its standard 1-form $\eta = dz - \sum_{i=1}^{n} y_i dx_i$. As a manifold, H^{2n+1} is just $\mathbb{R}^{2n+1}$, which can be realized in terms of $(n+2) \times (n+2)$ nilpotent matrices of the form

$$A = \begin{pmatrix} 1 & a_1 & \cdots & a_n & c \\ 0 & 1 & 0 & \cdots & b_1 \\ \vdots & & \ddots & \cdots & \vdots \\ 0 & \cdots & 0 & 1 & b_n \\ 0 & \cdots & 0 & 0 & 1 \end{pmatrix}, \tag{12}$$

where $a_i, b_i, c \in \mathbb{R}$, $i = 1, \ldots, n$. Then a global system of coordinates x_i, y_i, z for H^{2n+1} is defined by $x_i(A) = a_i$, $y_i(A) = b_i$, $z(A) = c$. Standard calculations show that we have a basis for the left-invariant 1-forms on H^{2n+1}, which consists of

$$\left\{dx_i\,,\ dy_i\,,\ dz - \sum\nolimits_{i=1}^{n} x_i dy_i\right\}.$$

Consider the discrete subgroup Γ of H^{2n+1} defined by the matrices of the form given in (12) with integer entries. The quotient manifold

$$M := \Gamma \backslash H^{2n+1}$$

is compact. The 1-forms dx_i, dy_i and $dz - \sum_{i=1}^{n} x_i dy_i$ descend to 1-forms α_i, β_i and γ, respectively, on M. We note that $\{\alpha_i, \beta_i, \gamma\}$ is a basis for the 1-forms on M. Let $\{X_i, Y_i, Z\}$ be the basis of vector fields on M that is dual to the basis $\{\alpha_i, \beta_i, \gamma\}$. Define the almost contact metric structure (η, ξ, ϕ, g) on M by

$$\eta = \gamma\,,\quad \xi = Z\,,\quad \phi(X_i) = Y_i\,,\quad \phi(Y_i) = -X_i\,,$$

$$\phi(\xi) = 0\,,\quad g = \gamma^2 + \sum\nolimits_{i=1}^{n}((\alpha_i)^2 + (\beta_i)^2)\,.$$

Then one can check that (η , ξ , ϕ , g) is a regular Sasakian structure on M. Moreover, M is non-formal (this is checked in [4]).

An example of non-simply connected *formal* Sasakian manifold is also constructed in [4]. It is a non-trivial circle bundle over $B^4 = (L_3/\Gamma) \times S^1$, where L_3 is the simply connected non-nilpotent solvable Lie group and Γ is a co-compact lattice. It is checked that M^5 is formal by calculating the model of the constructed Boothby-Wang fibration.

Example 2 *(Non-formal simply connected case)* A series of examples of non-formal simply connected Sasakian manifolds is given by the following result [4].

Theorem 16 *For every $n \geq 3$, there exists a simply connected compact regular Sasakian manifold of dimension $2n+1$, which is non-formal. More precisely, there is a non-trivial 3-sphere bundle over $(S^2)^{n-1}$, which is a non-formal simply connected compact regular Sasakian manifold.*

Vanishing of higher order Massey products of Sasakian manifolds.

Here, we show that the higher Massey products rule out the existence of Sasakian structures on a compact manifold.

Theorem 17 ([4]) *Let M be a compact Sasakian manifold. Then, all the higher order Massey products for M are zero.*

Sketch of proof. Let $(\mathcal{A} = \bigoplus_{i=0}^{2n} A^i, 0)$ be a graded differential algebra with a non-zero element $\omega \in A^2$. For every $0 \leq k \leq n$, define the Lefschetz map

$$L_\omega : A^{n-k} \longrightarrow A^{n+k}, \quad \beta \longmapsto \beta \cdot \omega^{n-k}.$$

We say that $\mathcal{A}$ satisfies the hard Lefschetz property if L_ω is an isomorphism for every $0 \leq k \leq n$.

Proposition 11 *Let $(\mathcal{A} = \bigoplus_{i=0}^{2n} A^i,\ 0)$ be a graded differential commutative algebra, and let $\omega \in A^2$ be a nondegenerate element, such that the hard Lefschetz property with respect to ω holds. Consider the elementary extension $(\mathcal{A} \otimes \bigwedge(y),\ d)$ of $(\mathcal{A},\ 0)$, where $\deg(y) = 1$ and $dy = \omega$. Then the higher order Massey products $\langle a_1, \ldots, a_m \rangle$, where $a_i \in H^*(\mathcal{A} \otimes \bigwedge(y),\ d)$ and $m \geq 4$, all vanish.*

Having Proposition 11, we continue as follows. We know that Massey products on a manifold M can be computed by using any model for M. Now, let $(M, \eta , \xi , \phi , g)$ be a compact Sasakian manifold of dimension $2n+1$. Put $\mathcal{A} = H_B^*(M)$ for the basic cohomology of M. From Theorem 4 we know that $(\mathcal{A} \otimes \bigwedge(x),\ D)$ with $|x| = 1$, $D(\mathcal{A}) = 0$ and $Dx = [d\eta]_B$, is a model for M.

Since $(M, \eta , \xi , \phi , g)$ is a Sasakian manifold, there is the orthogonal decomposition of the tangent bundle TM of M

$$TM = \mathcal{D} \oplus \mathcal{L},$$

where $\mathcal{L}$ is the trivial line subbundle generated by ξ, and the transversal foliation $\mathcal{D}$ is Kähler with respect to $\omega = [d\eta]_B \in H_B^2(M)$. Then, we can apply El Kacimi's basic hard Lefschetz Theorem for transversally Kähler foliations $\mathcal{F}$ on a compact manifold N such that $H_B^{cod\,\mathcal{F}}(N) \neq 0$ (see [20] for a more complete statement of the theorem). This implies that $(\mathcal{A}, 0)$ satisfies the hard Lefschetz property with respect to ω and so, by Proposition 11, all higher Massey products of the model $(\mathcal{A} \otimes \bigwedge(x), D)$ are zero.

7.3 K-Contact Non-Sasakian Manifolds

Now, we answer the question posed in [5]: are there *simply-connected* K-contact manifolds, which do not carry Sasakian structures? Note that this is an analogue of the classical question about the existence of symplectic manifolds that do not admit a Kähler structure (see [47]). It is interesting to note that such manifold do exist in each odd dimension. However, the methods of constructing such manifolds are quite different: in dimensions > 7 the problem is settled by pure rational homotopy theory [4, 26], in dimension 7 one needs a more refined approach, using rational homotopy theory combined with symplectic surgery [24] and Donaldson's hyperplane theorem [1]. The result can be found in [38], and we present the proof in order to show the difference.

Examples in dimensions dimension > 7

Theorem 18 ([4]) *Let M be a simply connected compact symplectic manifold of dimension $2k$ with an integral symplectic form ω. Assume that the quadruple Massey product in $H^*(M)$ is non-zero. There exists a sphere bundle*

$$S^{2m+1} \longrightarrow E \longrightarrow M\,,$$

for $m+1 > k$, such that the total space E is K-contact, but E does not admit any Sasakian structure.

The proof of this theorem is based on Theorem 17. One shows that the non-vanishing quadruple Massey products survive in the algebraic model of E, which rules out Sasakian structure on E. On the other hand, Lerman's theorem [32] ensures the existence of a K-contact structure. By dimensional reasons, we obtain the lower bound for the dimension of the K-contact non-Sasakian manifold $\dim E > 7$. An earlier paper [26] contains the same result proved without using higher order Massey products but based on a cohomological calculation together with the result from [48].

Examples in dimension 7.

The dimension 7 deserves some attention, since the rational homotopy theory still plays an important role. However, it is not sufficient anymore, and one needs to use symplectic surgery and Donaldson's hyperplane theorem in symplectic geometry (which we do not discuss here referring to [1, 24, 25]).

Theorem 19 *There exist 7-dimensional compact simply connected K-contact manifolds, which do not admit a Sasakian structure.*

Let us sketch the proof (we just shorten the proof of [38]).

Proposition 12 *There exists a simply connected 6-dimensional symplectic manifold* (M, ω) *such that* $\dim\ker(L_\omega : H^2(M) \to H^4(M))$ *is odd.*

Sketch of proof of Proposition 12. Gompf constructs in [24, Theorem 7.1] an example of a simply connected 6-dimensional symplectic manifold (M, ω) which does not satisfy the hard Lefschetz property, that is, the Lefschetz map $L_\omega : H^2(M) \to H^4(M)$ is not an isomorphism. If $\dim\ker L_\omega$ is already odd then we have finished.

So let us suppose that $\dim\ker L_\omega$ is even. Take a cohomology class $a \in H^2(M)$, which belongs to the kernel of L_ω. In [14, Lemma 2.4] Cavalcanti proves that given a symplectic manifold (M, ω) as above satisfying that there exists a symplectic surface $S \hookrightarrow M$ with $\langle a, [S]\rangle \neq 0$, then there is another 6-dimensional symplectic manifold (M', ω') (the symplectic blow-up of M along S) satisfying

$$\dim\ker(L_{\omega'} : H^2(M') \to H^2(M')) = \dim\ker(L_\omega : H^2(M) \to H^2(M)) - 1.$$

The symplectic blow-up construction of M along S is known [47]. It is proved that the fundamental groups $\pi_1(M') \cong \pi_1(M)$ (here M' denotes the blown-up manifold). Hence, M' is simply connected. This means that the simply connected 6-dimensional symplectic manifold M' satisfies that

$\dim\ker(L_{\omega'} : H^2(M') \to H^4(M'))$ is odd, as required.

It remains to find $S \hookrightarrow M$ as required. The cohomology class a is non-zero, so there exists $b \in H^4(M, \mathbb{Z})$ such that $a \cup b \neq 0$. It is easy to see that there is a rank 2 complex vector bundle $E \to M$ with $c_1(E) = 0$ and $c_2(E) = 2b$. This corresponds to the fact that the map $[M, B\,\mathrm{SU}(2)] \to H^4(M, \mathbb{Z})$ given by the second Chern class exhausts $2\,H^4(M, \mathbb{Z})$ (we omit the proof of this fact, referring to [38]).

Next, take the rank 2 bundle $E \to M$ just constructed. Assume that $[\omega]$ is a an integral cohomology class (which can always be done by perturbing ω slightly to make it rational and multiplying it by a large integer). Let $L \to M$ be the line bundle with first Chern class $c_1(L) = [\omega]$. We now use the asymptotically holomorphic techniques introduced by Donaldson in [19]. Specifically, the result of [1] guarantees the existence of a suitable large $k \gg 0$ and a section of $E \otimes L^{\otimes k}$, whose zero locus is a symplectic manifold (in fact, an asymptotically holomorphic manifold). This zero locus $S \subset M$ is a symplectic surface, and the cohomology class defined by S is $c_2(E \otimes L^{\otimes k}) = c_2(E) + 2kc_1(L) = 2b + 2k[\omega]$. Therefore, $\langle a, [S]\rangle = \langle a, 2b + 2k[\omega]\rangle = 2\langle a, b\rangle \neq 0$, as required. □

We will call the manifold obtained in Proposition 12 the *Gompf-Cavalcanti manifold*, since it is constructed using the R. Gompf [24] surgery technique together with the symplectic blow-up of G. Cavalcanti [14]. Note, however, that this is not a unique one, but a family of manifolds.

Completion of proof. We show the existence of simply connected compact K-contact non-Sasakian manifolds in dimension 7 by proving that the Boothby-Wang fibration over a Gompf-Cavalcanti manifold is K-contact but non-Sasakian.

Let (M, ω) be a Gompf-Cavalcanti manifold as given by Proposition 12. We can assume that $[\omega]$ is an integral cohomology class. Let

$$S^1 \to E \to M$$

be the associated Boothby-Wang fibration. Now we need to prove that E cannot carry Sasakian structures. There is an exact sequence

$$H_2(M) \to H_1(S^1) = \mathbb{Z} \to H_1(E) \to 0$$

from the Serre spectral sequence. The map $H_2(M) \to \mathbb{Z}$ is cupping with $[\omega] \in H^2(M)$. Using $[\omega]$ integral cohomology class and primitive, yields that $H_2(M) \to \mathbb{Z}$ is surjective; hence, $H_1(E) = 0$. Using the long homotopy exact sequence and the Gysin exact sequence, one obtains that $\pi_1(E)$ is abelian. Therefore, E is simply connected and

$$b^3(E) = b^3(M) + \dim(\ker L_\omega : H^2(M) \to H^4(M)).$$

As M is a 6-manifold, we conclude that $b^3(M)$ is even (by Poincaré duality, the intersection pairing on $H^3(M)$ is an antisymmetric non-degenerate bilinear form; hence, the dimension of $H^3(M)$ is even). By construction, $\dim(\ker L_\omega : H^2(M) \to H^4(M))$ is odd, so $b^3(E)$ is odd. As the third Betti number of a 7-dimensional Sasakian manifold has to be even (Theorem 1), we get that E cannot admit a Sasakian structure. □

Remark 2 In the context of works [3, 4, 38], we should also mention [12, 13], which are devoted to the analogues of the hard Lefschetz property in the Sasakian framework.

7.4 *Dimension 5*

This is the most difficult case. The solution to the problem of the existence of simply connected K-contact non-Sasakian manifolds in dimension 5 was a result of a series of works [17, 34–36] and required further development of the theory of Seifert fibrations over Kähler or symplectic cyclic 4-orbifolds with various types of singularities. This was described in Sect. 3. Therefore, we just formulate the main result obtained by V. Muñoz.

Theorem 20 ([35]) *There exists a Smale-Barden manifold that admits a K-contact structure but does not admit Sasakian structures.*

Acknowledgements This article is an extended version of the talk given at the conference *Differential geometric structures and applications* held at the University of Haifa in May 2023. The author is grateful to the organizers for the pleasant and creative atmosphere. The author was supported by the National Science Center, Poland, grant no. 2018/31/B/ST1/00053.

References

1. Auroux, D.: Asymptotically holomorphic families of symplectic submanifolds, Geom. Funct. Anal. 7 (1997), 971–995
2. Barth, W., Hulek, K., Peters, C. and Van de Ven, A.: *Compact Complex Surfaces*, Springer, 2004
3. Bazzoni, G., Biswas, I., Fernández, M., Muñoz, V. and Tralle, A.: Homotopic properties of Kähler orbifolds, in: S.G. Chiossi et. al (eds.) *Special Metrics and Group Actions in Geometry*, Springer INdAM Series 23, 2017, pp. 23–57
4. Biswas, I., M. Fernández, V. Muñoz, Tralle, A.: On formality of Sasakian manifolds, J. Topology 9 (2016), 161–180
5. Boyer, C., and Galicki, K.: *Sasakian Geometry*, Oxford, 2007
6. Boyer, C., Galicki, K. and Kollár, J.: Einstein metrics on spheres, Ann. Math. 162 (2005), 557–580
7. Boyer, C., Galicki, K. and Matzeu, P.: On η-Einstein Sasakian geometry, Commun. Math. Phys. 262 (2006), 177–208
8. Boyer, C., Galicki, K. and Mann, B.: The geometry and topology of 3-Sasakian manifolds, J. Reine Angew. Math. 455(1994), 183–220
9. Boyer, C., Galicki, K. and Nakamaye, M.: On positive Sasakian geometry, Geom. Dedicata 101 (2003), 93–102
10. Boyer, C. and Nakamaye, M.: On Sasaki-Einstein manifolds in dimension five, Geom. Dedicata 144 (2010), 141–156
11. Boothby, W.M. and Wang, H.C.: On contact manifolds, Annals Math. 68 (1958), 721–734
12. Cappelletti-Montano, B., de Nicola, A. and Yudin, I.: Hard Lefschetz theorem for Sasakian manifolds, J. Diff. Geom. 101 (2015), 47–66
13. Cappelletti-Montano, B., de Nicola, A., Marrero, J.C. and Yudin, I.: Examples of compact K-contact manifolds with no Sasakian metric, Intenat. J. Geom. Methods Phys. 11 (2014), art. no. 1460028
14. Cavalcanti, G.: The Lefschetz property, formality and blowing-up in symplectic geometry, Trans. Amer. Math. Soc. 359 (2007), 333–348
15. Cordero, L.A., Fernández, M. and de León, M.: Examples of compact almost contact manifolds admitting Sasakian and no cosymplectic structures, *Atti Sem. Mat. Fis. Univ. Modena* 34 (1985), 43–54
16. Cañas, A., Muñoz, V., Schütt, M. and Tralle, A.: Quasi-regular Sasakian and K-contact structures on Smale-Barden manifolds, Rev. Matem. Iberoam. 38 (2022), 1029–1050
17. Cañas, A., Muñoz, V., Rojo, J. and Viruel, A.: A K-contact simply connected 5-manifold with no semi-regular Sasakian structure, Publ. Math. 65 (2021), 615–651
18. Deligne, P., Griffiths, P., Morgan, J. and Sullivan, D.: Real homotopy theory of Kähler manifolds, Invent. Math. 29 (1975), 245–274
19. Donaldson, S.: Symplectic submanifolds and almost-complex geometry, J. Diff. Geom. 44 (1996) 666–705

20. El Kacimi Alaoui, A.: Operateurs transversalement elliptiques sur un feuilletage riemannien et applications, Compositio Math. 73 (1990), 57–106
21. Felix, Y., Oprea, J. and Tanré, D.: *Algebraic Models in Geometry*, Oxford Univ. Press, 2008
22. Felix, Y., Thomas, J.-C. and Halperin, S.: *Rational homotopy theory*, Springer, Berlin, 2002
23. Gómez, R.: A note on Smale manifolds and Lorentzian Sasaki-Einstein geometry, Bull. Math. Soc. Sci. Roumainie 59 (2016), 151–158
24. Gompf, R.: A new construction of symplectic manifolds, Annals Math. 142 (1995), 527–595
25. Gompf, R. and Stipsicz, A.: 4-*Manifolds and Kirby Calculus*, AMS, 2004
26. Hajduk, B.and Tralle, A.: On simply connected compact K-contact non-Sasakian manifolds, J. Fixed Point Theory Appl. 16 (2014), 229–241
27. Halperin, S.: Lectures on minimal models, Mém. Soc. Math. France **230**, 1983
28. Hartshorne, R.: *Algebraic Geometry*, Springer, 1977
29. Kobayashi, S.: Principal fibre bundles with the 1-dimensional toroidal group, Tôhoku Math. J. 8 (1956), 29–45
30. Kollár, J.: Einstein metrics on 5-dimensional Seifert bundles, J. Geom. Anal. 15 (2005), 445–476
31. Kollár, J.: Circle actions on simply connected 5-manifolds, Topology 45 (2006), 643–672
32. Lerman, E.: Contact fiber bundles, J. Geom. Phys. 49 (2004), 52–66
33. McDuff, D. and Salamon, D.: *Introduction to Symplectic Topology*, Oxford Univ. Press, 1998
34. Muñoz, V.: Gompf connected sum for orbifolds and K-contact Smale-Barden manifolds, Forum Mathematicum, 34 (2022), 197–223
35. Muñoz, V.: A Smale-Barden manifold admitting K-contact but not Sasakian structure, Arxiv: 2011.05783
36. Muñoz, V., Rojo, J.A. and Tralle, A.: Homology Smale-Barden manifolds with K-contact and Sasakian structures, Internat. Math. Res. Notices IMRN, Vol. 2020, (2020), 7397–7432
37. Muñoz, V. and Tralle, A.: On the classification of Smale-Barden manifolds with Sasakian structures, Comm. Contemporary Math., 24 (2022), art. no. 2150077
38. Muñoz, V. and Tralle, A.: Simply connected K-contact and Sasakian manifolds in dimension 7, Math. Z. 281 (2015), 457–470
39. Muñoz, V., Schütt, M. and Tralle, A.: Negative Sasakian structures on simply connected 5-manifolds, Math. Res. Lett. 29 (2022), 1827–1857
40. Nori, M.: Zariski's conjecture and related problems, Annales Sci. l'E.N.S. 4e série, 16 (1983), 305–344
41. Patra, D.S. and Rovenski, V.: On the rigidity of the Sasakian structure and characterization of cosymplectic manifolds, Differential Geometry and its Applications, 90 (2023) 102043
42. Park, J. and Won, J.: Simply connected Sasaki-Einstein rational homology spheres, Duke Math. J. 170 (2021), 1085–1112
43. Rukimbira, R.: Chern-Hamilton conjecture and K-contactness, Houston J. Math. 21 (1995), 709–718
44. Sasaki, S. and Hatakeyama, Y.: On differentiable manifolds with certain structures which are closely related to almost contact structure, II, Tôhoku Math. J. 13 (1961), 281–294
45. Teicher, M.: Hirzebruch surfaces: degenerations and related braid monodromy, Contemp. Math. 241 (1999), 305–326
46. Tievsky, A.: Analogues of Kähler geometry on Sasakian manifolds, Ph.D. Thesis, MIT, 2008
47. Tralle, A. and Oprea, J.: *Symplectic manifolds with no Kähler structure*, Springer, Berlin, 1997
48. Wang, Z. and Zafran, D.: A remark on the hard Lefschetz theorem for Kähler orbifolds, Proc. Amer. Math. Soc. 137 (2009), 2497–2501

Einstein-Type Metrics and Ricci-Type Solitons on Weak f-K-Contact Manifolds

Vladimir Rovenski

Abstract A weak metric f-structure (f, Q, ξ_i, η^i, g) $(i = 1, \dots, s)$ on a smooth manifold generalizes the metric f-structure, i.e., the linear complex structure on the contact distribution is replaced with a nonsingular skew-symmetric tensor. We study geometry of a weak f-K-contact structure, which is a weak f-contact structure, whose characteristic vector fields are Killing. We show that the distribution $\ker f$ of a weak f-contact manifold defines a $\mathfrak{g}$-foliation with an abelian Lie algebra, characterize weak f-K-contact manifolds among all weak metric f-manifolds by the property known for f-K-contact manifolds, and find when a Riemannian manifold endowed with a set of orthonormal Killing vector fields is a weak f-K-contact manifold. We prove that for $s > 1$, an Einstein weak f-K-contact manifold is Ricci flat and find sufficient conditions for a weak f-K-contact manifold with parallel Ricci tensor or with a generalized gradient Ricci soliton structure to be Ricci flat or a quasi Einstein manifold. We show positive definiteness of the Jacobi operators in the characteristic directions and use this to deform a weak f-K-contact structure to an f-K-contact structure. We define an η-Ricci soliton and η-Einstein structures on a weak metric f-manifold (which for $s = 1$, give the well-known structures on contact metric manifolds) and find sufficient conditions for a compact weak f-K-contact manifold with an η-Ricci soliton structure of constant scalar curvature to be η-Einstein.

Keywords Weak f-K-contact manifold · Killing vector field · Totally geodesic foliation · Einstein-type metric · Ricci-type soliton · ξ-curvature

Mathematics Subject Classifications (2010) 53C15, 53C25, 53D15

V. Rovenski (✉)
Department of Mathematics, University of Haifa, 3498838 Haifa, Israel
e-mail: vrovenski@univ.haifa.ac.il
URL: https://math.haifa.ac.il/ROVENSKI/rovenski.html

© The Author(s), under exclusive license to Springer Nature Switzerland AG 2024

V. Rovenski et al. (eds.), *Differential Geometric Structures and Applications*,
Springer Proceedings in Mathematics & Statistics 440,
https://doi.org/10.1007/978-3-031-50586-7_2

1 Introduction

Contact geometry is of growing interest due to its important role in mechanics in explaining physical phenomena. In addition, many recent articles have been motivated by the question: how interesting Ricci solitons—self-similar solutions of the Ricci flow equation, $\partial g(t)/\partial t = -2\,\mathrm{Ric}_{g(t)}$—can be for geometry of contact metric manifolds. Some of them consider conditions when a contact metric manifold $M(f, \xi, \eta, g)$, equipped with a Ricci-type soliton structure carries a canonical (e.g., constant curvature or Einstein-type) metric, e.g., [16, 17, 20, 21].

The Ricci soliton equation, where V is a vector field and $\lambda \in \mathbb{R}$,

$$(1/2)\,\pounds_V\, g + \mathrm{Ric} = \lambda\, g,$$

generalizes Einstein manifolds, $\mathrm{Ric} = \lambda\, g$. In the case that V is the gradient of a potential function σ one has a gradient Ricci soliton equation $\mathrm{Hess}_\sigma + \mathrm{Ric} = \lambda\, g$. Almost Ricci solitons (when $\lambda \in C^\infty(M)$, see [22]) are a generalization of quasi-Einstein manifolds (see [9]) and Ricci solitons. The following generalized Ricci soliton equation was studied in [21]:

$$(1/2)\,\pounds_V\, g - c_2\,\mathrm{Ric} = \lambda\, g - c_1\, V^\flat \otimes V^\flat, \tag{1}$$

where $c_1, c_2, \lambda \in \mathbb{R}$ and $V^\flat$ is the 1-form dual to the vector field V. If $V = \nabla\sigma$ in (1), then using the definition $\mathrm{Hess}_\sigma = \frac{1}{2}\,\pounds_{\nabla\sigma}\, g$, we get the *generalized gradient Ricci soliton* equation

$$\mathrm{Hess}_\sigma - c_2\,\mathrm{Ric} = \lambda\, g - c_1 d\sigma \otimes d\sigma\,, \tag{2}$$

and the tensor product notation $(d\sigma \otimes d\sigma)(Y, Z) = d\sigma(Y)\, d\sigma(Z)$ is used. Each equation above gives a generalization of Einstein manifolds. For different values of c_1, c_2 and λ, Eq. (1) generalizes Killing equation ($c_1 = c_2 = \lambda = 0$), equation for homotheties ($c_1 = c_2 = 0$), Ricci soliton equation ($c_1 = 0,\ c_2 = -1$), vacuum near-horizon geometry equation ($c_1 = 1,\ c_2 = 1/2$), e.g., [16]. Catino et al. [8] also introduced the notion of an *Einstein-type structure* on a manifold,

$$\beta\,\mathcal{L}_V\, g + \alpha\,\mathrm{Ric} = (\rho\,\mathrm{Scal} + \lambda)\, g + \mu\, V^\flat \otimes V^\flat$$

(with $\alpha, \beta, \mu, \rho \in \mathbb{R}$) unifying various cases studied recently, such as gradient Ricci solitons, and quasi-Einstein manifolds. There are no Einstein metrics on some compact manifolds, which motivated the study of generalizations of such metrics. A contact metric manifold is called *η-Einstein*, if

$$\mathrm{Ric} = \lambda\, g + \nu\,\eta \otimes \eta, \quad \lambda, \nu \in \mathbb{R}.$$

Under certain conditions, a Sasaki-Ricci flow (an odd-dimensional counterpart of the Kähler-Ricci flow) on a compact manifold converges to an η-Einstein metric on a contact metric manifold, see [29]. The notion of η-*Ricci soliton* on a contact metric manifold, see [11], generalizes Ricci solitons and η-Einstein metrics:

$$(1/2)\,\pounds_V\, g + \mathrm{Ric} = \lambda\, g + \nu\,\eta\otimes\eta,\quad \nu\in\mathbb{R}.$$

A metric f-structure on a $(2n+s)$-dimensional smooth manifold is a higher dimensional analog of a contact structure, defined by a (1,1)-tensor f of constant rank $2n$, which satisfies $f^3+f=0$, and orthonormal vector fields $\{\xi_i\}_{1\le i\le s}$ spanning ker f—the $2n$-dimensional characteristic distribution, see [4, 5, 14, 32–34]. Foliations of simple extrinsic geometry, i.e., vanishing second fundamental form of the leaves, appear on manifolds with degenerate differential forms and curvature-like tensors, e.g. [27]. The following problem was posed in [14]: find suitable structures on manifolds, which lead to totally geodesic foliations. The distribution ker f of an f-contact structure defines a totally geodesic foliation. An interesting case occurs when ker f is defined by a homomorphism of a s-dimensional Lie algebra $\mathfrak{g}$ to the Lie algebra of all vector fields on M, i.e., M admits a $\mathfrak{g}$-foliation, see [1]. In the presence of a compatible metric, such $\mathfrak{g}$-foliations are spanned by Killing vector fields, e.g., [7]. For a 1-dimensional Lie algebra, a $\mathfrak{g}$-foliation is generated by a nonvanishing vector field, and we get contact metric manifolds as well as K-contact and Sasakian ones, see [3].

An f-K-contact structure, see [18], i.e. an f-contact structure, whose characteristic vector fields are unit Killing vector fields, can be regarded as intermediate between a metric f-structure and S-structure (the Sasaki structure when $s=1$). The influence of constant length Killing vector fields on the geometry of Riemannian manifolds has been studied by several authors from different points of view, e.g., [2, 12]. The curvature of f-contact and f-K-contact manifolds was studied in [6, 30, 31]. Observe that there are no Einstein metrics on f-K-contact manifolds.

The study of "weak" f-contact structures on a smooth $(2n+s)$-dimensional manifold (i.e., the linear complex structure on the characteristic distribution is replaced with a nonsingular skew-symmetric tensor) was initiated in [24, 25, 28]. This generalizes a metric f-structure and its satellites, and allows a new look at the classical theory and new applications. In [24], we proved that the $\mathcal{S}$-structure is rigid, i.e., our weak $\mathcal{S}$-structure is the $\mathcal{S}$-structure, and a weak metric f-structure with parallel tensor f is a weak $\mathcal{C}$-structure. In [25], we obtained a topological obstruction (including the Adams number) to the existence of weak f-K-contact manifolds. In [28], we retracted weak structures with positive partial Ricci curvature onto the subspace of classical structures of the same type.

The article continues our study [24, 25, 28] of the geometry of weak f-contact manifolds. Following the approach in [23] for $s=1$, we study the sectional and Ricci curvature in the ξ-directions of a weak f-contact structure and its particular case—a weak f-K-contact structure. The characteristic distribution of a weak f-contact manifold defines a totally geodesic foliation with flat leaves. Our goal is to show that the weak structures can be a useful by studying unit Killing vector fields, totally geodesic foliations and ξ-sectional curvature on Riemannian manifolds, and that some results on f-K-contact manifolds can be extended to the case of weak f-K-contact manifolds. For example, we answer the questions of when a weak f-K-contact manifold (i) carries a generalized Ricci soliton structure or just a gradient quasi Einstein (e.g., an Einstein) metric, and (ii) carries an η-Ricci soliton structure or just an η-Einstein metric. The proofs use the properties of new tensors, as well as the constructions required in the classical case.

The article is organized as follows. Section 2, following the introductory Sect. 1, presents basics of weak metric f-manifolds. Section 3 presents basics of weak f-contact and weak f-K-contact manifolds and shows that these manifolds are endowed with totally geodesic foliations. We show that $\ker f$ of a weak f-contact manifold defines a $\mathfrak{g}$-foliation with an abelian Lie algebra. In Sect. 4, we characterize (Theorem 1) weak f-K-contact manifolds among all weak f-contact manifolds by the property $f = -\nabla\xi_i$ $(i \le s)$ known for f-K-contact manifolds, and find sufficient conditions (Theorem 2) under which when a Riemannian manifold endowed with a set of unit Killing vector fields is a weak f-K-contact manifold. In Sect. 5, for a weak f-K-contact manifold, we calculate the Ricci curvature in the ξ-directions and prove that the ξ-sectional curvature is positive (Proposition 6). This allows us to deform a weak f-K-contact structure to the f-K-contact structure (Theorem 4). We show that for $s > 1$, an Einstein weak f-K-contact manifold is Ricci flat and find (Theorem 3) sufficient conditions for a weak f-K-contact manifold with parallel Ricci tensor to be a Ricci flat manifold, or, for $s = 1$, an Einstein manifold. In Sect. 6, we find (Theorems 5 and 6) sufficient conditions for a weak f-K-contact manifold with a generalized gradient Ricci soliton to be a quasi Einstein or Ricci flat manifold, or, for $s = 1$, an Einstein manifold. In Sect. 7, we define an η-Ricci soliton structure and find sufficient conditions for a compact weak f-K-contact manifold with such a structure of constant scalar curvature to be η-Einstein (Theorem 7).

2 Preliminaries

Here, we survey the basics of a weak metric f-structure (see [24, 28]) as a higher dimensional analog of a weak almost contact structure (see [26]). We define weak f-contact and weak f-K-contact structures and show that these structures are endowed with totally geodesic foliations.

A *framed weak f-structure* on a smooth manifold M^{2n+s} is a set (f, Q, ξ_i, η^i), where f is a $(1,1)$-tensor, Q is a nonsingular $(1,1)$-tensor, ξ_i $(1 \le i \le s)$ are characteristic vector fields and η^i are dual 1-forms, satisfying

$$f^2 = -Q + \sum\nolimits_i \eta^i \otimes \xi_i, \quad \eta^i(\xi_j) = \delta^i_j, \quad Q\,\xi_i = \xi_i. \tag{3}$$

The forms η^i determine a $2n$-dimensional contact distribution $\mathcal{D} := \bigcap_i \ker \eta^i$. Assume that $\mathcal{D}$ is f-invariant (that is true for framed f-structure [4, 34], where $Q = \mathrm{id}_{TM}$),

$$fX \in \mathcal{D} \quad (X \in \mathcal{D}). \tag{4}$$

By (3) and (4), the distribution $\mathcal{D}$ is invariant for Q: $Q(\mathcal{D}) = \mathcal{D}$. Note that $f^3 + fQ = 0$. For a framed weak f-structure on a manifold M, the tensor f has rank $2n$, thus $\mathcal{D} = f(TM)$, and

$$f\,\xi_i = 0, \quad \eta^i \circ f = 0, \quad \eta^i \circ Q = \eta^i, \quad [Q,\, f] = 0.$$

A framed weak f-structure (f, Q, ξ_i, η^i) is called *normal* if the following tensor is zero:

$$N^{(1)}(X, Y) = [f, f](X, Y) + 2\sum\nolimits_i d\eta^i(X, Y)\,\xi_i, \quad X, Y \in \mathfrak{X}_M,$$

where the Nijenhuis torsion $[f, f]$ of f is given by

$$[f, f](X, Y) = f^2[X, Y] + [fX, fY] - f[fX, Y] - f[X, fY], \quad X, Y \in \mathfrak{X}_M, \tag{5}$$

and the exterior derivative of a differential form η^i is given by

$$d\eta^i(X, Y) = \frac{1}{2}\,\{X(\eta^i(Y)) - Y(\eta^i(X)) - \eta^i([X, Y])\}, \quad X, Y \in \mathfrak{X}_M. \tag{6}$$

Recall the following formulas with the Lie derivative $\pounds_Z$ in the Z-direction:

$$\begin{aligned}(\pounds_Z f)X &= [Z, fX] - f[Z, X] \quad (X, Y \in \mathfrak{X}_M),\\ (\pounds_{\xi_i} g)(X, Y) &= \xi_i(g(X, Y)) - g([\xi_i, X], Y) - g(X, [\xi_i, Y]).\end{aligned} \tag{7}$$

The following tensors $N^{(2)}_i$, $N^{(3)}_i$ and $N^{(4)}_{ij}$ are well known in the classical theory:

$$\begin{aligned}N^{(2)}_i(X, Y) &= (\pounds_{fX}\,\eta^i)(Y) - (\pounds_{fY}\,\eta^i)(X) \overset{(6)}{=} 2\,d\eta^i(fX, Y) - 2\,d\eta^i(fY, X),\\ N^{(3)}_i(X) &= (\pounds_{\xi_i} f)X = [\xi_i, fX] - f[\xi_i, X],\\ N^{(4)}_{ij}(X) &= (\pounds_{\xi_i}\,\eta^j)(X) = \xi_i(\eta^j(X)) - \eta^j([\xi_i, X]) \overset{(6)}{=} 2\,d\eta^j(\xi_i, X),\end{aligned}$$

where $\pounds_Z$ is the Lie derivative in the Z-direction, see [4, 34]. For $s = 1$, the above tensors reduce to the following tensors on (weak) almost contact manifolds:

$$N^{(2)}(X, Y) - (\pounds_{fX}\,\eta)Y - (\pounds_{fY}\,\eta)X, \quad N^{(3)} = \pounds_\xi\, f, \quad N^{(4)} = \pounds_\xi\, \eta.$$

Remark 1 *(see* [23]*)* Let $M^{2n+s}(f, Q, \xi_i, \eta^i)$ be a framed weak f-manifold. Consider the product manifold $\bar M = M^{2n+s} \times \mathbb{R}^s$, where $\mathbb{R}^s$ is a Euclidean space with a basis $\partial_1, \ldots, \partial_s$, and define tensors $\bar f$ and $\bar Q$ on $\bar M$ putting

$$\bar{f}(X, \sum_i a^i\partial_i) = (fX - \sum_i a^i\xi_i, \sum_j \eta^j(X)\partial_j),$$
$$\bar{Q}(X, \sum_i a^i\partial_i) = (QX, \sum_i a^i\partial_i),$$

where $a_i \in C^\infty(M)$. Hence, $\bar{f}(X,0) = (fX,0)$, $\bar{Q}(X,0) = (QX,0)$ for $X \in \mathcal{D}$, $\bar{f}(\xi_i,0) = (0,\partial_i)$, $\bar{Q}(\xi_i,0) = (\xi_i,0)$ and $\bar{f}(0,\partial_i) = (-\xi_i,0)$, $\bar{Q}(0,\partial_i) = (0,\partial_i)$. Then it is easy to verify that $\bar{f}^2 = -\bar{Q}$. The tensors $N^{(1)}, N_i^{(2)}, N_i^{(3)}, N_{ij}^{(4)}$ appear when we derive the integrability condition $[\bar{f},\bar{f}] = 0$ of $\bar{f}$ and express the normality condition $N^{(1)} = 0$ of a framed weak f-structure on M.

If there is a Riemannian metric g on M such that

$$g(fX, fY) = g(X, QY) - \sum_i \eta^i(X)\,\eta^i(Y), \quad X, Y \in \mathfrak{X}_M, \tag{8}$$

then (f, Q, ξ_i, η^i, g) is called a *weak metric f-structure* on M, and g is called a *compatible* metric. Note that $X = X^\top + X^\perp$, where $X^\top = \sum_i \eta^i(X)\,\xi_i$ is the projection of the vector $X \in TM$ on $\ker f$. A framed weak f-manifold $M(f, Q, \xi_i, \eta^i)$ endowed with a compatible Riemannian metric g is called a *metric weak f-manifold* and is denoted by $M(f, Q, \xi_i, \eta^i, g)$. Putting $Y = \xi_i$ in (8) and using $Q\,\xi_i = \xi_i$, we get, as in the classical theory, $\eta^i(X) = g(X, \xi_i)$. In particular, ξ_i is g-orthogonal to $\mathcal{D}$ for any compatible metric g. Thus, the orthogonal complement $\mathcal{D}^\perp$ of $\mathcal{D}$ is the characteristic (or, Reeb) distribution $\ker f$. Some sufficient conditions for the existence of a compatible metric on a weak almost contact manifold are given in [28].

Using the Levi-Civita connection ∇ of g, one can rewrite (5) as

$$[f, f](X, Y) = (f\nabla_Y f - \nabla_{fY} f)X - (f\nabla_X f - \nabla_{fX} f)Y. \tag{9}$$

For a weak metric f-structure, the tensor f is skew-symmetric and Q is positive definite and self-adjoint, see [24, 28]:

$$g(fX, Y) = -g(X, fY), \quad g(QX, Y) = g(X, QY), \quad X, Y \in \mathfrak{X}_M.$$

The fundamental 2-form Φ on $M(f, Q, \xi_i, \eta^i, g)$ is defined by

$$\Phi(X, Y) = g(X, fY), \quad X, Y \in \mathfrak{X}_M.$$

Recall the co-boundary formula for exterior derivative d on a 2-form Φ,

$$d\Phi(X, Y, Z) = \frac{1}{3}\big\{X\,\Phi(Y, Z) + Y\,\Phi(Z, X) + Z\,\Phi(X, Y) - \Phi([X, Y], Z) - \Phi([Z, X], Y) - \Phi([Y, Z], X)\big\}. \tag{10}$$

3 Weak f-Contact and Weak f-K-Contact Structures

Here, we define "weak" structures that generalize f-contact and f-K-contact structures.

Definition 1 A *weak f-contact structure* is a weak metric f-structure satisfying

$$\Phi = d\eta^1 = \ldots = d\eta^s, \tag{11}$$

thus, $d\Phi = 0$. A weak f-contact structure is called a *weak f-K-contact structure* if all characteristic vector fields ξ_i are Killing, i.e.

$$\begin{aligned}(\pounds_{\xi_i} g)(X, Y) &:= \xi_i(g(X, Y)) - g([\xi_i, X], Y) - g(X, [\xi_i, Y]) \\ &= g(\nabla_X \xi_i, Y) + g(\nabla_Y \xi_i, X) = 0.\end{aligned} \tag{12}$$

For a weak f-contact structure and all i and j, by (11), we get

$$d\eta^j(\xi_i, X) = \Phi(\xi_i, X) = 0.$$

For a weak f-contact structure, the distribution $\mathcal{D}$ is non-integrable (has no integral $2n$-dimensional submanifolds) since for any nonzero $X \in \mathcal{D}$ we get

$$g([X, fX], \xi_i) = 2\, d\eta^i(fX, X) = g(fX, fX) > 0.$$

The relationships between the different classes of weak manifolds (considered in this article) is summarized in the diagram well known for classical structures:

$$\left|\begin{array}{c}\text{metric weak}\\ f\text{-manifold}\end{array}\right| \xrightarrow{\Phi=d\eta^i} \left|\begin{array}{c}\text{weak}\\ f\text{-contact}\end{array}\right| \xrightarrow{\xi_i\ \text{-Killing}} \left|\begin{array}{c}\text{weak}\\ f\text{-K}-\text{contact}\end{array}\right| \xrightarrow{N^{(1)}=0} \left|\begin{array}{c}\text{weak}\\ \mathcal{S}\text{-manifold}\end{array}\right|.$$

For $s = 1$, we get the following diagram:

$$\left|\begin{array}{c}\text{weak almost}\\ \text{contact metric}\\ \text{manifold}\end{array}\right| \xrightarrow{\Phi=d\eta} \left|\begin{array}{c}\text{weak contact}\\ \text{metric}\\ \text{manifold}\end{array}\right| \xrightarrow{\xi\ \text{-Killing}} \left|\begin{array}{c}\text{weak}\\ \text{K}-\text{contact}\\ \text{manifold}\end{array}\right| \xrightarrow{N^{(1)}=0} \left|\begin{array}{c}\text{weak}\\ \text{Sasakian}\\ \text{manifold}\end{array}\right|.$$

Remark 2 In [13, 24, 30], the f-contact structure is called the almost $\mathcal{S}$-structure, and the normal f-contact structure is called the $\mathcal{S}$*-structure*. A weak metric f-structure with the properties $d\Phi = 0$ and $N^{(1)} = 0$ is called a weak $\mathcal{K}$-structure; in this case, $\xi_1, \ldots, \xi_s$ are Killing vector fields, see [24]; thus, we get a weak f-K-contact structure.

Proposition 1 (see Theorem 2.2 in [24]) *For a weak f-contact structure, the tensors* $N_i^{(2)}$ *and* $N_{ij}^{(4)}$ *vanish; moreover,* $N_i^{(3)}$ *vanishes if and only if* ξ_i *is a Killing vector field. In particular, for a weak f-K-contact structure, the tensors* $N_i^{(2)}$, $N_i^{(3)}$ *and* $N_{ij}^{(4)}$ *vanish.*

A "small" (1,1)-tensor $\widetilde{Q} = Q - \mathrm{id}_{TM}$ measures the difference between a weak f-contact structure and a contact one, and $\widetilde{Q} = 0$ means the classical f-contact geometry. We have

$$[\widetilde{Q}, f] = 0, \quad \widetilde{Q}\,\xi_i = 0, \quad \eta^i \circ \widetilde{Q} = 0.$$

Proposition 2 (see Corollary 2.1 in [24]) *For a weak f-contact structure we get*

$$\begin{aligned} 2\,g((\nabla_X f)Y, Z) &= g(N^{(1)}(Y, Z), fX) + 2\,g(fX, fY)\,\bar\eta(Z) - 2\,g(fX, fZ)\,\bar\eta(Y) \\ &\quad + N^{(5)}(X, Y, Z), \end{aligned} \tag{13}$$

where $\bar\eta = \sum_i \eta^i$, and a skew-symmetric w.r.t. Y and Z tensor $N^{(5)}(X, Y, Z)$ is given by

$$\begin{aligned} N^{(5)}(X, Y, Z) &= (fZ)\,(g(X, \widetilde{Q}Y)) - (fY)\,(g(X, \widetilde{Q}Z)) + g([X, fZ], \widetilde{Q}Y) \\ &\quad - g([X, fY], \widetilde{Q}Z) + g([Y, fZ] - [Z, fY] - f[Y, Z], \widetilde{Q}X). \end{aligned}$$

Taking $X = \xi_i$ in (13), we get

$$2\,g((\nabla_{\xi_i} f)Y, Z) = N^{(5)}(\xi_i, Y, Z), \quad 1 \le i \le s. \tag{14}$$

Recall that a distribution $\widetilde{\mathcal{D}} \subset TM$ is *totally geodesic* if and only if $\nabla_X Y + \nabla_Y X \in \widetilde{\mathcal{D}}$ for any vector fields $X, Y \in \widetilde{\mathcal{D}}$ – this is the case when any geodesic of M that is tangent to $\widetilde{\mathcal{D}}$ at one point is tangent to $\widetilde{\mathcal{D}}$ at all its points, e.g., [27, Sect. 1.3.1]. An integrable totally geodesic distribution determines a totally geodesic foliation.

Let $\mathfrak{g}$ be a Lie algebra of dimension s. We say that a foliation $\mathcal{F}$ of dimension s on a smooth connected manifold M is a $\mathfrak{g}$-*foliation* if there exist complete vector fields $\xi_1, \ldots, \xi_s$ on M, which restricted to each leaf of $\mathcal{F}$ form a parallelism of this submanifold which is Lie algebra isomorphic to $\mathfrak{g}$, see, for example, [28]. We show that the distribution $\ker f$ of a weak f-contact manifold defines a $\mathfrak{g}$-foliation with an abelian Lie algebra $\mathfrak{g}$.

Corollary 1 *For a weak f-contact structure, the distribution* $\ker f$ *is integrable and defines a $\mathfrak{g}$-foliation with totally geodesic flat leaves.*

Proof Recall that $[\xi_i, \xi_j] = \nabla_{\xi_i} \xi_j - \nabla_{\xi_j} \xi_i$. Taking $Y = \xi_j$ in (14), we get

$$\nabla_{\xi_i} \xi_j = 0, \qquad 1 \le i, j \le s. \tag{15}$$

By (15), $\ker f$ of a weak f-contact structure defines a $\mathfrak{g}$-foliation with an abelian Lie algebra (that is $[\xi_i, \xi_j] = 0$) with totally geodesic (that is $\nabla_{\xi_i} \xi_j = 0$) flat leaves (that is $R_{\xi_i, \xi_j}\, \xi_k = 0$). □

Remark 3 Only one new tensor $N^{(5)}$ (vanishing at $\widetilde{Q} = 0$), which supplements the sequence of well-known tensors $N^{(i)}$, $i = 1, 2, 3, 4$, is needed to study the weak f-contact structure. For particular values of the tensor $N^{(5)}$ we get

$$
\begin{aligned}
N^{(5)}(X,\xi_i,Z) &= -N^{(5)}(X,Z,\xi_i) = g(N_i^{(3)}(Z),\ \widetilde{Q}X),\\
N^{(5)}(\xi_i,Y,Z) &= g([\xi_i, fZ], \widetilde{Q}Y) - g([\xi_i, fY], \widetilde{Q}Z),\\
N^{(5)}(\xi_i,\xi_j,Y) &= N^{(5)}(\xi_i,Y,\xi_j) = 0. \qquad (16)
\end{aligned}
$$

Definition 2 ([24]) Framed weak f-structures (f, Q, ξ_i, η^i) and (f', Q', ξ_i, η^i) on a smooth manifold M are said to be *homothetic* if the following is valid for some real $\lambda > 0$:

$$
f = \sqrt{\lambda}\, f', \quad Q\,|_{\mathcal{D}} = \lambda\, Q'|_{\mathcal{D}}. \qquad (17)
$$

Two weak metric f-structures (f, Q, ξ_i, η^i, g) and $(f', Q', \xi_i, \eta^i, g')$ on M are said to be *homothetic* if they satisfy conditions (17) and

$$
g|_{\mathcal{D}} = \lambda^{-\frac{1}{2}}\, g'|_{\mathcal{D}}, \quad g(\xi_i, \cdot) = g'(\xi_i, \cdot). \qquad (18)
$$

Lemma 1 (see [24]) *Let (f, Q, ξ_i, η^i) be a framed weak f-structure such that $Q\,|_{\mathcal{D}} = \lambda\,\mathrm{id}_{\mathcal{D}}$ for some real $\lambda > 0$. Then the following is true:*
- *(f', ξ_i, η^i) is a framed f-contact structure, where f' is given by $(17)_1$.*
- *If (f, Q, ξ_i, η^i, g) is a weak f-contact structure and $(17)_1$ and (18) are valid, then (f', ξ_i, η^i, g') is an f-contact structure.*

4 Killing Vector Fields of Unit Length

Here, we characterize weak f-K-contact manifolds among all weak f-contact manifolds and find conditions under which a Riemannian manifold endowed with a set of unit Killing vector fields is a weak f-K-contact manifold.

The tensor $N_i^{(3)}$ is important for f-contact manifolds. Therefore, we define for a weak f-contact manifold the tensor field frame $h = (h_1, \ldots, h_s)$, where

$$
h_i = (1/2)\, N_i^{(3)} = (1/2)\, \pounds_{\xi_i} f.
$$

By Proposition 1, $h_i = 0$ if and only if ξ_i is a Killing vector field. First, we calculate

$$
\begin{aligned}
(\pounds_{\xi_i} f)X &\overset{(7)}{=} \nabla_{\xi_i}(fX) - \nabla_{fX}\,\xi_i - f(\nabla_{\xi_i} X - \nabla_X\,\xi_i)\\
&= (\nabla_{\xi_i} f)X - \nabla_{fX}\,\xi_i + f\nabla_X\,\xi_i. \qquad (19)
\end{aligned}
$$

Taking $X = \xi_i$ in (19) and using $\nabla_{\xi_i}\xi_j = 0$, see (15), and $(\nabla_{\xi_i} f)\,\xi_j = 0$, see (14) with $Y = \xi_j$, we get the equality $h_i\,\xi_j = 0$. From (19), using $g(\nabla_X\,\xi_i, \xi_j) = 0$ and $g((\nabla_{\xi_i} f)\,\xi_j, Z) = \frac{1}{2} N^{(5)}(\xi_i, \xi_j, Z) = 0$, see [24], we conclude that $\eta_j \circ h_i = 0$.

For an f-contact structure, the linear operator h_i is self-adjoint, trace-free and anti-commutes with f, i.e., $h_i f + f h_i = 0$, see [13]. We generalize this result for a weak f-contact structure.

Proposition 3 (see [24, 25]) *For a weak f-contact structure (f, Q, ξ_i, η^i, g), the tensor h_i and its conjugate tensor h_i^* satisfy*

$$\begin{aligned} g((h_i - h_i^*)X, Y) &= (1/2)\, N^{(5)}(\xi_i, X, Y), \quad X, Y \in TM, \\ h_i f + f\, h_i &= -(1/2)\, \pounds_{\xi_i} \widetilde{Q}, \\ h_i Q - Q\, h_i &= (1/2)[\, f,\, \pounds_{\xi_i} \widetilde{Q}\,]. \end{aligned}$$

The following corollary generalizes the known property of f-contact manifolds.

Corollary 2 *Let a weak f-contact manifold satisfy $\pounds_{\xi_i} \widetilde{Q} = 0$, then*

$$\operatorname{trace} h_i = 0.$$

Proof If $h_i X = \lambda X$, then using $h_i f = -f h_i$ (by the assumptions and Proposition 3), we get $h_i f X = -\lambda f X$. Thus, if λ is an eigenvalue of h_i, then $-\lambda$ is also an eigenvalue of h_i; hence, $\operatorname{trace} h_i = 0$. □

Lemma 2 *On a weak f-K-contact manifold $M^{2n+s}(f, Q, \xi_i, \eta^i, g)$, we obtain $N^{(1)}(\xi_i, \cdot) = 0$ and*

$$N^{(5)}(\xi_i, \cdot, \cdot) = N^{(5)}(\cdot, \xi_i, \cdot) = 0, \tag{20}$$

$$\pounds_{\xi_i} \widetilde{Q} = \nabla_{\xi_i} \widetilde{Q} = 0, \tag{21}$$

$$\nabla_{\xi_i} f = 0. \tag{22}$$

Proof From (9) with $Y = \xi_i$, since $f\, \xi_i = 0$, we get

$$[f, f](X, \xi_i) = f(\nabla_{\xi_i} f)X + \nabla_{fX}\, \xi_i - f\, \nabla_X\, \xi_i, \quad X \in \mathfrak{X}_M. \tag{23}$$

By (23) and $d\eta^j(\xi_i, \cdot) = \Phi(\xi_i, \cdot) = 0$ we have

$$N^{(1)}(\xi_i, X) = [f, f](X, \xi_i) = f^2[X, \xi_i] - f[fX, \xi_i] = f N_i^{(3)}(X) = 0.$$

By $(16)_1$, using $N_i^{(3)} = 0$ we get $N^{(5)}(\cdot, \xi_i, \cdot) = N^{(5)}(\cdot, \cdot, \xi_i) = 0$. Then, by Proposition 3 with $h_i := \frac{1}{2} N_i^{(3)} = 0$, we get $N^{(5)}(\xi_i, \cdot, \cdot) = 0$, $\pounds_{\xi_i} Q = 0$ and

$$g(Q\, \nabla_X\, \xi_i, Z) = g(fZ, QX) - (1/2)\, N^{(5)}(X, \xi_i, fZ) = g(fZ, QX).$$

We use $[f, Q] = 0$ and $Q = \widetilde{Q} + \mathrm{id}_{TM}$ to obtain $\nabla_{\xi_i} \widetilde{Q} = 0$:

$$(\pounds_{\xi_i} Q)X = [\xi_i, QX] - Q[\xi_i, X] = (\nabla_{\xi_i} Q)X + [f, Q]X = (\nabla_{\xi_i} Q)X = (\nabla_{\xi_i} \widetilde{Q})X.$$

This completes the proof of (20) and (21). From (14) and (20) we get (22). □

In the next theorem, we characterize weak f-K-contact manifolds among all weak f-contact manifolds by the following well known property of f-K-contact manifolds, see [4, 18]:

$$\nabla\,\xi_i = -f, \quad 1 \le i \le s. \tag{24}$$

Theorem 1 *A weak f-contact manifold is weak f-K-contact if and only if (24) holds.*

Proof Let a weak f-contact manifold satisfy (24). Due to the skew symmetry of f, we obtain

$$(\pounds_{\xi_i}\, g)(X, Y) = g(\nabla_X\,\xi_i, Y) + g(\nabla_Y\,\xi_i, X) = -g(fX, Y) - g(fY, X) = 0,$$

thus, ξ_i are Killing vector fields.

Conversely, let our manifold be weak f-K-contact. By (13) with $Y = \xi_i$, using (20), we get

$$g((\nabla_X\, f)\,\xi_i, Z) = g(f^2 X, Z).$$

Hence, $(\nabla_X\, f)\,\xi_i = f^2 X$. From this and

$$0 = \nabla_X\,(f\,\xi_i) = (\nabla_X\, f)\,\xi_i + f\,\nabla_X\,\xi_i,$$

we obtain $f(\nabla_X\,\xi_i + fX) = 0$. Since $\nabla_X\,\xi_i + fX \in \mathcal{D}$ and f is invertible when restricted on $\mathcal{D}$, we get $\nabla_X\,\xi_i = -fX$ that proves (24). □

Corollary 3 (see [23]) *A weak contact manifold* $M^{2n+1}(f, Q, \xi, \eta, g)$ *is weak K-contact if and only if* $\nabla\,\xi = -f$ *holds.*

Denote by $R_{X,Y}Z = (\nabla_X\nabla_Y - \nabla_Y\nabla_X - \nabla_{[X,Y]})Z$ the curvature tensor. The mapping $R_{\xi_i} : X \mapsto R_{X,\,\xi_i}\,\xi_i$ ($\xi \in \ker f,\ \|\xi\| = 1$) will be called the *Jacobi operator* in the ξ_i-direction, e.g., [27]. For a weak f-contact manifold, by Proposition 2, we get $R_{\xi_i}(X) \in \mathcal{D}$. Recall that a Riemannian manifold with a unit Killing vector field and the property $R_\xi(X) = X$ ($X \perp \xi$) is a K-contact manifold, e.g., [34, p. 268].

We generalize this result and [23, Theorem 2] in the following.

Theorem 2 *A Riemannian manifold* (M^{2n+s}, g) *with orthonormal Killing vector fields* ξ_i ($i = 1, \ldots, s$) *such that* $d\,\eta^1 = \ldots = d\,\eta^s$ *(where* η^i *is the 1-form dual to* ξ_i*) and the Jacobi operators* R_{ξ_i} ($i \le s$) *are positive definite on the distribution* $\mathcal{D} = \bigcap_i \ker \eta^i$ *is a weak f-K-contact manifold with the following structural tensors:*

$$f = -\nabla\,\xi_i, \quad QX = R_{\xi_i}(X) \quad (X \in \mathcal{D}, \ \ i = 1, \ldots, s).$$

Proof Put $f = -P\nabla\,\xi_i$ and $\Phi(X, Y) = g(X, fY)$, where $P : TM \to \mathcal{D}$ is the orthoprojector, that does no depend on i. Since ξ_i are Killing vector fields, we obtain the property (11):

$$d\eta^i(X, Y) = (1/2)\,(g(\nabla_X\,\xi_i, Y) - g(\nabla_Y\,\xi_i, X)) = -g(\nabla_Y\,\xi_i, X) = g(X, fY).$$

Set $QX = R_{\xi_i}(X)$ for some i and all $X \in \mathcal{D}$. Since ξ_i is a unit Killing vector field, we get $\nabla_{\xi_i}\xi_i = 0$ and $\nabla_X\nabla_Y\,\xi_i - \nabla_{\nabla_X Y}\,\xi_i = R_{X,\xi_i}Y$, see [34]. Thus, $f\,\xi_i = 0$ and

$$f^2X = \nabla_{\nabla_X \xi_i} \xi_i = R_{\xi_i, X}\, \xi_i = -R_{\xi_i}(X) = -QX, \quad X \in \mathcal{D}.$$

By the conditions, the Jacobi operator R_{ξ_i} does no depend on i and the tensor Q is positive definite on $\mathcal{D}$. Thus, f restricted to $\mathcal{D}$ has constant rank $2n$. Put $Q\,\xi_i = \xi_i$ $(i = 1, \ldots, s)$. Therefore, (3) and (8) are valid. □

Corollary 4 (see [23]) *A Riemannian manifold (M^{2n+1}, g) with a unit Killing vector field ξ such that R_ξ is positive definite on $\mathcal{D} = \{X \in TM : g(X, \xi) = 0\}$ is a weak K-contact manifold with the structural tensors $\eta(X) = g(X, \xi)$, $f = -\nabla\,\xi$ and $QX = R_\xi(X)$ for $X \in \mathcal{D}$.*

If a plane contains unit vectors: $\xi \in \ker f$ and $X \in \mathcal{D}$, then its sectional curvature is called *ξ-sectional curvature*. The ξ-sectional curvature of an f-contact manifold is an example of *mixed sectional curvature* of an almost product manifold, for example, [27]. It is well known that the ξ-sectional curvature of an f-K-contact manifold is constant equal to 1.

Proposition 4 *A weak f-K-contact structure (f, Q, ξ_i, η^i, g) with constant mixed sectional curvature, $K(\xi_i, X) = \lambda > 0$ for some $\lambda = const \in \mathbb{R}$ and all $X \in \mathcal{D}$, is homothetic to an f-K-contact structure (f', ξ_i, η^i, g') after the transformation (17)–(18) with $Q' = \mathrm{id}$.*

Proof Note that $K(\xi_i, X) = \lambda$ $(X \in \mathcal{D})$ if and only if $R_{\xi_i}(X) = \lambda X$ $(X \in \mathcal{D})$. By $QX = R_{\xi_i}(X)$ $(X \in \mathcal{D})$, see Theorem 2, we get $QX = \lambda X$ $(X \in \mathcal{D})$. By Lemma 1(ii), (f', ξ_i, η^i, g') is an f-contact structure. Using (12), we get $(\pounds_{\xi_i} g')(\xi_i, \cdot) = 0$ and

$$(\pounds_{\xi_i} g')(X, Y) = \lambda(\pounds_{\xi_i} g)(X, Y), \quad X, Y \in \mathcal{D}.$$

By $\pounds_{\xi_i} g = 0$, we get $\pounds_{\xi_i} g' = 0$; so (f', ξ_i, η^i, g') is an f-K-contact structure. □

Example 1 By Theorem 2, we can search for examples of weak f-K-contact (not f-K-contact) manifolds among Riemannian manifolds of positive sectional curvature that admit a set of $s \ge 1$ orthonormal Killing vector fields. Such example with $s = 1$ is given in [23] using idea of [12, p. 5]. Indeed, let M be a convex hypersurface (ellipsoid) with induced metric g of $\mathbb{R}^{2n+2}$,

$$M = \Big\{(u_1, \ldots, u_{2n+2}) \in \mathbb{R}^{2n+2} : \sum\nolimits_{i=1}^{n+1} u_i^2 + a \sum\nolimits_{i=n+2}^{2n+1} u_i^2 = 1\Big\},$$

where $0 < a = const \ne 1$ and $n \ge 1$. The sectional curvature of (M, g) is positive. It follows that

$$\xi = (-u_2, u_1, \ldots, -u_{n+1}, u_n, -\sqrt{a}\, u_{n+3}, \sqrt{a}\, u_{n+2}, \ldots, -\sqrt{a}\, u_{2n+2}, \sqrt{a}\, u_{2n+1})$$

is a Killing vector field on $\mathbb{R}^{2n+2}$, whose restriction to M has unit length. Since ξ is tangent to M, i.e., M is invariant under the flow of ξ, ξ is a unit Killing vector field on (M, g).

5 The Curvature in the Characteristic Direction

Denote by $\mathrm{Ric}^\sharp$ the Ricci operator of (M, g) associated with the Ricci tensor Ric and given by

$$\mathrm{Ric}(X, Y) = g(\mathrm{Ric}^\sharp X, Y), \quad X, Y \in \mathfrak{X}_M.$$

Since $K(\xi_i, \xi_j) = 0$ for a weak f-contact manifold, the Ricci curvature in the ξ_j-direction is given by

$$\mathrm{Ric}(\xi_j, \xi_j) = \sum\nolimits_{i=1}^{2n} g(R_{e_i, \xi_j}\, \xi_j, e_i),$$

where (e_i) is any local orthonormal basis of $\mathcal{D}$. In the next proposition, we generalize some particular properties of f-K-contact manifolds for weak f-K-contact manifolds.

Lemma 3 *For a weak f-K-contact manifold $M^{2n+s}(f, Q, \xi_i, \eta^i, g)$, we get for all i, j:*

$$R_{\xi_i, X} = \nabla_X f \quad (X \in \mathfrak{X}_M), \tag{25}$$

$$R_{X, \xi_i}\, \xi_j = -f^2 X \quad (X \in \mathfrak{X}_M), \tag{26}$$

$$\mathrm{Ric}^\sharp \xi_i = \mathrm{div}\, f, \tag{27}$$

$$\mathrm{Ric}(\xi_i, \xi_j) = \mathrm{trace}\, Q = 2n + \mathrm{trace}\, \widetilde{Q}. \tag{28}$$

Proof Using (24), we derive

$$\begin{aligned} R_{Z,X}\, \xi_i &= \nabla_Z(\nabla_X \xi_i) - \nabla_X(\nabla_Z \xi_i) - \nabla_{[Z,X]}\, \xi_i) \\ &= \nabla_X(fZ) - \nabla_Z(fX) + f([Z, X]) = (\nabla_X f)Z - (\nabla_Z f)X. \end{aligned} \tag{29}$$

Note that $(\nabla_X \Phi)(Y, Z) = g((\nabla_X f)Z, Y) = -g((\nabla_X f)Y, Z)$. Using condition $d\Phi = 0$ and (10), we get

$$(\nabla_X \Phi)(Y, Z) + (\nabla_Y \Phi)(Z, X) + (\nabla_Z \Phi)(X, Y) = 0. \tag{30}$$

From (29), using (30) and skew-symmetry of Φ, we get (25):

$$\begin{aligned} g(R_{\xi_i, X}\, Y, Z) &= g(R_{Y,Z}\, \xi_i, X) \overset{(29)}{=} (\nabla_Z \Phi)(X, Y) + (\nabla_Y \Phi)(Z, X) \\ &\overset{(30)}{=} -(\nabla_X \Phi)(Y, Z) = g((\nabla_X f)Y, Z). \end{aligned}$$

By (25) with $Y = \xi_j$, using $f\, \xi_j = 0$ and (24), we get (26):

$$R_{\xi_i, X}\, \xi_j = (\nabla_X f)\, \xi_j = -f \nabla_X \xi_j = f^2 X.$$

For weak f-K-contact manifolds, by (25) and (22), we get the equality (31):

$$\mathrm{Ric}(\xi_j, X) = \sum\nolimits_{k=1}^{2n} g((\nabla_{e_k} f)\, X, e_k) = g(\operatorname{div} f, X), \tag{31}$$

where (e_k) is any local orthonormal basis of $\mathcal{D}$. By (31), we get the equality (27). Using $\mathrm{Ric}(\xi_i, \xi_j) = \operatorname{trace}\,(X \to R_{X,\xi_i}\xi_j)$ and $f\xi_i = 0$, we get for all i, j,

$$\mathrm{Ric}(\xi_i, \xi_j) \overset{(26)}{=} -\sum\nolimits_{k=1}^{2n} g(f^2 e_k, e_k) \overset{(3)}{=} \sum\nolimits_{k=1}^{2n} g(Q e_k, e_k).$$

By the above and the equality $\operatorname{trace} Q = 2n + \operatorname{trace} \widetilde{Q}$, (28) is true. □

Proposition 5 *There are no Einstein weak f-K-contact manifolds $M^{2n+s}(f, Q, \xi_i, \eta^i, g)$ with $s > 1$.*

Proof For a weak f-K-contact manifold with $s > 1$ and $\tilde\xi = (\xi_1 + \xi_2)/\sqrt{2}$, we get

$$\mathrm{Ric}(\tilde\xi, \tilde\xi) = (1/2) \sum\nolimits_{i,j=1}^{2} \mathrm{Ric}(\xi_i, \xi_j) \overset{(28)}{=} 2 \operatorname{trace} Q. \tag{32}$$

If the manifold is an Einstein manifold, then we get $\mathrm{Ric}(\tilde\xi, \tilde\xi) = \mathrm{Ric}(\xi_i, \xi_i) = \operatorname{trace} Q$. Comparing with (32) gives a contradiction: $\operatorname{trace} Q = 0$. □

Proposition 6 *For a weak f-K-contact manifold $M^{2n+s}(f, Q, \xi_i, \eta^i, g)$, the ξ-sectional curvature is positive:*

$$K(\xi_i, X) = g(QX, X) > 0 \quad (X \in \mathcal{D},\ \|X\| = 1),$$

and for the Ricci curvature we get $\mathrm{Ric}(\xi_i, \xi_j) > 0$ *for all* $1 \le i, j \le s$.

Proof For any unit vectors $\xi \in \ker f$, $X \in \mathcal{D}$, from (26), we get $K(\xi, X) = g(fX, fX) > 0$. From (28), using (3) and non-singularity of f on $\mathcal{D}$, we get

$$\mathrm{Ric}(\xi_i, \xi_j) = \operatorname{trace} Q = -\sum\nolimits_{k=1}^{2n} g(f^2 e_k, e_k) = \sum\nolimits_{k=1}^{2n} g(f e_k, f e_k) > 0,$$

where (e_k) is a local orthonormal frame of $\mathcal{D}$, thus the second statement is valid. □

The following theorem generalizes [34, Proposition 5.1] and [23, Theorem 4].

Theorem 3 *A weak f-K-contact manifold $M^{2n+s}(f, Q, \xi_i, \eta^i, g)$ with conditions* $\operatorname{trace} Q = const$ *and* $(\nabla \mathrm{Ric})(\xi_i, \cdot) = 0$ $(i \le s)$ *is an Einstein manifold and* $s = 1$.

Proof Differentiating (28) and using (24) and the conditions, we have

$$0 = \nabla_Y (\mathrm{Ric}(\xi_i, \xi_i)) = (\nabla_Y \mathrm{Ric})(\xi_i, \xi_i) + 2\,\mathrm{Ric}(\nabla_Y \xi_i, \xi_i) = -2\,\mathrm{Ric}(fY, \xi_i),$$

hence $\mathrm{Ric}(Y, \xi_i) = (\operatorname{trace} Q)\, \eta^i(Y)$. Differentiating this, then using

$$X(\eta^i(Y)) = g(\nabla_X \xi_i, Y) = -g(fX, Y) + g(\nabla_X Y, \xi_i)$$

and assuming $\nabla_X Y = 0$ at $x \in M$, gives

$$\begin{aligned}
&(\,\mathrm{trace}\ Q)\, g(fY, X) = \nabla_X\, (\mathrm{Ric}(Y, \xi_i))\\
&= (\nabla_X \mathrm{Ric})(Y, \xi_i) + 2\,\mathrm{Ric}(Y, \nabla_X\, \xi_i) = -2\,\mathrm{Ric}(Y, fX).
\end{aligned}$$

Thus, $\mathrm{Ric}(Y, fX) = (\,\mathrm{trace}\ Q)\, g(Y, fX)$. Hence, $\mathrm{Ric}(X, Y) = (\,\mathrm{trace}\ Q)\, g(X, Y)$ for any vector fields X and Y on M, which means that (M, g) is an Einstein manifolds and $s = 1$. □

Corollary 5 (see [23]) *A weak K-contact manifold* $M^{2n+1}(f, Q, \xi, \eta, g)$ *with* $(\nabla\,\mathrm{Ric})(\xi, \cdot) = 0$ *and* trace $Q = const$ *is an Einstein manifold of scalar curvature* $(2\,n + 1)$ trace Q.

Remark 4 For f-contact manifolds, we have

$$\sum\nolimits_{i=1}^{2n} (\nabla_{e_i} f)\, e_i = 2\,n\,\bar{\xi}, \tag{33}$$

see [13, Proposition 2.6]. For f-K-contact manifolds, (27) and (33) give $\mathrm{Ric}^\sharp(\xi_i) = 2\,n\,\bar{\xi}$ $(i = 1, \ldots, s)$ and $\mathrm{Ric}(\xi_i, \xi_i) = 2\,n$. Can one generalize this property of Ric and (33) for weak f-contact manifolds ?

Define the skew-symmetric operator $T^\sharp_\xi$ $(\xi \in \ker f)$ by

$$g(T^\sharp_\xi X, Y) = g(T(X, Y), \xi), \quad X, Y \in \mathcal{D},$$

where $T(X, Y) = (1/2)(\nabla_X Y - \nabla_Y X)^\perp$ is the integrability tensor of $\mathcal{D}$.

The self-adjoint partial Ricci curvature tensor, see [28], is given by

$$\mathrm{Ric}^\perp(X) = \sum\nolimits_i (R_{X^\perp, \xi_i}\, \xi_i)^\perp,$$

and for an f-K-contact manifold, we have $\mathrm{Ric}^\perp = s\,\mathrm{id}^\perp$. By Proposition 6, the tensor $\mathrm{Ric}^\perp$ of a weak f-K-contact manifold is positive definite. In [28], using the flow of metrics on a $\mathfrak{g}$-foliation, a deformation retraction of a weak almost $\mathcal{S}$-structures with positive partial Ricci curvature onto the subspace of the aforementioned classical structures is constructed. In the following theorem, using Proposition 6 and the method of [28], we show that a weak f-K-contact structure can be transformed into an f-K-contact structure.

Theorem 4 *Let* $M^{2n+s}(f(0), Q_0, \xi_i, \eta^i, g_0)$ *be a weak f-K-contact manifold. Then there exist metrics* g_t $(t \in \mathbb{R})$ *such that each* $(f(t), Q_t, \xi_i, \eta^i, g_t)$ *is a weak f-K-contact structure on M with structural tensors defined on* $\mathcal{D}$ *as*

$$Q_t = (1/s)\,\mathrm{Ric}^\perp_t, \quad f(t)\,|_{\mathcal{D}} = T^\sharp_{\xi_i}(t). \tag{34}$$

Moreover, g_t converges exponentially fast, as $t \to -\infty$, to a limit metric $\hat g$ with $\mathrm{Ric}^\perp_{\hat g} = s\,\mathrm{id}^\perp$ that provides an f-K-contact structure on M.

Proof The partial Ricci curvature tensor of a weak f-K-contact manifold is positive definite. Thus, the result is similar to [28, Theorem 1]. Consider the "partial Ricci flow" of metrics on M,

$$\partial_t\, g_t = -2\,(\mathrm{Ric}^\perp)^\flat_{g_t} + 2\,s\,g_t^\perp, \tag{35}$$

where the (0, 2)-tensor $g^\perp$ is defined by $g^\perp(X, Y) := g(X^\perp, Y^\perp)$. For $t = 0$, we get on $\mathcal{D}$:

$$\mathrm{Ric}^\perp = -\sum\nolimits_i (T^\sharp_{\xi_i})^2 = -\sum\nolimits_i \phi_i^2 = s\,Q,$$

and find $T^\sharp_{\xi_i}\,\mathrm{Ric}^\perp = \mathrm{Ric}^\perp\,T^\sharp_{\xi_i}$, see also [28]. Thus, $\sum_i T^\sharp_{\xi_i}\,\mathrm{Ric}^\perp\,T^\sharp_{\xi_i} = -(\mathrm{Ric}^\perp)^2$. By the above, we obtain the following ODE:

$$\partial_t\,\mathrm{Ric}^\perp = 4\,\mathrm{Ric}^\perp(\mathrm{Ric}^\perp - s\,\mathrm{id}^\perp).$$

By the theory of ODE's, there exists a unique solution $\mathrm{Ric}^\perp_t$ for $t \in \mathbb{R}$; hence, a solution g_t of (35) exists for $t \in \mathbb{R}$ and it is unique. Observe that $(f(t), \xi_i, \eta^i, Q_t)$ with $f(t)$, Q_t given in (34) is a weak f-K-contact structure on (M, g_t). By uniqueness of a solution, the flow (35) preserves the directions of eigenvectors of $\mathrm{Ric}^\perp$, and each eigenvalue $\mu_i > 0$ of $\mathrm{Ric}^\perp$ satisfies the following ODE with $\mu_i(0) > 0$:

$$\dot\mu_i = 4\mu_i\,(\mu_i - s).$$

This ODE has the following solution (a function $\mu_i(t)$ on M for $t \in \mathbb{R}$):

$$\mu_i(t) = \frac{\mu_i(0)\,s}{\mu_i(0) + \exp(4\,s\,t)(s - \mu_i(0))}$$

with $\lim_{t\to-\infty} \mu_i(t) = s$. Thus, $\lim_{t\to-\infty} \mathrm{Ric}^\perp(t) = s\,\mathrm{id}^\perp$. Let $\{e_i(t)\}$ be a g_t-orthonormal frame of $\mathcal{D}$ of eigenvectors associated with $\mu_i(t)$, we then get $\partial_t e_i = (\mu_i - s)e_i$. Since $e_i(t) = z_i(t)\,e_i(0)$ with $z_i(0) = 1$, then $\partial_t \log z_i(t) = \mu_i(t) - s$. By the above, $z_i(t) = (\mu_i(t)/\mu_i(0))^{1/4}$. Hence,

$$g_t(e_i(0), e_j(0)) = z_i^{-1}(t) z_j^{-1}(t)\, g_t(e_i(t), e_j(t)) = \delta_{ij}(\mu_i(0)\mu_j(0)/(\mu_i(t)\mu_j(t)))^{1/4}.$$

As $t \to -\infty$, g_t converges to the metric $\hat g$ given on the frame $\{e_i(0)\}$ by $\hat g(e_i(0), e_j(0)) = \delta_{ij}\sqrt{\mu_i(0)/s}$. □

Corollary 6 *Let $M^{2n+1}(f(0), Q_0, \xi, \eta, g_0)$ be a weak K-contact manifold. Then there exist metrics g_t ($t \in \mathbb{R}$) such that each $(f(t), Q_t, \xi, \eta, g_t)$ is a weak K-contact structure on M with structural tensors $Q_t = \mathrm{Ric}^\perp_t$ and $f(t)\,|_{\mathcal{D}} = T^\sharp_\xi(t)$, defined on*

$\mathcal{D}$. Moreover, g_t converges exponentially fast, as $t \to -\infty$, to a limit metric $\hat{g}$ with $\mathrm{Ric}^{\perp}_{\hat{g}} = \mathrm{id}^{\perp}$ *that provides a K-contact structure on M.*

Remark 5 By Corollary 6 and results on K-contact manifolds, e.g., [3, p. 40], on a compact weak K-contact manifold M^{2n+1}, ξ has at least $n+1$ closed orbits; and if ξ has exactly $n+1$ closed orbits, then M^{2n+1} is covered by a sphere.

6 Weak f-K-Contact Structure as a Generalized Ricci Soliton

Proposition 5 motivates to consider *quasi Einstein manifolds*, introduced in [10] by the condition

$$\mathrm{Ric} = a\, g + b\, \mu \otimes \mu$$

for all vector fields X, Y, where a and $b \neq 0$ are real scalars, and μ is a 1-form of unit norm, and later this structure has been studied by several geometers. If μ is the differential of a function, then we get a *gradient quasi Einstein manifold.*

The following generalization of (2) was given in [20]:

$$\mathrm{Hess}_{\sigma_1} - c_2\, \mathrm{Ric} = \lambda\, g - c_1 d\sigma_2 \otimes d\sigma_2 \tag{36}$$

for some $\sigma_1, \sigma_2 \in C^\infty(M)$ and real constants c_1, c_2 and λ. For $\sigma_1 = \sigma_2$, (36) reduces to (2). First, we formulate some lemmas.

Lemma 4 (see Lemma 3.1 in [16] or Lemma 4 in [23] for $s = 1$) *For a weak f-K-contact manifold $M^{2n+s}(f, Q, \xi_i, \eta^i, g)$ the following holds:*

$$(\pounds_{\xi_i}(\pounds_X\, g))(Y, \xi_i) = g(X, Y) + g(\nabla_{\xi_i}\nabla_{\xi_i} X, Y) + Y g(\nabla_{\xi_i} X, \xi_i)$$

for any $i \le s$ and all smooth vector fields X, Y with Y orthogonal to $\ker f$.

Proof This uses $\nabla_{\xi_i}\xi_i = 0$ and (26), and is the same as for [23, Lemma 4]. □

Lemma 5 (see, for example, [16]) *Let $(M; g)$ be a Riemannian manifold and σ be a smooth function on M. Then the following holds for every vector fields ξ and Y on M :*

$$\pounds_\xi(d\sigma \otimes d\sigma)(Y, \xi) = Y(\xi(\sigma))\,\xi(\sigma) + Y(\sigma)\,\xi(\xi(\sigma)).$$

Recall that the Ricci curvature of any f-K-contact manifold satisfies the following condition:

$$\mathrm{Ric}(X, \xi) = 0 \quad (X \in \mathcal{D},\ \xi \in \ker f). \tag{37}$$

Lemma 6 (see Lemma 3.3 in [16] or Lemma 6 in [23] for $s = 1$) *Let a weak f-K-contact manifold $M^{2n+s}(f, Q, \xi_i, \eta^i, g)$ satisfy (37) and admit the generalized gradient Ricci soliton structure. Then*

$$\nabla_{\xi_i} \nabla\sigma = (\lambda + c_2 \operatorname{trace} Q)\, \xi_i - c_1 \xi_i(\sigma)\, \nabla\sigma.$$

Proof By (28) and (37) we get

$$\lambda\, \eta^i(Y) + c_2 \operatorname{Ric}(\xi_i, Y) = (\lambda + c_2 \operatorname{trace} Q)\, \eta^i(Y). \tag{38}$$

Using (2) and (38), we get

$$\operatorname{Hess}_\sigma(\xi_i, Y) = -c_1 \xi_i(\sigma)\, g(\nabla\sigma, Y) + (\lambda + c_2 \operatorname{trace} Q)\, \eta^i(Y). \tag{39}$$

Thus, (39) and the condition (2) for the Hessian complete the proof. □

The following theorem generalizes [23, Theorem 4] and [16, Theorem 3.1].

Theorem 5 *Let a weak f-K-contact manifold $M^{2n+s}(f, Q, \xi_i, \eta^i, g)$ with the property* trace $Q = const$ *satisfy the generalized gradient Ricci soliton equation (2) with $c_1(\lambda + c_2 \operatorname{trace} Q) \neq -1$:*

$$\operatorname{Hess}_\sigma - c_2 \operatorname{Ric} = \lambda\, g - c_1 d\sigma \otimes d\sigma\, .$$

Suppose that (37) is true. Then $\sigma = const$. Moreover, if $c_2 \neq 0$, then the manifold is an Einstein one and $s = 1$.

Proof Let $Y \perp \ker f$. Then by Lemma 4 with $X = \nabla f$, we obtain

$$2\, (\pounds_{\xi_i}(\operatorname{Hess}_\sigma))(Y, \xi_i) = Y(\sigma) + g(\nabla_{\xi_i} \nabla_{\xi_i} \nabla\sigma, Y) + Y g(\nabla_{\xi_i} \nabla\sigma, \xi_i). \tag{40}$$

Using Lemma 6 in (40) and the properties $\nabla_{\xi_i} \xi_i = 0$ and $g(\xi_i, \xi_i) = 1$, yields

$$\begin{aligned} 2\, (\pounds_{\xi_i}(\operatorname{Hess}_\sigma))(Y, \xi_i) &= Y(\sigma) + a\, g(\nabla_{\xi_i} \xi_i, Y) \\ &\quad - c_1 g(\nabla_{\xi_i}(\xi_i(\sigma)\nabla\sigma), Y) + a\, Y(g(\xi_i, \xi_i)) - c_1 Y(\xi_i(\sigma)^2) \\ &= Y(\sigma) - c_1 g(\nabla_{\xi_i}(\xi_i(\sigma)\nabla\sigma), Y) - c_1 Y(\xi_i(\sigma)^2), \end{aligned} \tag{41}$$

where $a := \lambda + c_2 \operatorname{trace} Q$. Using Lemma 6 with $Y \in \mathcal{D}$, from (41) it follows that

$$2\, (\pounds_{\xi_i}(\operatorname{Hess}_\sigma))(Y, \xi_i) = Y(\sigma) - c_1 \xi_i(\xi_i(\sigma))\, Y(\sigma) + c_1^2 \xi_i(\sigma)^2 Y(\sigma) - c_1 Y(\xi_i(\sigma)^2). \tag{42}$$

Since ξ_i is a Killing vector field, thus $\pounds_{\xi_i} g = 0$, this implies $\pounds_{\xi_i} \operatorname{Ric} = 0$. Using the above fact and applying the Lie derivative to Eq. (2), gives

$$2\, (\pounds_{\xi_i}(\operatorname{Hess}_\sigma))(Y, \xi_i) = -2\, c_1 (\pounds_{\xi_i}(d\sigma \otimes d\sigma))(Y, \xi_i). \tag{43}$$

Using (42), (43) and Lemma 5, we obtain

$$Y(\sigma)\big(1 + c_1\xi_i(\xi_i(\sigma)) + c_1^2\,\xi_i(\sigma)^2\big) = 0. \tag{44}$$

By Lemma 6, we get

$$c_1\xi_i(\xi_i(\sigma)) = c_1\,\xi_i(g(\xi_i, \nabla\sigma)) = c_1 g(\xi_i, \nabla_{\xi_i}\nabla\sigma) = c_1 a - c_1^2\,\xi_i(\sigma)^2. \tag{45}$$

Using (44) in (45), we get $Y(\sigma)(c_1 a + 1) = 0$. This implies $Y(\sigma) = 0$ provided by $c_1 a + 1 \neq 0$. Hence, $\nabla\sigma$ is parallel to $\ker f$. Taking the covariant derivative of $\nabla\sigma = \sum_i \xi_i(\sigma)\,\xi_i$ and using (24) and $\bar\xi = \sum_i \xi_i$, we obtain

$$g(\nabla_X \nabla\sigma, Y) = \sum\nolimits_i X(\xi_i(\sigma))\,\eta^i(Y) - \bar\xi(\sigma)\,g(\sigma X, Y), \quad X, Y \in \mathfrak{X}_M.$$

From this, by symmetry of Hess_σ, i.e. $g(\nabla_X \nabla\sigma, Y) = g(\nabla_Y \nabla\sigma, X)$, we get $\bar\xi(\sigma)\,g(\sigma X, Y) = 0$. For $Y = \sigma X$ with some $X \neq 0$, since $g(\sigma X, \sigma X) > 0$, we get $\bar\xi(\sigma) = 0$. Replacing (ξ_i) with another orthonormal frame from $\ker f$ preserves the weak f-K-contact structure and allows reaching any direction $\bar\xi$ in $\ker f$. So $\nabla\sigma = 0$, i.e. $\sigma = const$. Thus, from (2) and $c_2 \neq 0$ it follows the claim. □

Corollary 7 (see [23]) *Let a weak K-contact manifold $M(f, Q, \xi, \eta, g)$ with* trace $Q = const$ *satisfy the generalized gradient Ricci soliton equation (2) with* $c_1(\lambda + c_2\,\mathrm{trace}\,Q) \neq -1$. *Suppose that condition* $\mathrm{Ric}(\xi, X) = 0$ $(X \perp \xi)$ *is true. Then $\sigma = const$. Furthermore, if $c_2 \neq 0$, then the manifold is an Einstein one.*

The following theorem contains 3 cases and generalizes (and uses) Theorem 5.

Theorem 6 *Let a weak f-K-contact manifold $M^{2n+s}(f, Q, \xi_i, \eta^i, g)$ satisfy the generalized gradient Ricci soliton equation (36) with $c_1 a \neq -1$, where $a = \lambda + c_2\,\mathrm{trace}\,Q$. Suppose that conditions* trace $Q = const$ *and (37) are true. Then $\tilde\sigma = \sigma_1 + c_1 a\sigma_2$ is constant and*

$$-\,c_1 a\,\mathrm{Hess}_{\sigma_2} = -c_1\,d\sigma_2 \otimes d\sigma_2 + c_2\,\mathrm{Ric} + \lambda\,g. \tag{46}$$

Furthermore,

1. if $c_1 a \neq 0$, then (46) reduces to $\mathrm{Hess}_{\sigma_2} = \frac{1}{a}\,d\sigma_2 \otimes d\sigma_2 - \frac{c_2}{c_1 a}\,\mathrm{Ric} - \frac{\lambda}{c_1 a}\,g$. By Theorem 5, if $c_1 a \neq -1$, then $\sigma_2 = const$; moreover, if $c_2 \neq 0$, then (M, g) is an Einstein manifold and $s = 1$.

2. if $a = 0$ and $c_1 \neq 0$, then (46) reduces to $0 = c_2\,\mathrm{Ric} - c_1\,d\sigma_2 \otimes d\sigma_2 + \lambda\,g$. If $c_2 \neq 0$ and $\sigma_2 \neq const$, then we get a gradient quasi Einstein manifold.

3. if $c_1 = 0$, then (46) reduces to $0 = c_2\,\mathrm{Ric} + \lambda\,g$, and for $c_2 \neq 0$ we get an Einstein manifold and $s = 1$.

Proof Similarly to Lemma 6, we get

$$\nabla_{\xi_i}\nabla\sigma_1 = a\,\xi_i - c_1\xi_i(\sigma_2)\,\nabla\sigma_2. \tag{47}$$

Using (47) and Lemmas 4 and 5, and slightly modifying the proof of Theorem 5, we find that the vector field $\nabla\tilde\sigma$ is parallel to $\ker f$, where $\tilde\sigma = \sigma_1 + c_1 a\sigma_2$. As in the proof of Theorem 5, we get $d\tilde\sigma = 0$, i.e., $d\sigma_1 = -c_1 a\, d\sigma_2$. Using this in (36), we get (46). Analyzing (46), we obtain the required 3 cases of the theorem. □

7 η-Ricci Solitons on Compact Weak f-K-Contact Manifolds

A weak metric f-manifold $M^{2n+s}(f, Q, \xi_i, \eta^i, g)$ is said to be *η-Einstein*, if

$$\mathrm{Ric} = \alpha\, g + \beta \sum\nolimits_i \eta^i \otimes \eta^i + (\alpha+\beta) \sum\nolimits_{i<j} \eta^i \otimes \eta^j \tag{48}$$

for some functions $\alpha, \beta \in C^\infty(M)$. By an *$\eta$-Ricci soliton* we mean here a weak f-contact manifold, whose Ricci tensor satisfies

$$(1/2)\, \pounds_V\, g + \mathrm{Ric} = \lambda\, g + \nu \sum\nolimits_i \eta^i \otimes \eta^i + (\lambda+\nu) \sum\nolimits_{i<j} \eta^i \otimes \eta^j \tag{49}$$

for some functions $\lambda, \nu \in C^\infty(M)$ and some smooth vector field V on M.

Remark 6 For $s=1$, (48) and (49) give the well-known definitions on contact metric manifolds. Let $s=1$ and $Q=\mathrm{id}$. Then from (48) we get an η-Einstein structure $\mathrm{Ric} = \alpha\, g + \beta\, \eta \otimes \eta$, e.g., [17]; and from (49) we get an η-Ricci soliton $\frac12\, \pounds_V\, g + \mathrm{Ric} = \lambda\, g + \beta\, \eta \otimes \eta$ on an almost contact metric manifold, e.g., [17]. Equation (49) gives a generalized η-Ricci soliton $\frac12\, \pounds_V\, g + \mathrm{Ric} = \lambda\, g + \nu\, \eta \otimes \eta$ on an almost contact metric manifold, e.g., [17]; if also $\nu = 0$, then (49) gives a Ricci almost soliton; moreover, for $\lambda = const$ it is a Ricci soliton.

For weak f-K-manifolds and $V = \bar\xi$, the notion of η-Ricci soliton reduces to η-Einstein structure. For this reason, in this section we study an η-Ricci soliton on a weak f-K-contact manifold.

We recall the generalized Pohozaev-Schoen identity due to Gover-Orsted [15].

Lemma 7 *Let (M^n, g) be a compact Riemannian manifold without boundary. If E is a divergence free symmetric (0,2)-tensor and V is a vector field on M, then*

$$\int_M V(\mathrm{trace}_g\, E)\, \mathrm{d}\ \mathrm{vol} = \frac{n}{2} \int_M g(E^0, \pounds_V\, g)\, \mathrm{d}\ \mathrm{vol}, \tag{50}$$

where $E^0 = E - \frac1n\, (\mathrm{trace}_g\, E)\, g$ and d vol *is the volume form of g.*

Denote by r the scalar curvature of a Riemannian manifold (M,g). The constancy of r plays an important role for the triviality of a compact generalized η-Ricci soliton, see [17]. We extend this result for weak f-K-contact manifolds.

Theorem 7 *Let a compact weak f-K-contact manifold $M^{2n+s}(f, \xi_i, \eta^i, Q, g)$ satisfy $r = const$ and* trace $Q = const$. *If (g, V, λ, ν) represents an η-Ricci soliton, then M is η-Einstein.*

Proof We take the trace of (48) to get the scalar curvature $r = (2n + s)\,\alpha + s\,\beta$. On the other hand, using (28) in (48), we obtain $\alpha + \beta =$ trace Q. From these it follows that

$$\alpha = \frac{r - s\,\text{trace}\,Q}{2n}, \quad \beta = \frac{(2n + s)\,\text{trace}\,Q - r}{2n}. \tag{51}$$

Define a traceless symmetric (0,2)-tensor E as

$$E = \text{Ric} - \frac{r - s\,\text{trace}\,Q}{2n}\,g + \frac{r - (2n + s)\,\text{trace}\,Q}{2n}\sum\nolimits_i \eta^i \otimes \eta^i$$
$$-(\,\text{trace}\,Q)\sum\nolimits_{i<j} \eta^i \otimes \eta^j. \tag{52}$$

We express (49), using (52), as

$$(1/2)\,\pounds_V\,g + E = \big(\lambda - \frac{r-s\,\text{trace}\,Q}{2n}\big)\,g + \big(\nu + \frac{r-(2n + s)\,\text{trace}\,Q}{2n}\big)\sum\nolimits_i \eta^i \otimes \eta^i$$
$$+ (\lambda + \nu - \text{ trace}\,Q)\sum\nolimits_{i<j} \eta^i \otimes \eta^j. \tag{53}$$

Taking covariant derivative of (52) with X and using (24) and trace $Q = const$, gives

$$(\nabla_X E)(Y, Z) = (\nabla_X \text{Ric})(Y, Z) - \frac{X(r)}{2n}\big(g(Y, Z) - \sum\nolimits_i \eta^i(Y)\,\eta^i(Z)\big) \tag{54}$$
$$+ \Big((s - 1)\,\text{trace}\,Q + \frac{(2n + s)\,\text{trace}\,Q - r}{2n}\Big)\big(\bar\eta(Z)\,g(fX, Y) + \bar\eta(Y)\,g(fX, Z)\big).$$

Taking the trace of (54) over X and Z, and since $r = const$, $\text{div}_g\,\text{Ric} = \frac{1}{2}\,d\,r$ (twice-contracted 2nd Bianchi identity, [19, p. 135]), and trace $f = 0$, we get the equality $\text{div}_g\,E = 0$:

$$(\text{div}_g\,E)(Y) = (\text{div}_g\,\text{Ric})(Y)$$
$$+ \frac{s\,(2n + 1)\,\text{trace}\,Q - r}{2n}\big(\bar\eta(e_i)\,g(fe_i, Y) + \bar\eta(Y)\,g(fe_i, e_i)\big) = 0.$$

Since the tensor E is traceless, we get $E^0 = E$ and $V(\,\text{trace}_g E) = 0$. Thus, applying (50), we find $\int_M g(E, \pounds_V\,g)\,\text{d vol} = 0$. Applying (53) and (28) in the above integral formula, and using equalities $g(E, g) = \text{ trace}_g E = 0$ and $g(E, \eta^i \otimes \eta^j) = E(\xi_i, \xi_j) = 0$, we obtain

$$\int_M g(E, E)\,\text{d vol} = 0. \tag{55}$$

From (55) we get $E = 0$. Thus, our manifold is η-Einstein, see (48), where α, β satisfy (51). □

The following consequence of Theorem 7 generalizes [17, Theorem 3.1(i)].

Corollary 8 *Let a compact weak* K*-contact manifold* $M^{2n+1}(f, \xi, \eta, Q, g)$ *satisfy* $r = const$ *and* trace $Q = const$. *If* (g, V, λ, ν) *represents a generalized* η*-Ricci soliton on* M*, then* M *is* η*-Einstein.*

Notes and Comments. It is shown that a weak f-contact structure is a useful tool for studying Killing vector fields, totally geodesic foliations and mixed sectional curvature on Riemannian manifolds. Some results for f-contact and f-K-contact structures [4, 34] have been extended to certain weak structures. Conditions are found under which (i) a weak f-K-contact manifold with parallel Ricci tensor or with a generalized gradient Ricci soliton structure is a quasi Einstein or an Einstein manifold; (ii) a compact weak f-K-contact manifold with constant scalar curvature and an η-Ricci soliton structure is η-Einstein. Based on applications of the weak f-contact structure in contact geometry considered in the article, we expect that this structure will be fruitful in theoretical physics as well.

In conclusion, we pose several questions. Is the condition "the ξ-sectional curvature is positive" sufficient for a weak f-contact manifold to be weak f-K-contact? Does a weak f-contact manifold of dimension greater than 3 have some positive ξ-sectional curvature? Is a compact weak f-K-contact Einstein manifold an $\mathcal{S}$-manifold? When a given weak f-K-contact manifold is a mapping torus (see [18]) of a manifold of lower dimension? When a weak f-contact manifold equipped with a Ricci-type soliton structure, carries a canonical (for example, with constant sectional curvature or Einstein-type) metric? One could answer these questions by generalizing some deep results on f-contact manifolds for weak f-contact manifolds.

References

1. Alekseevsky, D. and Michor, P.: Differential geometry of $\mathfrak{g}$-manifolds, Differential Geom. Appl. 1995, 5, 371–403
2. Berestovskij, V. N. and Nikonorov, Yu. G.: Killing vector fields of constant length on Riemannian manifolds. Sib. Math. J. 49, No. 3 (2008), 395–407
3. Blair, D. E.: *Riemannian geometry of contact and symplectic manifolds*, 2nd edition, Springer-Verlag, New York, 2010
4. Blair, D. E.: Geometry of manifolds with structural group $U(n) \times O(s)$, J. Diff. Geom. 4 (1970), 155–167
5. Brunetti, L. and Pastore, A.M.: S-manifolds versus indefinite S-manifolds and local decomposition theorems. International Electronic J. of Geometry, 2016, 9:1, 1–8
6. Cabrerizo, J. L., Fernández, L. M. and Fernández, M.: The curvature tensor fields on f-manifolds with complemented frames. An. Stiint. Univ. Al. I. Cuza Iasi, 36 (1990), 151–161
7. Cairns, G.: A general description of totally geodesic foliations, Tohoku Math. J. 1986, 38, 37–55
8. Catino, G., Mastrolia, P., Monticelli, D. and Rigoli, M.: On the geometry of gradient Einstein-type manifolds, Pac. J. Math. 286 (1) (2017), 39–67
9. Cao, H. D.: Recent progress on Ricci soliton. In: "Recent Advances in Geometric Analysis", Adv. Lect. Math., 11 (2009), 1–38

10. Chaki, M. C. and Maity, R. K.: On quasi Einstein manifolds, Publ. Math. Debrecen, 57 (2000), 297–306
11. Cho, J. T. and Kimura, M.: Ricci solitons and real hypersurfaces in complex space form, Tohoku Math. J. 61 (2) (2009), 205–212
12. Deshmukh, S. and Belova, O.: On Killing vector fields on Riemannian manifolds, Mathematics, 9, 259 (2021), 1–17
13. Duggal, K. L., Ianus, S. and Pastore, A. M.: Maps interchanging f-structures and their harmonicity. Acta Appl. Math. 67 (2001), 91–115
14. Falcitelli, M., Ianus, S. and Pastore, A.: *Riemannian Submersions and Related Topics*, World Scientific, 2004
15. Gover, A. R. and Orsted, B.: Universal principles for Kazdan-Warner and Pohozaev-Schoen type identities, Commun. Contemp. Math. 15 (4) (2013), Art. ID 1350002
16. Ghosh, G. and De, U. C.: Generalized Ricci soliton on K-contact manifolds, Math. Sci. Appl. E-Notes, 8 (2020), 165–169
17. Ghosh, A.: K-contact and (k, μ)-contact metric as a generalized η-Ricci soliton. Math. Slovaca 73, No. 1, 185–194 (2023)
18. Goertsches, O. and Loiudice, E.: On the topology of metric f-K-contact manifolds, Monatshefte für Mathematik, 192 (2020), 355–370
19. Kobayashi, S. and Nomizu, K.: *Foundations of differential geometry*, Vols. I, II, USA, Interscience Publishers, New York–London–Sydney, 1963, 1969
20. Mohammed Cherif, A., Zegga, K. and Beldjilali, G.: On the generalised Ricci solitons and Sasakian manifolds. Communications in Mathematics, 30, Issue 1 (2022) 119–123
21. Nurowski, P. and Randall, M.: Generalised Ricci solitons, J. Geom. Anal., 26 (2016), 1280–1345
22. Pigola S., Rigoli M., Rimoldi M. and Setti A. G.: Ricci almost solitons. Ann. Scuola Norm. Sup. Pisa Cl. Sci. 10 (4) (2011), 757–799
23. Rovenski, V.: Generalized Ricci solitons and Einstein metrics on weak K-contact manifolds. Communications in Analysis and Mechanics, 2023, Vol. 15 (2), 177–188
24. Rovenski, V.: Metric structures that admit totally geodesic foliations. J. Geom. (2023) 114:32
25. Rovenski, V.: On the splitting tensor of the weak f-contact structure. Symmetry 2023, 15(6), 1215. https://doi.org/10.3390/sym15061215
26. Rovenski, V. and Patra, D. S.: On the rigidity of the Sasakian structure and characterization of cosymplectic manifolds, Differential Geometry and its Applications, 90 (2023) 102043
27. Rovenski, V. and Walczak, P. G.: *Extrinsic geometry of foliations*, Progress in Mathematics, vol. 339, Birkhäuser, Cham, 2021
28. Rovenski, V. and Wolak, R.: New metric structures on $\mathfrak{g}$-foliations, Indagationes Mathematicae, 33 (2022), 518–532
29. Smoczyk, K., Wang, G., Zhang, Y.: The Sasaki-Ricci flow. Int. J. Math. 21(7), 951–969 (2010)
30. Di Terlizzi, L.: On the curvature of a generalization of contact metric manifolds, Acta Math. Hung. 110, No. 3 (2006), 225–239
31. Di Terlizzi, L.: Correction to the paper "On the curvature of a generalization of a contact metric manifolds", Acta Math. Hung. 124, No. 4 (2009), 399–401
32. Di Terlizzi, L., Pastore, A.M. and Wolak, R.: Harmonic and holomorphic vector fields on an f-manifold with parallelizable kernel. An. Stiint. Univ. Al. I. Cuza Iausi, Ser. Noua, Mat. 2014, 60, No. 1, 125–144
33. Yano, K.: On a structure f satisfying $f^3 + f = 0$, Technical Report No. 12, University of Washington, 1961
34. Yano, K.: and Kon, M. *Structures on Manifolds*, Vol. 3 of Series in Pure Math. World Scientific Publ. Co., Singapore, 1985

Weak β-Kenmotsu Manifolds and η-Ricci Solitons

Dhriti Sundar Patra and Vladimir Rovenski

Abstract Weak contact metric structures on a smooth manifold have been recently introduced by V. Rovenski and R. Wolak in 2022. In this paper, we define a new structure of this kind called a weak β-Kenmotsu structure (that generalizes the notion by K. Kenmotsu with $\beta = 1$ and its extension for $\beta \neq 0$ by Z. Olszak). We show that a weak β-Kenmotsu manifold is locally the warped product $(-\varepsilon, \varepsilon) \times_\sigma \bar{M}$, where $(\partial_t \sigma)/\sigma = \beta \neq 0$, and $(\bar{M}, \bar{g})$ is equipped with a parallel skew-symmetric (1,1)-tensor $\bar{\phi}$ such that $\bar{\phi}^2$ is negative definite. Then, we show that an η-Einstein weak β-Kenmotsu manifold with $\beta = const \neq 0$ admitting an η-Ricci soliton structure is an Einstein manifold. Finally, we prove that a weak β-Kenmotsu manifold and admitting an η-Ricci soliton structure, whose non-zero potential vector field is weak contact or is collinear to ξ, is an Einstein manifold.

Keywords Warped product · Weak β-Kenmotsu manifold · Einstein manifold · η-Einstein manifold · η-Ricci soliton · Weak contact vector field

1 Introduction

The notion of a warped product is popular in geometry as well as in general relativity, e.g., [2]. Some solutions of Einstein field equations are warped products and some spacetime models, e.g., Schwarzschild, Robertson-Walker and Reissner-Nordström spacetimes, are warped products (see [4] for more details). In [10], Z. Olszak characterized in terms of warped products a special class of almost contact metric manifolds, known as β-Kenmotsu manifolds for $\beta = const > 0$ (Kenmotsu man-

D. S. Patra
Department of Mathematics, Indian Institute of Technology, Sangareddy, Hyderabad 502285, India
e-mail: dhriti@math.iith.ac.in

V. Rovenski (✉)
Department of Mathematics, University of Haifa, 3498838 Haifa, Israel
e-mail: vrovenski@univ.haifa.ac.il

© The Author(s), under exclusive license to Springer Nature Switzerland AG 2024

V. Rovenski et al. (eds.), *Differential Geometric Structures and Applications*,
Springer Proceedings in Mathematics & Statistics 440,
https://doi.org/10.1007/978-3-031-50586-7_3

ifolds when $\beta = 1$, see [14]). Namely, the warped product $\mathbb{R} \times_\sigma \bar{M}$ (of the real line $\mathbb{R}$ and a Kähler manifold $(\bar{M}, \bar{g})$) with the metric $g = dt^2 \oplus \sigma^2 \bar{g}$ and the warping function $\sigma > 0$ given on an open interval $(-\varepsilon, \varepsilon)$ and satisfying

$$(\partial_t \sigma)/\sigma = \beta, \tag{1}$$

admits a β-Kenmotsu structure, i.e., (ξ, η, φ) with the properties $\eta(\xi) = 1$ and

$$(\nabla_X \varphi)Y = \beta\{g(\varphi X, Y)\,\xi - \eta(Y)\,\varphi X\};$$

conversely, every point of a β-Kenmotsu manifold has a neighbourhood, which is locally a warped product $(-\varepsilon, \varepsilon) \times_\sigma \bar{M}$ of an open interval and a Kähler manifold, where σ satisfies (1).

There are no Einstein metrics on some compact manifolds, which motivates the study of generalizations of such metrics. Some authors studied the question: when a contact metric manifold $M(\varphi, \xi, \eta, g)$, equipped with a Ricci-type soliton structure carries an Einstein-type metric, e.g., [6]. In [3], Cho-Kimura generalized the notion of Ricci soliton to *η-Ricci soliton*, defined on a contact metric manifold $M^{2n+1}(\varphi, \xi, \eta, g)$ by the following equation:

$$\frac{1}{2}\,\pounds_V\, g + \mathrm{Ric} + \lambda\, g + \mu\, \eta \otimes \eta = 0, \tag{2}$$

where Ric is the Ricci tensor, $\pounds_V$ is the Lie derivative in the V-direction, V is the potential vector field on M and λ, μ are real constants. The tensor product notation $(\eta \otimes \eta)(X, Y) = \eta(X)\,\eta(Y)$ was also used. For $\mu = 0$, (2) defines a Ricci soliton; moreover, if $V = 0$, then (2) gives an Einstein manifold. A contact metric manifold $M^{2n+1}(\varphi, \xi, \eta, g)$ is called *η-Einstein*, if its Ricci tensor has the following form, e.g., [6]:

$$\mathrm{Ric} = a\, g + b\, \eta \otimes \eta, \tag{3}$$

where $a, b \in C^\infty(M)$, see [16, p. 285] for $a, b \in \mathbb{R}$. Note that a and b in (3) can be non-constant for an η-Einstein Kenmotsu manifold M^{2n+1}, $n > 1$, see [14].

In [12, 13], we introduced new metric structures on a smooth manifold that generalize the almost contact, Sasakian, cosymplectic, etc. metric structures. Such so-called "weak" structures (i.e., the linear complex structure on the contact distribution is replaced by a nonsingular skew-symmetric tensor) allow us to take a fresh look at the theory of classical structures and find new applications.

In this paper (consisting of an introduction and four sections), we define a new structure of this kind called a weak β-Kenmotsu structure, where $\beta \in C^\infty(\mathbb{R})$ is positive, and derive its fundamental properties. In Sect. 3, we provide a geometrical interpretation of a weak β-Kenmotsu structure and prove (Theorem 1) that a weak β-Kenmotsu manifold is locally a warped product of a real line and a Riemannian manifold equipped with a skew-symmetric (1,1)-tensor, whose square is negative definite. In Sects. 4 and 5, we study the question: *Are there weak β-Kenmotsu mani-*

folds, whose metrics are η-Ricci solitons? In Theorem 3 (Theorem 6 for $\dim M = 3$), we prove that an η-Einstein weak β-Kenmotsu manifold with $\beta = const \neq 0$ equipped with an η-Ricci soliton structure is an Einstein manifold. We supplement this result with the cases, when the nonzero potential vector field (of an η-Ricci soliton) is a weak contact vector field (Theorem 4), or is collinear to ξ (Theorem 5).

2 Preliminaries: Weak Almost Contact Manifolds

A *weak almost contact structure* on a smooth manifold M^{2n+1} is a set (φ, Q, ξ, η), where φ is a $(1,1)$-tensor, ξ is a vector field (called Reeb vector field) and η is a dual 1-form, satisfying

$$\varphi^2 = -Q + \eta \otimes \xi, \quad \eta(\xi) = 1, \tag{4}$$

see [12, 13], where Q is a nonsingular $(1,1)$-tensor field such that

$$Q\,\xi = \xi. \tag{5}$$

By (4), η defines a $2n$-dimensional distribution $\mathcal{D} := \ker \eta$. Assume that $\mathcal{D}$ is φ-invariant, i.e.,

$$\varphi X \in \mathcal{D}, \quad \forall X \in \mathcal{D}, \tag{6}$$

as in the classical theory [1], where $Q = \mathrm{id}_{TM}$. By (6) and (4a), $\mathcal{D}$ is invariant for Q: $Q(\mathcal{D}) = \mathcal{D}$. If there is a Riemannian metric g on M such that

$$g(\varphi X, \varphi Y) = g(X, Q\,Y) - \eta(X)\,\eta(Y), \quad X, Y \in \mathfrak{X}_M, \tag{7}$$

then $(\varphi, Q, \xi, \eta, g)$ is called a *weak almost contact metric structure* on M. A weak almost contact manifold $M(\varphi, Q, \xi, \eta)$ with a compatible Riemannian metric g called a *weak almost contact metric manifold* and is denoted by $M(\varphi, Q, \xi, \eta, g)$. Setting $Y = \xi$ in (7), we get as in the classical theory, $\eta(X) = g(X, \xi)$.

Proposition 1 *For a weak almost contact structure on M, the tensor φ has rank $2n$ and*

$$\varphi\,\xi = 0, \quad \eta \circ \varphi = 0, \quad \eta \circ Q = \eta, \quad [Q, \varphi] = 0. \tag{8}$$

For a weak almost contact metric structure, φ is skew-symmetric and Q is self-adjoint.

The *fundamental 2-form* Φ on $M(\varphi, Q, \xi, \eta, g)$ is defined by

$$\Phi(X, Y) = g(X, \varphi Y), \quad X, Y \in \mathfrak{X}_M.$$

A weak almost contact structure (φ, Q, ξ, η) on a manifold M will be called *normal* if the following tensor $N^{(1)}$ is identically zero:

$$N^{(1)}(X, Y) = [\varphi, \varphi](X, Y) + 2\,d\eta(X, Y)\,\xi, \quad X, Y \in \mathfrak{X}_M.$$

where $d\eta(X, Y) = \frac{1}{2}\{X(\eta(Y)) - Y(\eta(X)) - \eta([X, Y])\}$. The Nijenhuis torsion $[\varphi, \varphi]$ of φ is given by

$$[\varphi, \varphi](X, Y) = \varphi^2[X, Y] + [\varphi X, \varphi Y] - \varphi[\varphi X, Y] - \varphi[X, \varphi Y], \quad X, Y \in \mathfrak{X}_M. \tag{9}$$

Recall that the Levi-Civita connection ∇ of a Riemannian metric g is given by

$$\begin{aligned} 2\,g(\nabla_X Y, Z) &= X\,g(Y, Z) + Y\,g(X, Z) - Z\,g(X, Y) \\ &\quad + g([X, Y], Z) + g([Z, X], Y) - g([Y, Z], X), \end{aligned}$$

and has the properties

$$[X, Y] = \nabla_X\,Y - \nabla_Y\,X, \quad X\,g(Y, Z) = g(\nabla_X\,Y, Z) + g(Y, \nabla_X\,Z).$$

Thus, (9) can be written in terms of ∇ as

$$[\varphi, \varphi](X, Y) = (\varphi\nabla_Y\,\varphi - \nabla_{\varphi Y}\,\varphi)X - (\varphi\nabla_X\,\varphi - \nabla_{\varphi X}\,\varphi)Y; \tag{10}$$

in particular, since $\varphi\,\xi = 0$,

$$[\varphi, \varphi](X, \xi) = \varphi(\nabla_\xi\,\varphi)X + \nabla_{\varphi X}\,\xi - \varphi\,\nabla_X\,\xi \quad (X \in \mathfrak{X}_M).$$

Define a "small" (1,1)-tensor $\tilde{Q} = Q - \mathrm{id}_{TM}$. Note that, see also (8),

$$[\tilde{Q}, \varphi] = 0, \quad \eta \circ \tilde{Q} = 0, \quad \tilde{Q}\,\xi = 0.$$

3 Geometry of Weak β-Kenmotsu Manifolds

In the following definition, we generalize the notion of β-Kenmotsu manifolds, for example, [5].

Definition 1 A weak almost contact metric manifold $M^{2n+1}(\varphi, Q, \xi, \eta, g)$ will be called a *weak β-Kenmotsu manifold* if

$$(\nabla_X\,\varphi)Y = \beta\{g(\varphi X, Y)\,\xi - \eta(Y)\,\varphi X\} \quad (X, Y \in \mathfrak{X}_M), \tag{11}$$

where β is a smooth function on M such that $\{\beta = 0\}$ is a nowhere dense set; and if $n = 1$, then we also require $d\beta \wedge \eta = 0$. If $\beta \equiv 1$, then $M^{2n+1}(\varphi, Q, \xi, \eta, g)$ is called a *weak Kenmotsu manifold*.

Taking $X = \xi$ in (11) and using $\varphi\,\xi = 0$, we get $\nabla_\xi\,\varphi = 0$.

Lemma 1 *The following formulas are true for a weak β-Kenmotsu manifold $M^{2n+1}(\varphi, Q, \xi, \eta, g)$:*

$$\nabla_X \xi = \beta\{X - \eta(X)\,\xi\} \quad (X \in \mathfrak{X}_M), \tag{12}$$

$$d\eta(\xi, \cdot) = 0. \tag{13}$$

Proof Taking $Y = \xi$ in (11) and using $g(\varphi X, \xi) = 0$ and $\eta(\xi) = 1$, we get $\varphi(\nabla_X \xi - \beta X) = 0$. Since φ is non-degenerate on $\mathcal{D}$ and has rank $2n$, we get $\nabla_X \xi - \beta X = \alpha\,\xi$. The scalar product with ξ gives $0 = g(\nabla_X \xi, \xi) = \beta\, g(X, \xi) - \alpha$; hence, $\alpha = \beta\, g(X, \xi)$. This proves (12). Taking $X = \xi$ in (12), we get $\nabla_\xi \xi = 0$, i.e., ξ is a geodesic vector field. By this and the definition of $d\eta$, we get (13). □

Recall the co-boundary formula for exterior derivative d on a 2-form Φ,

$$\begin{aligned} d\Phi(X, Y, Z) = \frac{1}{3}\big\{&X\,\Phi(Y, Z) + Y\,\Phi(Z, X) + Z\,\Phi(X, Y) \\ &- \Phi([X, Y], Z) - \Phi([Z, X], Y) - \Phi([Y, Z], X)\big\}. \end{aligned} \tag{14}$$

Proposition 2 *A weak β-Kenmotsu manifold $M^{2n+1}(\varphi, Q, \xi, \eta, g)$ with $n > 1$ is a weak almost contact metric manifold satisfying $N^{(1)} = 0$, $d\eta = 0$ and $3\,d\Phi = 2\,\beta\,\eta \wedge \Phi$.*

Proof Using (12), we obtain

$$(\nabla_X \eta)Y = Xg(\xi, Y) - g(\xi, \nabla_X Y) = g(\nabla_X \xi, Y) \overset{(12)}{=} \beta\{g(X, Y) - \eta(X)\,\eta(Y)\} \tag{15}$$

for all $X, Y \in \mathfrak{X}_M$. By (15), $(\nabla_X \eta)Y = (\nabla_Y \eta)X$ is true. Thus, we obtain

$$0 = (\nabla_X \eta)Y - (\nabla_Y \eta)X \overset{(15)}{=} -\beta\, g([X, Y], \xi)$$

for $X, Y \in \ker\eta$, that means integrability of the distribution $\ker\eta$, or equivalently, $d\eta(X, Y) = 0$ for all $X, Y \in \ker\eta$. By this and (13), we find $d\eta = 0$. Using (11) and (14) and assuming $\nabla X = \nabla Y = \nabla Z = 0$ at a point $x \in M$, we get

$$3\,d\Phi = 2\,\beta\{\eta(X)g(\varphi Z, Y) + \eta(X)g(\varphi X, Z) + \eta(X)g(\varphi Y, X)\}.$$

On the other hand, we have

$$\eta \wedge \Phi = \eta(X)g(\varphi Z, Y) - \eta(X)g(\varphi Z, X) + \eta(X)g(\varphi Y, X).$$

Thus, $3\,d\Phi = 2\,\beta\,\eta \wedge \Phi$ is valid. Using (10) and (11), we obtain $[\varphi, \varphi] = 0$; hence, the normality condition $N^{(1)} = 0$. □

The condition $d\beta \wedge \eta = 0$ follows from (11) if $\dim M > 3$, and it does not hold in general if $\dim M = 3$. Indeed, by Proposition 2, we get $0 = 3\,d^2\Phi = 2\,d\beta \wedge \eta \wedge \Phi$, which gives the assertion.

Definition 2 A Riemannian manifold $(\bar M, \bar g)$ equipped with a skew-symmetric (1,1)-tensor $\bar\phi$ such that the tensor $\bar\phi^2$ is negative definite will be called a *weak Kählerian manifold*, if $\bar\nabla\bar\phi = 0$, where $\bar\nabla$ is the Levi-Civita connection of $\bar g$.

Many authors investigated the problem of finding skew-symmetric parallel 2-tensors (different from a complex structure) on Riemannian spaces and proved that some spaces do not admit such tensors (for example, [9]), or classified such tensors (for example, [8]).

Example 1 Let $(\bar M, \bar g)$ be a Riemannian manifold. A *warped product* $\mathbb{R} \times_\sigma \bar M$ with a real line $\mathbb{R}$ (called the *base*) is the product $M = \mathbb{R} \times \bar M$ with the metric $g = dt^2 \oplus \sigma^2(t)\,\bar g$, where $\sigma > 0$ is a smooth function on $\mathbb{R}$, called a warping function. Set $\xi = \partial_t$ and denote by $\mathcal{D}$ the distribution on M orthogonal to ξ. The Levi-Civita connections, ∇ of g and $\bar\nabla$ of $\bar g$, are related as follows:
(i) $\nabla_\xi\, \xi = 0$.
(ii) $\nabla_X\, \xi = \nabla_\xi\, X = \xi(\log\sigma)X$ for $X \in \mathcal{D}$.
(iii) $\pi_{1*}(\nabla_X Y) = -g(X, Y)\,\xi(\log\sigma)$, where $\pi_1 : M \to \mathbb{R}$ is the orthoprojector.
(iv) $\pi_{2*}(\nabla_X Y)$ is the lift of $\bar\nabla_X Y$, where $\pi_2 : M \to \bar M$ is the orthoprojector.
Submanifolds $\{t\} \times \bar M$ (called the *fibers*) are totally umbilical, i.e., the Weingarten operator $A_\xi = -\nabla\xi$ on $\mathcal{D}$ is conformal with the factor $-\xi(\log\sigma)$, see (iv). Note that σ is constant along the fibers; thus, $X(\sigma) = \xi(\sigma)\,\eta(X)$ for all $X \in \mathfrak{X}_M$.

Let $(\bar M, \bar g, \bar\phi)$ be a weak Kählerian manifold, see Definition 2 (e.g., a 4-dimensional Lie group, whose metric Lie algebra belongs to [8, Table 4]) and $\sigma \in C^\infty(\mathbb{R}_+)$, $\partial_t\sigma \ne 0$. Then the warped product $\mathbb{R} \times_\sigma \bar M$ is a weak β-Kenmotsu manifold with $\beta = (\partial_t\sigma)/\sigma$. Indeed, define tensors on $M = \mathbb{R} \times \bar M$,

$$\varphi = \begin{pmatrix} 0 & 0 \\ 0 & \bar\phi \end{pmatrix}, \quad Q = \begin{pmatrix} 1 & 0 \\ 0 & -\bar\phi^2 \end{pmatrix}, \quad \xi = \begin{pmatrix} \partial_t \\ 0 \end{pmatrix}, \quad \eta = (dt, 0), \quad g = \begin{pmatrix} dt^2 & 0 \\ 0 & \sigma^2\bar g \end{pmatrix}. \tag{16}$$

Using (16), we show that (4)–(5) and (7) are valid. Note that

$$X(\beta) = 0 \quad (X \perp \xi). \tag{17}$$

If $X, Y \in \mathcal{D}$, then

$$\pi_{2*}((\nabla_X\, \varphi)Y) \overset{(4)}{=} (\bar\nabla_X\, \bar\varphi)Y = 0, \quad \pi_{1*}((\nabla_X\, \varphi)Y) \overset{(3)}{=} -\beta\, g(X, \varphi Y).$$

If $X \in \mathcal{D}$ and $Y = \xi$, then

$$(\nabla_X\, \varphi)\xi = -\varphi\nabla_X\, \xi \overset{(2)}{=} -\beta\, \varphi X.$$

If $X = \xi$ and $Y \in \mathcal{D}$, then

$$(\nabla_\xi\,\varphi)Y \overset{(2)}{=} \beta(\varphi Y) - \varphi(\beta Y) = 0.$$

Also, using (1), we get $(\nabla_\xi\,\varphi)\xi = 0$. By the above, (11) is true. In particular, if $(\partial_t\sigma)/\sigma \equiv 1$, then $\mathbb{R}\times_\sigma \bar M$ is a weak Kenmotsu manifold.

Theorem 1 *Every point of a weak β-Kenmotsu manifold $M(\varphi, Q, \xi, \eta, g)$ has a neighborhood U that is a warped product $(-\varepsilon, \varepsilon)\times_\sigma \bar M$, where $(\partial_t\sigma)/\sigma = \beta$, $(\bar M, \bar g, \bar\phi)$ is a weak Kählerian manifold, and the structure $(\xi, \eta, \varphi, Q, g)$ is given on U as (16). Thus, (17) is valid.*

Proof We show integrability of $\mathcal{D}$: if $X, Y \in \mathcal{D}$, then using $d\eta = 0$ (see Proposition 2), we get $\eta([X, Y]) = 0$. Next, by $\nabla_\xi\,\xi \perp \xi$ and $(\nabla_\xi\,\varphi)\,\xi = 0$, we get $\nabla_\xi\,\xi = 0$; thus, ξ is a geodesic vector field. By (12), the Weingarten operator $A_\xi = -\nabla\xi$ on $\mathcal{D}$ is conformal, i.e., $\mathcal{D}$ is totally umbilical. Thus, any point of M has a warped product neighborhood $(-\varepsilon, \varepsilon)\times_\sigma \bar M$, where $(\partial_t\sigma)/\sigma = \beta$, and (16) holds. To show $\bar\nabla\bar\phi = 0$ we find $(\bar\nabla_X\,\bar\varphi)Y \overset{(4)}{=} \pi_{2*}((\nabla_X\,\varphi)Y) = 0$ for $X, Y \in \mathcal{D}$. □

Remark 1 From Example 1 and Theorem 1, we obtain the following generalization of [10, Theorem 2.3]. A weak almost contact metric manifold $M(\phi, Q, \xi, \eta, g)$ is a weak β-Kenmotsu manifold if and only if the following conditions are valid:

(a) ξ-trajectories are geodesics, i.e., $\nabla_\xi\,\xi = 0$, and φ is ξ-invariant, i.e., $\pounds_\xi\,\varphi = 0$,
(b) the distribution $\ker\eta$ is integrable and defines a totally umbilical foliation $\mathcal{F}$ of codimension one, whose leaves have constant mean curvature,
(c) a weak structure $(\tilde g, \tilde\phi)$ induced on any leaf $\tilde M \in \mathcal{F}$ is weak Kählerian.

Example 2 Let $(\bar M, \bar g, \bar\phi)$ be a weak Kählerian manifold and $\sigma(t) = c\,e^t$ ($c = const \ne 0$) a function on a line $\mathbb{R}$. Then the warped product manifold $M = \mathbb{R}\times_\sigma \bar M$ has a weak almost contact metric structure which satisfies (11) with $\beta \equiv 1$.

Corollary 1 *Any point of a weak Kenmotsu manifold (M, g) has a neighborhood that is a warped product $(-\varepsilon, \varepsilon)\times_\sigma \bar M$, where $\sigma = c\,e^t$ ($c = const \ne 0$) and $(\bar M, \bar g, \bar\phi)$ is a weak Kählerian manifold.*

The curvature tensor of g is given by $R_{X,Y} = [\nabla_X, \nabla_Y] - \nabla_{[X,Y]},\ X, Y \in \mathfrak{X}_M$. The Ricci operator $\mathrm{Ric}^\sharp$ associated with the Ricci tensor is given by the equality

$$\mathrm{Ric}(X, Y) = g(\mathrm{Ric}^\sharp X, Y), \quad X, Y \in \mathfrak{X}_M.$$

Lemma 2 (see [5, 10, 14] for $\beta = 1$) *The following formulas hold on weak β-Kenmotsu manifolds:*

$$R_{X,Y}\,\xi = (\xi(\beta) + \beta^2)(\eta(X)Y - \eta(Y)X) \quad (X, Y \in \mathfrak{X}_M), \tag{18}$$

$$\mathrm{Ric}^\sharp\,\xi = -2n(\xi(\beta) + \beta^2)\xi. \tag{19}$$

Proof First taking covariant derivative of (12) along $Y \in \mathfrak{X}_M$, we get

$$\nabla_Y \nabla_X \xi = Y(\beta)\,(X - \eta(X)\xi) + \beta\,\big\{\nabla_Y X - \eta(\nabla_Y X)\xi - \beta\, g(X,Y)\xi - \beta\,\eta(Y)X + 2\beta\,\eta(X)\eta(Y)\xi\big\}.$$

Repeated application of (12) and the foregoing equation in the Riemann curvature tensor R of the Riemannian manifold (M, g), we get

$$R_{X,Y}\,\xi = X(\beta)(Y - \eta(Y)\xi) - Y(\beta)(X - \eta(X)\xi) + \beta^2\big\{\eta(X)Y - \eta(Y)X\big\}. \quad (20)$$

As $X(\beta) = \xi(\beta)\eta(X)$, the required result (18) follows from (20). By (18) with $Y = \xi$, we get

$$R_{X,\xi}\,\xi = (\xi(\beta) + \beta^2)(\eta(X)\,\xi - X), \quad (X \in \mathfrak{X}_M).$$

From this and $g(\mathrm{Ric}^\sharp\,\xi, X) = \sum_i g(R_{e_i,X}\,\xi, e_i)$, where (e_i) is a local orthonormal basis of $\mathcal{D}$, we get (19). □

The following formulas are known, e.g., [5, Eqs. (6), (7)]:

$$(\nabla_X \pounds_V\, g)(Y, Z) = g((\pounds_V \nabla)(X, Y), Z) + g((\pounds_V \nabla)(X, Z), Y), \quad (21)$$

$$(\pounds_V\, R)_{X,Y} Z = (\nabla_X \pounds_V \nabla)(Y, Z) - (\nabla_Y \pounds_V \nabla)(X, Z). \quad (22)$$

The scalar curvature of g is given by $r = \mathrm{trace}_g\, \mathrm{Ric}$.

The following lemma generalizes [5, Lemma 2] with $\beta \equiv 1$.

Lemma 3 *The following formulas hold on weak β-Kenmotsu manifolds:*

$$(\nabla_\xi\, \mathrm{Ric}^\sharp)X = -2\beta\, \mathrm{Ric}^\sharp X - 2(\xi(\beta^2) + 2\,n\beta^3)X + 2(1-n)\,\xi(\beta^2)\eta(X)\,\xi, \quad (23)$$

$$\xi(r) = -2\,\beta\,(r + 2\,n(2\,n+1)\beta^2) - 6\,n\,\xi(\beta^2), \quad (24)$$

for all $X \in \mathfrak{X}_M$; in particular, if $\beta = const$, then

$$(\nabla_\xi\, \mathrm{Ric}^\sharp)X = -2\,\beta\, \mathrm{Ric}^\sharp X - 4n\,\beta^3 X.$$

Proof This is similar to the proof of [5, Lemma 2]. Using (12), we get

$$(\pounds_\xi\, g)(Y, Z) = g(\nabla_Y\,\xi, Z) + g(\nabla_Z\,\xi, Y) = 2\,\beta(g(Y,Z) - \eta(Y)\,\eta(Z)). \quad (25)$$

Taking covariant derivative of (25) in the X-direction and using (12) gives

$$(\nabla_X \pounds_\xi\, g)(Y, Z) = 2\,\beta^2\big(2\,\eta(X)\,\eta(Y)\,\eta(Z) - g(X,Y)\,\eta(Z) - g(X,Z)\,\eta(Y)\big)$$

for all $X, Y, Z \in \mathfrak{X}_M$. Using this in (21) with $V = \xi$, we obtain

$$g((\pounds_\xi \nabla)(X, Y), Z) + g((\pounds_\xi \nabla)(X, Z), Y) = 2\,\beta^2\big(2\,\eta(X)\,\eta(Y)\,\eta(Z) - g(X,Y)\,\eta(Z) - g(X,Z)\,\eta(Y)\big). \quad (26)$$

By a combinatorial computation, we find

$$\begin{aligned} & g((\pounds_\xi \nabla)(Y, Z), X) + g((\pounds_\xi \nabla)(Y, X), Z) \\ &= 2\,\beta^2\big(2\,\eta(X)\,\eta(Y)\,\eta(Z) - g(Y, Z)\,\eta(X) - g(Y, X)\,\eta(Z)\big), \\ & g((\pounds_\xi \nabla)(Z, X), Y) + g((\pounds_\xi \nabla)(Z, Y), X) \\ &= 2\,\beta^2\big(2\,\eta(X)\,\eta(Y)\,\eta(Z) - g(Z, X)\,\eta(Y) - g(Z, Y)\,\eta(X)\big). \end{aligned}$$

Subtracting (26) from the sum of the last two equations gives

$$(\pounds_\xi \nabla)(Y, Z) = 2\,\beta^2\big(\eta(Y)\,\eta(Z) - g(Y, Z)\big)\,\xi \quad (Y, Z \in \mathfrak{X}_M). \tag{27}$$

Taking covariant derivative of (27) in the X-direction and using (12) gives

$$\begin{aligned} (\nabla_X \pounds_\xi \nabla)(Y, Z) = 2\,\beta^3\big\{\big(& g(X, Y)\,\eta(Z) + g(X, Z)\,\eta(Y) + g(Y, Z)\,\eta(X) \\ & - 3\,\eta(X)\,\eta(Y)\,\eta(Z)\big)\,\xi + (\eta(Y)\,\eta(Z) - g(Y, Z))\,X\big\} \\ & + 2\,\eta(X)\,\xi(\beta^2)\big(\eta(Y)\,\eta(Z) - g(Y, Z)\big)\,\xi. \end{aligned}$$

Using this in (22) with $V = \xi$, we obtain

$$\begin{aligned} (\pounds_\xi\, R)_{X,Y} Z = 2\,\beta^3\big(& (g(X, Z) - \eta(X)\,\eta(Z))\,Y - (g(Y, Z) - \eta(Y)\,\eta(Z))\,X\big) \\ & + 2\,\xi(\beta^2)\big(g(X, Z)\,\eta(Y) - g(Y, Z)\,\eta(X)\big)\,\xi. \end{aligned} \tag{28}$$

Contracting (28) over X, we deduce

$$(\pounds_\xi\, \mathrm{Ric})(Y, Z) = 2\,(\xi(\beta^2) + 2\,n\beta^3)(\eta(Y)\,\eta(Z) - g(Y, Z)). \tag{29}$$

Taking the Lie derivative of equality $\mathrm{Ric}(Y, Z) = g(\mathrm{Ric}^\sharp\, Y, Z)$, we obtain

$$(\pounds_\xi\, \mathrm{Ric})(Y, Z) = (\pounds_\xi\, g)(\mathrm{Ric}^\sharp\, Y, Z) + g((\pounds_\xi\, \mathrm{Ric}^\sharp)Y, Z). \tag{30}$$

On the other hand, replacing Y by $\mathrm{Ric}^\sharp\, Y$ in (25) and using (19), we obtain

$$\begin{aligned} (\pounds_\xi\, g)(\mathrm{Ric}^\sharp\, Y, Z) &= 2\,\beta\big(g(\mathrm{Ric}^\sharp\, Y, Z) - g(\mathrm{Ric}^\sharp\, \xi, Y)\,\eta(Z)\big) \\ &= 2\,\beta\big(g(\mathrm{Ric}^\sharp\, Y, Z) + 2\,n(\xi(\beta) + \beta^2)\,\eta(Y)\,\eta(Z)\big). \end{aligned} \tag{31}$$

Applying (30) and (31) in (29) we get

$$(\pounds_\xi\, \mathrm{Ric}^\sharp)Y = -2\,\beta\,\mathrm{Ric}^\sharp\, Y - 2\,(\xi(\beta^2) + 2\,n\,\beta^3)Y + 2\,(1 - n)\,\xi(\beta^2)\,\eta(Y)\,\xi. \tag{32}$$

Using (12), we calculate

$$
\begin{aligned}
(\pounds_\xi \operatorname{Ric}^\sharp)Y &= \pounds_\xi(\operatorname{Ric}^\sharp Y) - \operatorname{Ric}^\sharp \pounds_\xi Y\\
&= \nabla_\xi(\operatorname{Ric}^\sharp Y) - \nabla_{\operatorname{Ric}^\sharp Y}\,\xi - \operatorname{Ric}^\sharp \nabla_\xi Y + \operatorname{Ric}^\sharp \nabla_Y \xi\\
&= (\nabla_\xi \operatorname{Ric}^\sharp)Y - \operatorname{Ric}^\sharp Y - 2n\,\eta(Y)\,\xi + \operatorname{Ric}^\sharp Y + 2n\,\eta(Y)\,\xi = (\nabla_\xi \operatorname{Ric}^\sharp)Y.
\end{aligned}
$$

Using this in (32), gives (23). Contracting (23), we get (24). □

The following theorem and corollary generalize [5, Theorem 1 and Corollary 1] with $\beta \equiv 1$.

Theorem 2 *Let $M^{2n+1}(\varphi, Q, \xi, \eta, g)$ be a weak β-Kenmotsu manifold with $\beta = const \neq 0$. If $\nabla_\xi \operatorname{Ric}^\sharp = 0$, then (M, g) is an Einstein manifold with scalar curvature $r = -2\,n(2n+1)\,\beta^2$.*

Proof By Lemma 3, $\operatorname{Ric}^\sharp X = -2\,n\,\beta^2 X$. Since $\dim M > 2$, M is Einstein. □

A 3-dimensional Einstein manifold has constant sectional curvature. Thus, we get the following

Corollary 2 *Let $M(\varphi, Q, \xi, \eta, g)$ be a 3-dimensional weak β-Kenmotsu manifold with $\beta = const \neq 0$. If $\nabla_\xi \operatorname{Ric}^\sharp = 0$, then M has sectional curvature $-\beta^2$.*

4 η-Ricci Solitons on Weak β-Kenmotsu Manifolds

Here, we study η-Ricci solitons on weak β-Kenmotsu manifolds and generalize some results in [11].

Lemma 4 *Let $M^{2n+1}(\varphi, Q, \xi, \eta, g)$ be a weak β-Kenmotsu manifold. If g represents an η-Ricci soliton with the potential vector field V, then for all $X \in \mathfrak{X}_M$:*

$$
\begin{aligned}
(\pounds_V R)_{X,\xi}\,\xi &= -2\,\xi(\beta)\operatorname{Ric}^\sharp X + \big(4\,n\,\xi(\beta^3) + 2\,\xi(\xi(\beta^2))\big)X\\
&\quad -\big(8\,n\,\xi(\beta^3) + 2(n+1)\,\xi(\xi(\beta^2))\big)\,\eta(X)\,\xi;
\end{aligned}
$$

moreover, if $\beta = const \neq 0$, then $(\pounds_V R)_{X,\xi}\xi = 0$ (for $\beta = 1$ see [11, Lemma 3.1]).

Proof Taking the covariant derivative of (2) for $Z \in \mathfrak{X}_M$ and using (11), we get

$$
\begin{aligned}
&\tfrac{1}{2}\,(\nabla_Z \pounds_V\, g)(X, Y) = -(\nabla_Z \operatorname{Ric})(X, Y)\\
&-\beta\mu\{g(X, Z)\,\eta(Y) + g(Y, Z)\,\eta(X) - 2\,\eta(X)\,\eta(Y)\,\eta(Z)\}\ \ (X, Y \in \mathfrak{X}_M). \qquad (33)
\end{aligned}
$$

Recall the commutation formula with the tensor $\pounds_V \nabla$, see [15, p. 23],

$$
\begin{aligned}
&(\pounds_V \nabla_Z\, g - \nabla_Z \pounds_V\, g - \nabla_{[V,Z]}\, g)(X, Y)\\
&= -g((\pounds_V \nabla)(Z, X), Y) - g((\pounds_V \nabla)(Z, Y), X). \qquad (34)
\end{aligned}
$$

Since Riemannian metric is parallel, $\nabla g = 0$, it follows from (34) that

$$(\nabla_Z \pounds_V\, g)(X, Y) = g((\pounds_V \nabla)(Z, X), Y) + g((\pounds_V \nabla)(Z, Y), X). \tag{35}$$

Plugging it into (33), we obtain

$$\begin{gathered} g((\pounds_V \nabla)(Z, X), Y) + g((\pounds_V \nabla)(Z, Y), X) = -2(\nabla_Z \mathrm{Ric})(X, Y) \\ -2\,\beta\mu\{g(X, Z)\,\eta(Y) + g(Y, Z)\,\eta(X) - 2\,\eta(X)\,\eta(Y)\,\eta(Z)\} \end{gathered} \tag{36}$$

for all $X, Y, Z \in \mathfrak{X}_M$. Cyclically rearranging X, Y and Z in (36), we obtain

$$\begin{aligned} g((\pounds_V \nabla)(X, Y), Z) = (\nabla_Z \mathrm{Ric})(X, Y) - (\nabla_X \mathrm{Ric})(Y, Z) - (\nabla_Y \mathrm{Ric})(Z, X) \\ - 2\,\beta\mu\{g(X, Y)\,\eta(Z) - \eta(X)\,\eta(Y)\,\eta(Z)\}. \end{aligned} \tag{37}$$

Note that $X(\beta) = \eta(X)\,\xi(\beta)$. Thus,

$$X(\xi(\beta) + \beta^2) = \eta(X)\,\xi(\xi(\beta) + \beta^2).$$

Taking covariant derivative of (19) along $X \in \mathfrak{X}_M$ and using (11), we obtain

$$(\nabla_X \mathrm{Ric}^\sharp)\,\xi = -\beta\,\mathrm{Ric}^\sharp X - 2\,n\beta(\xi(\beta) + \beta^2)X - 2n\,\eta(X)\,\xi(\xi(\beta) + \beta^2)\,\xi. \tag{38}$$

Substituting $Y = \xi$ in (37) yields:

$$g((\pounds_V \nabla)(X, \xi), Z) = (\nabla_Z \mathrm{Ric})(X, \xi) - (\nabla_X \mathrm{Ric})(\xi, Z) - (\nabla_\xi \mathrm{Ric})(Z, X).$$

Applying (23) and (38) to this, we obtain

$$(\pounds_V \nabla)(X, \xi) = 2\,\beta\,\mathrm{Ric}^\sharp X + 2\,(\xi(\beta^2) + 2\,n\,\beta^3)X + 2\,(n-1)\,\xi(\beta^2)\,\eta(X)\,\xi \tag{39}$$

for any $X \in \mathfrak{X}_M$. Next, using (12) in the covariant derivative of (39) for Y, yields

$$\begin{gathered} (\nabla_Y(\pounds_V \nabla))(X, \xi) + \beta(\pounds_V \nabla)(X, Y) = 2\,\beta(\nabla_Y \mathrm{Ric}^\sharp)X \\ +2(\xi(\beta) + \beta^2)\,\eta(Y)\,\mathrm{Ric}^\sharp X + 2\big(\beta(\xi(\beta^2) + 2\,n\,\beta^3) - \xi(\xi(\beta^2) + 2\,n\,\beta^3)\big)\,\eta(Y)X \\ +2(n-1)\,\beta\,\xi(\beta^2)\eta(X)Y + 2(n-1)\big(\xi(\xi(\beta^2)) - 3\,\beta\,\xi(\beta^2)\big)\,\eta(X)\,\eta(Y)\,\xi \end{gathered}$$

for any $X \in \mathfrak{X}_M$. Plugging this in the following formula (see [15], p. 23):

$$(\pounds_V R)_{X,Y} Z = (\nabla_X(\pounds_V \nabla))(Y, Z) - (\nabla_Y(\pounds_V \nabla))(X, Z) \tag{40}$$

with $Z = \xi$, we deduce for all $X, Y \in \mathfrak{X}_M$:

$$
\begin{aligned}
(\pounds_V R)_{X,Y}\,\xi &= 2\,\beta\{(\nabla_X \operatorname{Ric}^\sharp)Y - (\nabla_Y \operatorname{Ric}^\sharp)X\}\\
&+2\,(\xi(\beta)+\beta^2)\{\eta(X)\operatorname{Ric}^\sharp Y - \eta(Y)\operatorname{Ric}^\sharp X\} + 2\big\{\beta(\xi(\beta^2)+2\,n\,\beta^3)\\
&-\xi(\xi(\beta^2)+2\,n\,\beta^3) - (n-1)\,\beta\,\xi(\beta^2)\big\}(\eta(X)Y - \eta(Y)X). \qquad (41)
\end{aligned}
$$

Substituting $Y=\xi$ in (41) gives

$$
\begin{aligned}
(\pounds_V R)_{X,\xi}\,\xi &= 2\,\beta\{(\nabla_X \operatorname{Ric}^\sharp)\xi - (\nabla_\xi \operatorname{Ric}^\sharp)X\}\\
&+2\,(\xi(\beta)+\beta^2)\{\eta(X)\operatorname{Ric}^\sharp \xi - \operatorname{Ric}^\sharp X\}\\
&+2\,\big\{\beta(\xi(\beta^2)+2\,n\,\beta^3) - \xi(\xi(\beta^2)+2\,n\,\beta^3) - (n-1)\,\beta\,\xi(\beta^2)\big\}(\eta(X)\xi - X),
\end{aligned}
$$

then using (19), (23) and (38), yields our result. □

Now, we consider an η-Einstein weak β-Kenmotsu metric as an η-Ricci soliton and characterize the Einstein metrics in such a wider class of metrics.

Lemma 5 *On an η-Einstein weak β-Kenmotsu manifold $M^{2n+1}(\varphi, Q, \xi, \eta, g)$, the expression of* $\operatorname{Ric}^\sharp$ *is the following:*

$$
\operatorname{Ric}^\sharp X = \Big(\xi(\beta)+\beta^2+\frac{r}{2n}\Big)X - \Big((2n+1)(\xi(\beta)+\beta^2)+\frac{r}{2n}\Big)\,\eta(X)\,\xi \quad (X\in\mathfrak{X}_M). \qquad (42)
$$

Proof First, tracing (3) gives $r=(2\,n+1)\,a+b$ and putting $X=Y=\xi$ in (3) and using (19), yields $a+b=-2\,n(\xi(\beta)+\beta^2)$. Therefore, $r=2\,n(a-\xi(\beta)-\beta^2)$ and

$$
a=\xi(\beta)+\beta^2+\frac{r}{2n},\quad b=-(2n+1)(\xi(\beta)+\beta^2)-\frac{r}{2n}.
$$

Hence, (3) gives the required result. □

Theorem 3 *Let $M^{2n+1}(\varphi, Q, \xi, \eta, g)$, $n>1$, be an η-Einstein weak β-Kenmotsu manifold with $\beta=const\neq 0$. If g represents an η-Ricci soliton with the potential vector field V, then (M,g) is an Einstein manifold with scalar curvature $r=-2\,n(2\,n+1)\,\beta^2$.*

Proof Taking covariant derivative of (42) for Y-direction and using (12), we get

$$
\begin{aligned}
(\nabla_Y \operatorname{Ric}^\sharp)X &= \tfrac{Y(r)}{2n}\big\{X-\eta(X)\xi\big\}\\
&-\Big((2n+1)\beta^2+\tfrac{r}{2n}\Big)\beta\big\{g(X,Y)\xi+\eta(X)(Y-2\eta(Y)\xi)\big\} \qquad (43)
\end{aligned}
$$

for all $X\in\mathfrak{X}_M$. Now, contracting (43) over Y, we get

$$
(n-1)\,X(r) = -\{\xi(r)+2n\beta(r+2n(2n+1)\beta^2)\}\,\eta(X).
$$

Using (24) in the previous equation, we have for $n > 1$,

$$X(r) = \xi(r)\,\eta(X), \tag{44}$$

i.e., $X(r) = 0$ $(X \perp \xi)$. Using (43) and (42) in (41), one can obtain

$$\begin{aligned} n\,(\pounds_V R)_{X,Y}\,\xi &= \beta\left\{X(r)(Y - \eta(Y)\xi) - Y(r)(X - \eta(X)\xi)\right\} \\ &+2\,\beta^2\,(r + 2n(2n+1)\beta^2)(\eta(X)Y - \eta(Y)X), \end{aligned} \tag{45}$$

as β is a constant. Using (44) in (45) and applying Lemma 4 gives $(\pounds_V R)_{X,Y}\,\xi = 0$ for all $X, Y \in \mathfrak{X}_M$. Therefore, we get

$$(\pounds_V\,\mathrm{Ric})(Y, \xi) = 0 \quad (Y \in \mathfrak{X}_M). \tag{46}$$

As β is constant, Eq. (19) gives $\mathrm{Ric}(Y, \xi) = -2n\,\beta\,\eta(Y)$. Taking its covariant derivative along V yields

$$(\pounds_V\,\mathrm{Ric})(Y, \xi) + \mathrm{Ric}(Y, \pounds_V\xi) = -2n\,\beta^2\,(\pounds_V\eta)(Y)$$

for all $X \in \mathfrak{X}_M$. Inserting (46) in the preceding equation we have

$$\mathrm{Ric}(Y, \pounds_V\xi) = -2n\,\beta^2\left\{(\pounds_V g)(Y, \xi) + g(X, \pounds_V\xi)\right\}. \tag{47}$$

In view of (2), (19) and (42), the Eq. (47) becomes

$$\begin{aligned} &\Big(\beta^2 + \frac{r}{2n}\Big)\,g(Y, \pounds_V\xi) - \Big((2n+1)\beta^2 + \frac{r}{2n}\Big)\,\eta(\pounds_V\xi)\,\eta(Y) \\ &\quad = -4n\,\beta^2\,(2n\beta^2 - \lambda - \mu)\eta(Y) - 2n\,\beta^2\,g(Y, \pounds_V\xi). \end{aligned} \tag{48}$$

Inserting $Y = \xi$ in (48) yields $\lambda + \mu = 2n\,\beta^2$. Thus, Lie-derivative of $g(\xi, \xi) = 1$ and (2) give us $\eta(\pounds_V\xi) = 0$. In view of these, Eq. (48) reduces to

$$\Big((2n+1)\beta^2 + \frac{r}{2n}\Big)\,\pounds_V\xi = 0. \tag{49}$$

Case I. Suppose $(2n+1)\beta^2 + \frac{r}{2n} = 0$. By (42), $\mathrm{Ric}^\sharp X = -2\,n\beta^2\,X$; thus, our (M, g) is an Einstein manifold with scalar curvature $r = -2\,n(2\,n+1)\,\beta^2$.

Case II. Suppose $(2n+1)\beta^2 + \frac{r}{2n} \neq 0$ on some open set $\mathcal{O}$ of M. Then $\pounds_V\xi = [V, \xi] - 0$ on $\mathcal{O}$, see (49). It follows that

$$\nabla_\xi\,V = \nabla_V\,\xi = \beta\{V - \eta(V)\,\xi\}, \tag{50}$$

where we have used (12). Next, we recall the well known formula, see [15]:

$$(\pounds_V \nabla)(X, Y) = \nabla_X \nabla_Y V - \nabla_{\nabla_X Y} V + R_{V,X} Y. \tag{51}$$

Replacing Y by ξ in (51) and using (12), (18) and (50), we get $(\pounds_V \nabla)(X, \xi) = 0$. Further, inserting (42) in (39), we get

$$(\pounds_V \nabla)(X, \xi) = 2\Big((2n+1)\beta^2 + \frac{r}{2n}\Big)\,(X - \eta(X)\xi).$$

Combining both, we get $(2n+1)\beta^2 + \frac{r}{2n} = 0$, – a contradiction. □

The next consequence of Theorem 3 is [11, Theorem 3.2].

Corollary 3 *Let $M^{2n+1}(\varphi, \xi, \eta, g)$, $n > 1$, be an η-Einstein Kenmotsu manifold. If g represents an η-Ricci soliton, then (M, g) is an Einstein manifold with scalar curvature $r = -2\,n(2n+1)$.*

Definition 3 *(see* [12]*)* A vector field X on a weak contact metric manifold $M^{2n+1}(\varphi, Q, \xi, \eta, g)$ is called a *weak contact vector field* (or, a *weak contact infinitesimal transformation*), if the flow of X preserves the form η, i.e., there exists a smooth function $\rho : M \to \mathbb{R}$ such that

$$\pounds_X\, \eta = \rho\, \eta, \tag{52}$$

and if $\rho = 0$, then the vector field X is said to be *strict weak contact vector field.*

We consider a weak β-Kenmotsu metric as an η-Ricci soliton, whose potential vector field V is weak contact, or V is collinear to ξ. First, we derive the following.

Lemma 6 *Let $M^{2n+1}(\varphi, Q, \xi, \eta, g)$ be a weak β-Kenmotsu manifold. If g represents an η-Ricci soliton with a potential vector field V, then $\lambda + \mu = 2n(\xi(\beta)+\beta^2)$.*

Proof For a weak β-Kenmotsu manifold, we get

$$(\pounds_V\, g)(\xi, \xi) = g(\xi, [V, \xi]) = 0. \tag{53}$$

Thus, using (2) in the Lie derivative of $g(\xi, \xi) = 1$, we obtain

$$\mathrm{Ric}(\xi, \xi) + \lambda + \mu = 0.$$

Then, using $\mathrm{Ric}(\xi, \xi) = -2\,n\,(\xi(\beta) + \beta^2)$, see (19), we achieve the required result. □

Recall the known formula (see [15, p. 23]):

$$(\pounds_V \nabla)(X, Y) = \pounds_V(\nabla_X Y) - \nabla_X\,(\pounds_V Y) - \nabla_{[V,X]} Y \quad (X, Y \in \mathfrak{X}_M). \tag{54}$$

Theorem 4 *Let $M^{2n+1}(\varphi, Q, \xi, \eta, g)$, $n > 1$, be a weak β-Kenmotsu manifold with $\beta = const \neq 0$. If g represents an η-Ricci soliton with a weak contact potential vector field V, then V is strict and (M, g) is an Einstein manifold with scalar curvature $r = -2\,n(2n+1)\beta^2$.*

Proof Taking Lie derivative of $\eta(X) = g(X, \xi)$ along V and using (2), (52) and Lemma 6, we obtain $\pounds_V\, \xi = \rho\, \xi$. By (53), using the equality $(\pounds_V\, g)(\xi, \xi) = -$pg $2\, g(\pounds_V\, \xi, \xi)$, we get $\eta(\pounds_V\, \xi) = 0$. Thus, $\rho = 0$. Therefore, $\pounds_V\, \xi = 0$ and V is strict. Also, (52) gives $\pounds_V\, \eta = 0$. Setting $Y = \xi$ in (54) and using (12) and the equality $(\pounds_V\, \eta)(X) = V(\eta(X)) - \eta([V, X])$, we find

$$\begin{aligned}(\pounds_V \nabla)(X, \xi) &= \beta \pounds_V (X - \eta(X)\xi) - \beta(\pounds_V X - \eta(\pounds_V X)\, \xi) \\ &= -\beta\{(\pounds_V\, \eta)(X)\xi + \eta(X)\pounds_V\, \xi\}. \end{aligned} \tag{55}$$

It follows from (55) that $(\pounds_V \nabla)(X, \xi) = 0$. Using (39), we prove the rest part of this theorem. □

Theorem 5 *Let $M^{2n+1}(\varphi, Q, \xi, \eta, g)$, $n > 1$ be a weak β-Kenmotsu manifold with $\beta = const \neq 0$. Suppose that g represents an η-Ricci soliton with a non-zero potential vector field V collinear to ξ: $V = \delta\, \xi$ for a smooth function $\delta \neq 0$ on M. Then $\delta = const$ and (M, g) is an Einstein manifold with scalar curvature $r = -2n(2n + \delta(\beta - 1))$.*

Proof Using (11) in the covariant derivative of $V = \delta\, \xi$ with any $X \in \mathfrak{X}_M$ yields

$$\nabla_X V = X(\delta)\, \xi + \delta\beta(X - \eta(X)\, \xi)$$

for any $X \in \mathfrak{X}_M$. Using this, the soliton equation (2) transforms into

$$2\, \mathrm{Ric}(X, Y) + X(\delta)\, \eta(Y) + Y(\delta)\, \eta(X) + 2(\delta\beta + \lambda)\, g(X, Y) - 2(\delta\beta - \mu)\, \eta(X)\, \eta(Y) = 0 \tag{56}$$

for all $X, Y \in \mathfrak{X}_M$. Inserting $X = Y = \xi$ in (56) and using (19) and Lemma 6, we get $\xi(\sigma) = 0$. It follows from (56) and (19) that $X(\delta) = 0$. Putting it into (56) gives

$$\mathrm{Ric} = -(\delta\beta + \lambda)\, g + (\delta\beta - \mu)\, \eta \otimes \eta. \tag{57}$$

This shows that (M, g) is an η-Einstein manifold, therefore, from Theorem 3 we conclude that (M, g) is an Einstein manifold. Thus, from (57) we have $\delta\beta = \mu$, therefore, $\delta + \lambda = 2n$ (follows from Lemma 6). This implies, using (57), that

$$\mathrm{Ric} = -(2n + \delta(\beta - 1))\, g;$$

hence, $r = -2n(2n + \delta(\beta - 1))$, as required. □

In [14], Kenmotsu found the following necessary and sufficient condition for a Kenmotsu manifold M to have constant φ-holomorphic sectional curvature H:

$$\begin{aligned} 4R_{X,Y}Z &= (H - 3)\{g(Y, Z)X - g(X, Z)Y\} \\ &+(H + 1)\{\eta(X)\, \eta(Z)Y - \eta(Y)\, \eta(Z)X + \eta(Y)g(X, Z)\xi - \eta(X)g(Y, Z)\xi \\ &+ g(X, \varphi Z)\varphi Y - g(Y, \varphi Z)\varphi X + 2g(X, \varphi Y)\varphi Z\}, \quad X, Y, Z \in \mathfrak{X}_M. \end{aligned} \tag{58}$$

The following corollary of Theorem 3 illustrates Lemma 6 and gives an example of a weak β-Kenmotsu manifold that admits an η-Ricci soliton.

Corollary 4 *Let a metric g of a warped product $\mathbb{R}\times_\sigma N^{2n}$, where $n>1$, $(\partial_t\sigma)/\sigma=\beta=const\neq 0$ and N is a weak Kählerian manifold, represent an η-Ricci soliton with the potential vector field V. Then g has constant φ-holomorphic sectional curvature $H=-\beta^2$.*

Proof Idea of the following construction is that if we choose the potential vector field $V=n\,\xi$, i.e., $\delta=n$, then $\mathbb{R}\times_\sigma N^{2n}$ is an η-Ricci soliton for $\lambda=n>1$ and $\mu=n$. It follows from (58) that

$$4\,\mathrm{Ric}^\sharp X=((2n-3)H-3(2n+1))X-(2n-3)(H+1)\,\eta(X)\xi \tag{59}$$

for any $X\in\mathfrak{X}_M$. Also, a warped product $\mathbb{R}\times_\sigma N^{2n}$, where $(\partial_t\sigma)/\sigma=\beta$ on the real line $\mathbb{R}$, and N^{2n} is a weak Kählerian manifold, admits the weak β-Kenmotsu structure. Thus, it follows from (59) that $\mathbb{R}\times_\sigma N^{2n}$ is an η-Einstein weak β-Kenmotsu manifold, and by Theorem 3, we conclude that it is an Einstein manifold. It follows from (3) and (59) that $b=0$, i.e., $H=-\beta^2$; hence, $a=-2\,n$. □

5 3-Dimensional Weak β-Kenmotsu Manifolds

Here, we complete Theorem 3 by studying the 3-dimensional case.

Lemma 7 *A 3-dimensional weak β-Kenmutsu manifold $M(\varphi,Q,\xi,\eta,g)$ with $\beta=const\neq 0$ represents an η-Einstein manifold.*

Proof The curvature tensor of a 3-dimensional Riemannian manifold has well-known form

$$\begin{aligned}R_{X,Y}Z=g(Y,Z)\,\mathrm{Ric}^\sharp X-g(X,Z)\,\mathrm{Ric}^\sharp Y+g(\mathrm{Ric}^\sharp Y,Z)X-g(\mathrm{Ric}^\sharp X,Z)Y\\-\tfrac{r}{2}\big(g(Y,Z)X-g(X,Z)Y\big)\end{aligned} \tag{60}$$

for all $X,Y,Z\in\mathfrak{X}_M$. Replacing $Y=Z=\xi$ in (60) and using $\mathrm{Ric}^\sharp\xi=-2n\beta^2\xi$, see (19), we get (as β is a constant)

$$\mathrm{Ric}^\sharp X=\Big(\beta^2+\frac{r}{2}\Big)X-\Big(3\,\beta^2+\frac{r}{2}\Big)\eta(X)\xi\quad(X\in\mathfrak{X}_M) \tag{61}$$

– the η-Einstein structure (3). Note that (61) generalizes [7, Lemma 1]. □

Using Lemma 7 and the proof of Theorem 3, we obtain the following.

Corollary 5 *If a 3-dimensional weak β-Kenmutsu manifold $M^3(\varphi, Q, \xi, \eta, g)$ with $\beta = const \neq 0$ and $X(r) = 0$ $(X \perp \xi)$ represents an η-Ricci soliton (2), then (M, g) has constant sectional curvature $-\beta^2$.*

Proof According to Theorem 3 (as (44) assumed), we deduce $r = -6\beta^2$; thus, (61) gives $\mathrm{Ric}^\sharp X = -2\beta^2 X$ for any $X \in \mathfrak{X}_M$. It follows from (60) that (M, g) is of constant negative curvature $-\beta^2$. □

Lemma 8 *If a 3-dimensional weak β-Kenmutsu manifold $M(\varphi, Q, \xi, \eta, g)$ with $\beta = const \neq 0$ represents an η-Ricci soliton (2) with the potential vector field V, then for all $X \in \mathfrak{X}_M$,*

$$(i)\ 2\,(\pounds_V \nabla)(X, \xi) = -\xi(r)\,(X - \eta(X)\xi), \quad \text{where} \quad \xi(r) = -2\,\beta\,(r + 6\,\beta^2), \tag{62}$$

$$\begin{aligned}(ii)\ &(r + 6\,\beta^2)\,\big\{g(X, \pounds_V\xi) - \eta(X)\,\eta(\pounds_V\xi)\big\} + X(\xi(r)) + \xi(\xi(r))\,\eta(X)\\ &- 4\big\{\beta\,\xi(r) - 2\,\beta^2\,(\lambda - 2\,\beta^2)\big\}\,\eta(X) = 0.\end{aligned} \tag{63}$$

Proof Let e_i be an orthonormal frame of T_pM at some point $p \in M$. It is well known that (getting by contraction of Bianchi's second identity, e.g., [15, p. 14])

$$\sum\nolimits_i g((\nabla_X \mathrm{Ric}^\sharp)e_i, e_i) = X(r), \quad \sum\nolimits_i g((\nabla_{e_i} \mathrm{Ric}^\sharp)X, e_i) = \frac{1}{2}\,X(r).$$

Thus, taking $X = Y = e_i$ in (37), summing over i and using the symmetry property of $\mathrm{Ric}^\sharp$, we get

$$\sum\nolimits_i (\pounds_V \nabla)(e_i, e_i) = -2\,\beta\mu\,\xi. \tag{64}$$

Now, combining equations (2) and (61), we get

$$(\pounds_V g)(Y, Z) = -(2\lambda + r + 2\,\beta^2)\,g(Y, Z) - (2\mu - r - 6\,\beta^2)\,\eta(Y)\,\eta(Z). \tag{65}$$

Taking covariant derivative of (65) along X and using (12), we have

$$\begin{aligned}(\nabla_X \pounds_V g)(Y, Z) &= -X(r)(g(Y, Z) - \eta(Y)\,\eta(Z))\\ -\beta(2\mu - r - 6\,\beta^2)&\big\{g(X, Y)\,\eta(Z) + g(X, Z)\eta(Y) - 2\,\eta(X)\,\eta(Y)\,\eta(Z)\big\}.\end{aligned} \tag{66}$$

Using (35) in (66) and cyclically rearranging X, Y and Z, we obtain

$$\begin{aligned}2\,g((\pounds_V \nabla)(X, Y), Z) &= -X(r)\,(g(Y, Z) - \eta(Y)\,\eta(Z))\\ -Y(r)\,(g(Z, X) - \eta(X)\,\eta(Z)) &+ Z(r)\,(g(X, Y) - \eta(X)\,\eta(Y))\\ -2\,\beta\,(2\,\mu - r - 6\,\beta^2)&\big\{g(X, Y) - 2\,\eta(X)\,\eta(Y)\big\}\eta(Z).\end{aligned} \tag{67}$$

Next, taking $X = Y = e_i$ in (37) and summing over i, we get

$$\sum\nolimits_i (\pounds_V \nabla)(e_i, e_i) = \xi(r)\,\xi - 2\,\beta\,(2\,\mu - r - 6\,\beta^2)\xi, \tag{68}$$

where we used the symmetry property of $\mathrm{Ric}^\sharp$. Comparing (64) and (68), we get

$$\xi(r) = 2\,\beta\,(\mu - r - 6\,\beta^2). \tag{69}$$

Plugging this in (24) with $n = 1$ and $\beta = const$, we find that $\mu = 0$, as $\beta \neq 0$. Thus, our η-Ricci soliton reduces to a Ricci soliton and (67) with $Y = \xi$ gives us (62).

To prove (ii), we take the covariant derivative of (62) along Y and use (12):

$$\begin{aligned}2\,(\nabla_Y £_V \nabla)(X, \xi) + 2\beta\,(£_V \nabla)(X, Y) &= -Y(\xi(r))\,(X - \eta(X)\xi)\\ &+ \beta\,\xi(r)\,\big\{g(X, Y)\xi + \eta(X)Y - \eta(Y)X - \eta(X)\eta(Y)\xi\big\}.\end{aligned} \tag{70}$$

Repeated application of (70) in the formula (40) with $Z = \xi$, we obtain

$$\begin{aligned}2\,(£_V R)_{X,Y}\xi &= Y(\xi(r))\,(X - \eta(X)\xi) - X(\xi(r))\,(Y - \eta(Y)\xi)\\ &\quad -2\,\beta\,\xi(r)\,\big\{\eta(X)Y - \eta(Y)X\big\}.\end{aligned} \tag{71}$$

As β is a constant, (18) becomes

$$R_{X,Y}\,\xi = \beta^2\,\big\{\eta(X)Y - \eta(Y)X\big\}. \tag{72}$$

Taking the Lie derivative of (72) along V and using (12), we get

$$\begin{aligned}(£_V R)_{X,Y}\,\xi + R_{X,Y} £_V \xi &= \beta^2\{2(\lambda - 2\beta^2)\{\eta(Y)X - \eta(X)Y\}\\ &\quad + g(X, £_V\xi)Y - g(Y, £_V\xi)X\big\},\end{aligned} \tag{73}$$

where λ is defined in (2). Comparing (71) and (73), we get

$$\begin{aligned}2\,R_{X,Y} £_V \xi + Y(\xi(r))\,(X - \eta(X)\,\xi) - X(\xi(r))\,(Y - \eta(Y)\,\xi)\\ -\big\{2\,\beta\,\xi(r) - 4\,\beta^2\,(\lambda - 2\,\beta^2)\big\}\big\{\eta(X)Y - \eta(Y)X\big\}\\ = 2\,\beta^2\big\{g(X, £_V\xi)Y - g(Y, £_V\xi)X\big\}.\end{aligned} \tag{74}$$

Contracting (74) over Y, we have

$$\begin{aligned}2\,\mathrm{Ric}(X, £_V\xi) + X(\xi(r)) + \xi(\xi(r))\eta(X)\\ - \big\{4\beta\,\xi(r) - 8\,\beta^2\,(\lambda - 2\beta^2)\big\}\eta(X) + 4\beta^2 g(X, £_V\xi) = 0.\end{aligned} \tag{75}$$

Equation (65) with $Z = \xi$ gives us (recalling that $\mu = 0$)

$$(£_V g)(Y, \xi) = -2(\lambda - 2\,\beta^2)\,\eta(Y).$$

Thus, using (61), Eq. (75) becomes (63). □

The next theorem uses Lemma 8 and generalizes [7, Theorem 1].

Theorem 6 *If a 3-dimensional weak β-Kenmutsu manifold $M^3(\varphi, Q, \xi, \eta, g)$ with $\beta = const \neq 0$ represents an η-Ricci soliton (2) with the potential vector field V, then (M, g) has constant sectional curvature $-\beta^2$.*

Proof By $\xi(r) = -2\beta(r + 6\beta^2)$, see (62), we get $X(\xi(r)) = -2\beta X(r)$. Plugging this in (63), we find

$$(r + 6\beta^2)\, g(Y, \pounds_V \xi) = 2\beta \big\{Y(r) - \xi(r)\,\eta(Y)\big\}. \tag{76}$$

We claim that $r + 6\beta^2 = 0$; thus, M^3 has constant curvature $-\beta^2$.

Assume the contrary, i.e., $r + 6\beta^2 \neq 0$ on some open set $\mathcal{O}$ of M. Then (76) can be written as

$$\pounds_V \xi = \nabla f - \xi(f)\xi, \tag{77}$$

where $f = \frac{2\beta r}{r+6\beta^2}$. Setting $Y = \xi$ in (54) and using (12), (62) and (77), we get

$$\begin{aligned}&\beta\big\{X(f)\xi + \eta(X)\nabla f\big\} - 2\beta\,\xi(f)\,\eta(X)\,\eta(Y) + g(\nabla_X \nabla f, Y)\\ &- X(\xi(f))\,\eta(Y) = \big(\xi(f)\beta + \tfrac{1}{2}\,\xi(r)\big)\,(g(X, Y) - \eta(X)\,\eta(Y)),\end{aligned} \tag{78}$$

where we used $(\pounds_V g)(X, \xi) = 0$. Taking inner product of (78) with Y and anti-symmetrizing, yields

$$X(\xi(f))\,\eta(Y) = Y(\xi(f))\,\eta(X). \tag{79}$$

Now, taking covariant derivative of $f = \frac{2\beta r}{r+6\beta^2}$ along ξ using and (69) with $\mu = 0$, we get

$$\xi(f) = 2\beta\,(f - 2\beta).$$

Applying this in (79) gives us $2\beta\{X(f)\,\eta(Y) - Y(f)\,\eta(X)\} = 0$, which implies $\nabla f = \xi(f)\xi$ on $\mathcal{O}$. By the above, (77) yields the equality $\pounds_V \xi = 0$ on $\mathcal{O}$. This implies that

$$\nabla_\xi V = \nabla_V \xi = \beta\{V - \eta(V)\,\xi\}, \tag{80}$$

where we used (12). Replacing ξ by Y in (51) and using (12), (72) and (80), we get

$$(\pounds_V \nabla)(X, \xi) = -\beta\,\eta(\nabla_X V)\xi - \beta^2\big\{g(V, X) - \eta(X)\,\eta(V)\,\xi\big\}. \tag{81}$$

Further, using (80) in $(\pounds_V g)(X, \xi) = 0$, we get

$$\eta(\nabla_X V) = -g(\nabla_\xi V, X) = -\beta\big\{g(X, V)\xi - \eta(X)\,\eta(V)\,\xi\big\}.$$

Inserting this in (81) gives $(\pounds_V \nabla)(X, \xi) = 0$. It follows from (62) that $\xi(r) = 0$; therefore, (69) provides the equality $r + 6\,\beta^2 = 0$, a contradiction. Hence, the proof of our assertion. □

References

1. Blair, D.: *Riemannian geometry of contact and symplectic manifolds*, 2nd ed. Progress in Mathematics, vol. 203, Birkhäuser, 2010
2. Chen, B.-Y.: *Differential Geometry of Warped Product Manifolds and Submanifolds*, World Sci. Publ., 2017
3. Cho, J. and Kimura, M.: Ricci solitons and real hypersurfaces in a complex space form, Tohoku Math. J. 61(2), 205–212 (2009)
4. Ellis, G.F.R. and Hawking, S.W.: *The large scale structure of space-time*, vol. 1. Cambridge University Press, Cambridge, 1973
5. Ghosh, A.: Ricci soliton and Ricci almost soliton within the framework of Kenmotsu manifold, Carpathian Math. Publ., 11(1), 59–69, (2019)
6. Ghosh, A.: K-contact and (k, μ)-contact metric as a generalized η-Ricci soliton, Math. Slovaca 73, No. 1, 185–194 (2023)
7. Ghosh, A.: Kenmotsu 3-metric as a Ricci soliton. Chaos, Solitons & Fractals, 44 (2011), 647–650
8. Herrera, A.C.: Parallel skew-symmetric tensors on 4-dimensional metric Lie algebras. Revista de la Uni6n Matemática Argentina, **65**(2), 295–311, (2023)
9. Kiran Kumar, D.L. and Nagaraja, H.G.: Second order parallel tensor and Ricci solitons on generalized $(k; \mu)$-space forms, Mathematical Advances in Pure and Applied Sciences, 2019, Vol. 2, No. 1, 1–7
10. Olszak, Z.: Normal locally conformal almost cosymplectic manifolds, Publ. Math. Debrecen, 39(3-4) (1991), 315–323
11. Patra, D.S. and Rovenski V.: Almost η-Ricci solitons on Kenmotsu manifolds, European J. of Mathematics, 7 (2021), 1753–1766
12. Patra, D.S. and Rovenski, V.: On the rigidity of the Sasakian structure and characterization of cosymplectic manifolds, Differential Geometry and its Applications, 90 (2023) 102043
13. Rovenski, V. and Wolak, R.: New metric structures on $\mathfrak{g}$-foliations, Indagationes Mathematicae, 33 (2022), 518–532
14. Kenmotsu, K.: A class of almost contact Riemannian manifolds, Tôhoku Math. J., 24 (1972), 93–103
15. Yano, K.: *Integral formulas in Riemannian geometry*, Vol. 1. M. Dekker, 1970
16. Yano, K. and Kon, K.: *Structures on manifolds*, Series in Pure Mathematics, 3, World Scientific Pub. Co., Singapore, 1984

Twistor Bundles of Foliated Riemannian Manifolds

R. Mohseni and R. Wolak

Dedicated to Vladimir Rovenski on his 70th birthday.

Abstract In this paper, we present an overview of the theory of twistors on foliated manifolds, which has been developed in several papers of ours. We construct the twistor space of the normal bundle of a foliation. We demonstrate that the classical constructions of the twistor theory lead to foliated objects and permit us to formulate and prove foliated versions of some well-known results on holomorphic mappings. Since any orbifold can be understood as the leaf space of a suitably defined Riemannian foliation we obtain orbifold versions of the classical results as a simple consequence of the results on foliated mappings. Finally, we recall suspension construction, which is used to construct examples of transversely harmonic maps between foliated manifolds and discuss the twistor methods in this case.

Keywords Foliation · Leaf space · Orbifold · Transversely harmonic mapping · Transversely holomorphic mapping · Twistor space · Suspension construction

1 Introduction

The classical twistor theory was developed as an approach to the unification of general relativity and quantum mechanics [17]. Roughly speaking one has a double fibration which allows to have a correspondence between the "twistor space" and the "Minkowski space". On one side physics lives and on the other side we have the powerful tools of differential geometry and algebraic geometry. The initial goal of

R. Mohseni · R. Wolak (✉)
Jagiellonian University, Institute of Mathematics, ul. St. Lojasiewicza 4, 30-348 Krakow, Poland
e-mail: robert.wolak@uj.edu.pl

© The Author(s), under exclusive license to Springer Nature Switzerland AG 2024
V. Rovenski et al. (eds.), *Differential Geometric Structures and Applications*,
Springer Proceedings in Mathematics & Statistics 440,
https://doi.org/10.1007/978-3-031-50586-7_4

twistor theory has not yet been achieved to this day due to a number of difficulties, however, many applications of the theory in mathematics have arisen, one of which is in the theory of harmonic maps. Let $\phi\colon (M, g) \to (N, h)$ be a map of Riemannian manifolds, it is called harmonic if and only if

$$trace \nabla \phi_* = 0,$$

where ∇ is the connection on $T^*M \otimes \phi^{-1}TN$ induced by the Levi-Civita connections on M and N. Harmonic maps occur in many cases in mathematics and physics [3]. O'Brian and Rawnsley [16] noticed that for an oriented Riemannian manifold (M, g), its twistor space can be understood as the space of linear complex structures on the tangent space TM that are compatible with the metric g and orientation. They generalized this idea to any even-dimensional manifold. This twistor space has proven to be useful in constructing harmonic maps between Riemannian manifolds.

Let M be an oriented $2n$-dimensional Riemannian manifold and $\pi : J(M) \to M$, be the bundle such that at every point $x \in M$, the fiber $J_x(M)$ consists of all almost Hermitian structures in T_xM compatible with the orientation. The Levi-Civita connection ∇^M on M gives a decomposition $T_zJ(M) = H_z \oplus V_z$ ($z \in J(M)$, $\pi(z) = x$). The fibers of this bundle are isomorphic to $SO(2n)/U(n)$ and hence naturally complex. An almost complex structure on T_xM induces an almost complex structure J^H on H_z, hence, on the horizontal and vertical spaces we have J^H and J^V respectively and we can define $J^+ = J^H \oplus J^V$ and $J^- = J^H \oplus (-J^V)$ on $J(M)$. It is well-known that J^+ can be integrable but J^- is never integrable. If one drops the condition of compatibility with the orientation, then the fibers of the twistor space will be isomorphic to $O(2n)/U(n)$.

A unified description of all physical forces and phenomena has been an elusive task of modern physics. One of the hopeful mathematical tools has been the string theory in which points of a phase space are replaced by strings, so submanifolds. In the modern language, such a phase space is a foliated Riemannian manifold. Therefore in a series of joint papers, cf. [13–15], we have formulated and developed the theory of twistor spaces of foliated Riemannian manifolds. In Sect. 2, we introduce three twistor spaces $Z(M)$, $Z(M, \mathcal{F})$, $Z(\mathcal{F})$ on a foliated Riemannian manifold $(M, g, \mathcal{F})$, that are respectively the twistor space of the manifold M, the bundle of normal twistors and the bundle of tangent twistors. In Sect. 3, using these twistor spaces we provide some theorems on the foliated counterparts of twistor fibration, for more details see [13]. O'Brian and Rawnsley [16], realized that the definition of a twistor space as the space of all Hermitian structures on an even-dimensional manifold can be "too big" in many cases to work with. Therefore, they introduced the idea of a smaller twistor space called a generalized twistor space, which we discuss in more detail at the end of the section. As it was remarked by A. Haefliger et al. in [9], any orbifold can be realized as the leaf space of a foliation with compact leaves, we reformulate our theory for the leaf spaces of Riemannian foliations, hence obtaining similar results for orbifolds which we present in Sect. 4. In Sect. 5, we introduce the twistor space of a quaternion Kähler manifold, see [14] for more details.

This short note is an introduction to the theory of twistor spaces of foliated Riemannian manifolds formulated and developed by the authors in the three joint papers mentioned above, and it's supplemented by the last section on suspension construction.

2 Twistor Spaces on Foliated Manifolds

Let a Riemannian foliation $(M, \mathcal{F})$ be defined by a cocycle $\mathcal{U} = \{U_i, f_i, g_{ij}\}_{i,j \in I}$ that is modeled on a Riemannian manifold $(N, \bar{g})$ such that

1. $f_i : U_i \to N$ is a submersion with connected fibers;
2. $g_{ij} : f_j(U_i \cap U_j) \to f_i(U_i \cap U_j)$ are local isometries of $(N, \bar{g})$;
3. $f_i = g_{ij} f_j$ on $U_i \cap U_j$.

Denote by $Q = TM/T\mathcal{F}$ the normal bundle of the foliation $\mathcal{F}$. The natural action of the Lie algebra sheaf $(\mathcal{X}, \mathcal{F})$ of local vector fields tangent to the leaves of the foliation $\mathcal{F}$ defines the foliation $\mathcal{F}_Q$ of the manifold Q, which is of the same dimension as $\mathcal{F}$ and its leaves are covering spaces of the corresponding leaves of $\mathcal{F}$. Therefore, the tensor bundle $\bigotimes^2 Q^*$ is a foliated bundle with the induced foliation $\mathcal{F}_Q^2$ of the same dimension as $\mathcal{F}$. The normal bundle over the open set U_i is isomorphic to the pull-back $f_i^* TN$. This isomorphism sends traces of leaves of $\mathcal{F}_Q$ to the fibers of the submersion $df_i|_Q$. Moreover, it induces an isomorphism of the foliated bundles $\bigotimes^2 Q^*_{|U_i}$ and $f_i^* \otimes^2 TN^*$. Let g be a bundle-like metric on the manifold M, it induces a metric tensor g_Q on Q, i.e., a foliated mapping $g_Q\colon (M, \mathcal{F}) \to (Q^* \otimes Q^*, \mathcal{F}_Q^2)$. Under the same isomorphisms the tensor g_Q corresponds to $f_i^* \bar{g}$. Summing up, we can say that a Riemannian metric g_Q in the normal bundle Q is foliated if and only if for any foliated local sections X, Y of Q the function $g(X, Y)$ is basic, i.e., the function is constant on fibers of local submersions f_i defining the foliation, or, equivalently, its differential annihilates all vectors tangent to the leaves of the foliation.

At any point $x \in M$ we define the normal twistors as the set

$$Z(M, \mathcal{F})_x = \{J_x^Q : Q_x \to Q_x,\ (J^Q)^2 = -id_Q,\ g_Q(J^Q X,\ J^Q Y) = g_Q(X, Y),\ \forall X, Y \in \Gamma(Q)\} \tag{1}$$

and the bundle of normal twistors $Z(M, \mathcal{F})$ as the union $\bigcup_{x \in M} Z(M, \mathcal{F})_x$.

The cocycle definition of the foliation allows us to relate normal twistors on the foliated manifold $(M, \mathcal{F})$ to the twistors of the model manifold N, which will be called transverse twistors. In fact, at any point $x \in M$, the differential $d_x f_i$ defines a linear isomorphism $d_x f_i|_{Q_x} : Q_x \to T_{f_i(x)} N$, which in turn defines an isomorphism of the twistor spaces $Z(M, \mathcal{F})_x$ and $Z(N)_{f_i(x)}$, where $Z(N)_{f_i(x)}$ is the twistor space of N at the point $f_i(x) \in N$. Therefore, using any submersion f_i, one can define a submersion $Z(f_i)$ on the level of twistors

$$Z(f_i) : Z(M, \mathcal{F})|_{U_i} \to Z(N).$$

As a result, the bundle $Z_i(M, \mathcal{F}) := Z(M, \mathcal{F})|_{U_i}\ Z(g_{ij})$ is isomorphic to $f_i^* Z(N)$. Moreover, the cocycle $Z(\mathcal{U}) = \{Z_i(M, \mathcal{F}), Z(f_i), Z(k_{ij})\}$ defines a foliation $\mathcal{F}_Z$ on the manifold $Z(M, \mathcal{F})$, which is modeled on $Z(N)$ and the leaves of $\mathcal{F}_Z$ are covering spaces of leaves of $\mathcal{F}$. The manifold $Z(N)$ carries the natural almost Hermitian structure (h_N, J_N) and transformations $Z(g_{ij})$ are holomorphic isometries of the almost Hermitian structure of $Z(N)$. The normal bundle Q_Z of the foliation $\mathcal{F}_Z$, locally, is isomorphic to the pullback $Z(f_i)^* Z(N)$ as a foliated manifold. Therefore, the bundle Q_Z admits a natural Hermitian metric (h_Z, J_Z), which, in fact, is foliated and the submersions $Z(f_i)$ are transversely Hermitian for any extension of h_Z to a Riemannian metric on M. Moreover, the mappings $Z(f_i)$ are almost Hermitian submersions. Thus, the foliation $\mathcal{F}_Z$ on the manifold $Z(M, \mathcal{F})$ is transversely almost Hermitian for the canonical Riemannian metric and the metric compatible transverse almost complex structure J^Q. The normal Hermitian metric can be constructed directly without involving the transverse manifold and its twistor bundle.

The twistor bundle of a vector bundle can be constructed as in [2]. The normal bundle as described earlier is a foliated vector bundle over M with foliation $\mathcal{F}_Q$ and after performing the twistor construction a foliated twistor bundle will be obtained, which we denote by $Z(Q, \mathcal{F}_Q)$, moreover, using the model manifold N, the twistor space of Q can be constructed that is isomorphic to the pull-back bundle of $Z(N)$ and has a cocycle, which takes values in $Z(\mathcal{U})$. An important fact is that a twistor bundle of the normal bundle constructed in either way is isomorphic to another.

3 Foliated Harmonic Mappings

Let $(M, g, \mathcal{F})$ be a foliated Riemannian manifold with a foliation such that for obvious reasons we take the codimension of the foliation to be $m = 2q$. Denote by $Q = TM/T\mathcal{F}$ the normal bundle and the projection map by $\pi : TM \to Q$. The tangent bundle has the decomposition $TM = T\mathcal{F} \oplus T\mathcal{F}^\perp$, and identifying $T\mathcal{F}^\perp$ with Q, a metric g_Q is induced on Q. Foliation $\mathcal{F}$ is called a transversely almost Hermitian foliation, if it is equipped with a foliated metric g_Q and a transverse almost complex structure $(J^Q)^2 = -Id_Q$ such that they are compatible with each other, i.e. we have

$$g_Q(JX, JY) = g_Q(X, Y), \quad \forall X, Y \in \Gamma(Q).$$

Furthermore, let $\pi(X') = X$ and $\pi(Y') = Y$ for each $X', Y' \in \Gamma(TM)$, a normal Kähler 2-form w and its dual fundamental 2-vector F can be defined as follows:

$$w(X', Y') = g_Q(JX, Y), \tag{2}$$

$$g_Q(F, X \wedge Y) = g_Q(JX, Y) = -g_Q(X, JY). \tag{3}$$

where $F \in \Gamma(M, \bigwedge^2 Q)$. If the 2-form ω is a foliated section of $\bigwedge^2 Q^*$ or, equivalently, defines a basic 2-form on $(M, \mathcal{F})$, then the 2-vector F is a foliated section of $(\bigwedge^2 Q, \mathcal{F}_Q)$ and it corresponds to a 2-vector on the transverse manifold.

The relation between harmonic and holomorphic maps is a classic subject of research. Now we want to recall some of our results, which are the foliated analogs of known results for Riemannian manifolds. In order to do that, we first formulate foliated versions of the notions already available for Riemannian manifolds. The second author and late Jerzy Konderak in [10, 11] developed the theory of transversely harmonic mappings. Foliated (i.e. leaf-preserving) mappings preserving transverse (almost) complex structures have been studied by many authors, e.g., [6, 21].

Definition 1 Let $(M, \mathcal{F}, J^Q)$ and $(M', \mathcal{F}', J^{Q'})$ be two foliated Riemannian manifolds having transverse almost complex structures J^Q and $J^{Q'}$, respectively. A smooth foliated map $\phi : (M, \mathcal{F}) \to (M', \mathcal{F}')$ is *transversely holomorphic* if

$$\phi_*^T \circ J^Q = J^{Q'} \circ \phi_*^T,$$

where $\phi_*^T = \pi' \circ \phi_*$, is defined as the composition of the push-forward map ϕ_* and the projection $\pi' : TM' \to Q'$ and it is a foliated section of the bundle $Q^* \otimes \phi^{-1}Q' \to M$.

Let $\{E_a\}_{a=1,\dots,m}$ be a local orthonormal basis frame of the normal bundle Q and the transverse tension field of ϕ be defined as

$$\tau_b(\phi) = trace_Q \tilde{\nabla}\phi_*^T,$$

where $\tilde{\nabla}$ is the connection on the bundle $Q^* \otimes \phi^{-1}Q'$ induced by the normal Levi-Civita connection. $\tau(\phi)$ is a foliated section of the pull-back bundle $\phi^{-1}Q'$.

Definition 2 Let $\phi : (M, g, \mathcal{F}, J^Q) \to (M', g', \mathcal{F}', J^{Q'})$ be a foliated map between two transversely almost Hermitian foliated manifolds. It is called *transversely harmonic* if the transverse tension field $\tau_b(\phi)$ vanishes.

One can show that this definition is equivalent to the condition that for any cocycles defining the foliation, the induced mapping of the transverse manifolds is harmonic and it is exactly done in [10].

Let $Q^{\mathbb{C}} = Q \otimes \mathbb{C}$ denote the complexification of the normal bundle. It is straightforward to imagine that it is actually a foliated manifold and using the transverse almost complex structure J^Q from $(M, \mathcal{F})$, which maps foliated sections of the normal bundle to foliated sections of the normal bundle, we have the decomposition $Q^{\mathbb{C}} = Q^{1,0} \oplus Q^{0,1}$ that is also compatible with this foliation and

$$Q^{1,0} = \{X \in \Gamma Q^{\mathbb{C}} : J^Q X = iX\}, \quad Q^{0,1} = \{X \in \Gamma Q^{\mathbb{C}} : J^Q X = -iX\}.$$

The kth exterior power is decomposed as $(\bigwedge^k Q)^{\mathbb{C}} = \bigoplus_{p+q=k} Q^{p,q}$ written using the standard notation $Q^{p,q} \cong \bigwedge^p(Q^{1,0}) \otimes \bigwedge^q(Q^{0,1})$, where we also have the natural isomorphisms $(Q^{p,q})^* \cong Q^{\overline{p,q}} \cong Q^{q,p}$.

The second fundamental form of the subspace $Q^{1,0}$ in $Q^{\mathbb{C}}$ is denoted by α, it can be shown that it belongs to the subspace $Q^{0,2}$ and is related to $\nabla_X F$; here, ∇ is the Levi-Civita connection and $X \in Q^{\mathbb{C}}$. We have

$$Q^{0,2} \otimes (Q^*)^{\mathbb{C}} \cong Q^{0,2} \otimes (Q^{0,1} \oplus Q^{1,0}) \cong (Q^{0,1} \otimes Q^{0,2}) \oplus Q^{1,2}.$$

The covariant derivative of the fundamental 2-vector F, see (3), can be taken into two parts; namely, $\nabla F = D_1 F + D_2 F$ (cf. [18, Sect. 1]), where

$$D_1 F \in (Q^{1,0} \otimes Q^{2,0}) \oplus (Q^{0,1} \otimes Q^{0,2}), \quad D_2 F \in (Q^{1,2} \oplus Q^{2,1}).$$

The integrability of an almost complex structure J on the normal bundle Q can be defined as vanishing of its Nijenhuis tensor N_J, i.e.,

$$N_J(X, Y) := [J(X), J(Y)] - J([X, J(Y)] - J([J(X), Y]) - [X, Y] = 0,$$

for any sections $X, Y \in \Gamma Q$, where $[X, Y] = \pi[X', Y'], X', Y' \in \Gamma(TM)$. The normal almost complex structure J^Q is foliated and correspondingly it induces an almost complex structure J^N on the transverse manifold N. The foliated structure J^Q is integrable if and only if the induced (transverse) almost complex structure J^N is integrable. We mention here that we prefer the name "normal" to be used for the almost complex structure on the normal bundle and "transverse" for the one on the transverse manifold, however, some authors use transverse for the almost complex structure on the normal bundle.

Lemma 1 *The following holds:*

- $D_1 F = 0 \Longleftrightarrow \nabla_X(Q^{1,0}) \subset Q^{1,0}, \quad \forall X \in \Gamma(M, Q^{1,0})$,
- $D_2 F = 0 \Longleftrightarrow \nabla_X(Q^{1,0}) \subset Q^{1,0}, \quad \forall X \in \Gamma(M, Q^{0,1})$.

Definition 3 A transversely almost Hermitian foliation is called *transversely* $(1, 2)$-symplectic if $D_2 F = 0$ and *transversely cosymplectic* if $trace(D_2 F) = 0$.

It is obvious that "transversely $(1, 2)$-symplectic" implies "transversely cosymplectic". Now let us state the following theorem, that has its roots in A. Lichnerowicz's work, cf. [12].

Theorem 1 *Let $(M, g, \mathcal{F}, J^Q)$ and $(M', g', \mathcal{F}', J^{Q'})$ be transversely almost Hermitian foliated manifolds, that are, respectively, transversely cosymplectic and transversely $(1, 2)$-symplectic. Then any transversely holomorphic map $\phi : M \to M'$ is transversely harmonic.*

Theorem 2 *Let $(M, g, \mathcal{F}, J)$ be transversely cosymplectic and $(N, \mathcal{F}')$ a foliated Riemannian manifold with a normal twistor bundle $Z(N, \mathcal{F}')$, and we have the projection $\pi : Z(N, \mathcal{F}')$. Then, if $\psi : M \to Z(N, \mathcal{F}')$ is transversely J^--holomorphic, $\pi \circ \psi$ is transversely harmonic.*

Note that earlier we only specified the codimension of a foliated Riemannian manifold as $m = 2q$, now let us have a foliated Riemannian manifold $(M^{2n+2q}, \mathcal{F})$ and the foliation to be transversely almost Hermitian foliation of codimension $2q$, as before, and the dimension of the manifold is even. The tangent bundle of M is decomposed as $TM = L \oplus Q$, where $L = T\mathcal{F}$. Let J be an almost complex structure on TM, which preserves the tangent bundle to the leaves, i.e., $J(L) \subset L$. Restricting J to L defines an almost complex structure in L denoted by J^L. Furthermore, we have an almost complex structure J^Q in the normal bundle Q induced by J. If the almost complex structure J is compatible with the bundle-like metric g, then J decomposes as $J = J^L \oplus J^Q$ because $J(L) \subset L$ implies that $J(L^\perp) \subset L^\perp$. Similar to the normal twistor space defined as (1), using J^L we can define the tangent twistors as

$$Z(\mathcal{F})_x = \{J_x^L : L_x \to L_x,\ (J^L)^2 = -id_L,\ g_L(JX, JY) = g_L(X, Y),\ X, Y \in \Gamma(L)\},$$

where $x \in M$, g_L is the restriction of the Riemannian metric to the subbundle L and the bundle of tangent twistors $Z(\mathcal{F})$ as the union of $Z(\mathcal{F})_x$ over all $x \in M$.

Summing up, we showed that on a foliated Riemannian manifold $(M, \mathcal{F})$, there exist three twistor spaces, which we denote by $Z(M)$, $Z(M, \mathcal{F})$ and $Z(\mathcal{F})$ defined, respectively, using J, J^Q and J^L. Namely, $Z(M)$ is the twistor space of the foliated manifold $(M, \mathcal{F})$, $Z(M, \mathcal{F})$ is the bundle of normal twistors and $Z(\mathcal{F})$ is bundle of tangent twistors, moreover, we have $Z(M) \supset Z(\mathcal{F}) \times_M Z(M, \mathcal{F})$.

Next, we define the notion of a tangentially holomorphic map similar to the definition of a transversely holomorphic map given in Definition 1.

Definition 4 Let $(M, \mathcal{F})$ and $(M', \mathcal{F}')$ be two foliated Riemannian manifolds with tangential almost complex structures J^L and $J^{L'}$, respectively. A smooth foliated map $\phi : (M, \mathcal{F}) \to (M', \mathcal{F}')$ is called *tangentially holomorphic* if

$$\phi_*^L \circ J^L = J^{L'} \circ \phi_*^L,$$

where $\phi_*^L = \phi_*|_L$, is defined as the restriction of the push-forward mapping ϕ_* to the tangents to the foliation and is a section of the bundle $L^* \otimes \phi^{-1}L' \to M$.

Some of the known conditions, for which a transversely holomorphic map is transversely harmonic, can be stated as a theorem.

Theorem 3 *Let $(M, \mathcal{F})$ and $(M', \mathcal{F}')$ be almost Hermitian foliated manifolds and $\phi : M \to M'$ a transversely holomorphic map. Then ϕ is transversely harmonic if*

1. *$(M, \mathcal{F})$ is transversely cosymplectic and $(M', \mathcal{F}')$ is transversely $(1, 2)$-symplectic.*

2. *$(M, \mathcal{F})$ and $(M', \mathcal{F}')$ are transversely* $(1, 2)$*-symplectic.*
3. *$(M, \mathcal{F})$ and $(M', \mathcal{F}')$ are Riemannian manifolds with transversely Kähler foliations.*

If on a foliated manifold $(M, \mathcal{F})$, we take the open subsets of leaves of the foliation as the open subsets, then we obtain a new topology. Denote the set M together with this topology by $M_{\mathcal{F}}$. One can show that a smooth map $\phi : (M, \mathcal{F}) \to (M', \mathcal{F}')$ is foliated if and only if it induces a smooth map $\hat{\phi} : M_{\mathcal{F}} \to M'_{\mathcal{F}'}$. Denote by $\tau(\hat{\phi})$ the tension field of the map $\hat{\phi}$, then $\hat{\phi}$ is called *harmonic* if $\tau(\hat{\phi}) = 0$. Furthermore, ϕ is called *leaf-wise harmonic* if $\hat{\phi}$ is harmonic. One can formulate a counterpart of Theorems 1 and 2 using tangentially holomorphic and leaf-wise harmonic maps instead of transversely holomorphic and transversely harmonic maps and the twistor space will be the bundle of tangent twistors. Let us recall a corollary (cf. [11, Corollary 2.1]), which connects these three notions.

Corollary 1 *Let $\phi : (M, \mathcal{F}) \to (M', \mathcal{F}')$ be a smooth foliated mapping between two regular foliated Riemannian manifolds (i.e., the leaves the foliation are of the same dimension), with $\mathcal{F}$ being minimal and $\mathcal{F}'$ totally geodesic. Then the map ϕ is harmonic if and only if ϕ is transversely harmonic and leaf-wise harmonic.*

Let (N^{2n}, h) be an even-dimensional Riemannian manifold with Riemannian metric h and $J(N)$ be the bundle over N such that its fibers at each point $x \in N$ consist of all linear almost complex structures on the tangent space $T_x N$ compatible with the metric h. If we denote by $O(N)$ the orthonormal frame bundle of N, then $J(N) = O(N) \times_{O(2n)} O(2n)/U(n)$.

The Levi-Civita connection induces a horizontal distribution on $O(N)$ and, as a result, also on $J(N)$, which gives the splitting $TJ(N) = H \oplus V$, where H denotes the induced horizontal distribution and V is the vertical distribution with respect to $\pi : J(N) \to N$. Similarly, any linear connection ∇ on TN induces a horizontal distribution H^{∇} on $J(N)$. Using the fact that H is isomorphic to $\pi^{-1}TN$, one can obtain an almost complex structure on H denoted by J^H. The subbundle V is the tangent space through the fibers of π and since the fibers are complex manifolds, V inherits an almost complex structure denoted by J^V.

It is well-known [16], that on $J(N)$ there exist two almost complex structures

$$J^{\pm} = \begin{cases} J^H, & \text{on } H, \\ \pm J^V, & \text{on } V. \end{cases}$$

In some cases one needs to study smaller twistor spaces than $J(N)$, and this was done by O'Brian and Rawnsley [16]. They introduced a method of constructing such spaces as the following and called them *generalized twistor spaces*.

Let $\pi : Z \to N$ be a smooth submersion such that its fibers have a smoothly varying complex structure. Suppose that we have a fiber-preserving map $j : Z \to J(N)$ over the even-dimensional manifold N and that Z has a horizontal distribution denoted by H^Z. Each H^Z_z is isomorphic to $T_{\pi(z)}N$ and, therefore, is endowed with the endomorphism $j(z)$, which if summed together with the given complex structures on fibers (i.e. the vertical subspace in this case) gives Z an almost complex structure denoted by J^Z. This almost complex structure depends only on H^Z and j. Assume that on TN, there exists a linear connection ∇ and consequently a horizontal distribution H^∇ on $J(N)$ and also that the map j is holomorphic on each fiber. In addition, suppose that j maps H^Z into H^∇, i.e. $j_*(H^Z) \subset H^\nabla$, denoting the connection dependent almost complex structure on $J(N)$ by J^∇, one has $j_* \circ J^Z = J^\nabla \circ j_*$; hence, j is J^Z-holomorphic; moreover, integrability of J^Z is related to the torsion and curvature of ∇, see [16].

Let (N, h, J) be an almost Hermitian manifold, TN has a $U(n)$-structure and let $Z = G_k(TN)$ be the Grassmannian bundle of complex k-planes in TN. Denoting the bundle of unitary frames on N by $F(N)$, one has

$$Z = F(N) \times_{U(n)} G_k(\mathbb{C}^n).$$

For $w \in Z_x$ in the fibre at a point $x \in N$ one can define a linear almost complex structure on T_xN and hence obtain a map $j : Z \to J(N)$ as

$$j(w) = \begin{cases} J, & \text{on } w, \\ -J, & \text{on } w^\perp. \end{cases}$$

Now suppose that ∇ is a linear connection on TN and $\nabla h = \nabla J = 0$, then j gives a correspondence between the horizontal distributions H^Z and H^∇, once again one can check whether j is holomorphic on the fibers, which results in a relation between integrability of an almost complex structure J^Z defined similarly as before and the torsion and curvature tensors of ∇ satisfying certain conditions, see [16]. If N is oriented, the twistor space $J(N)$ has two connected components $J^\pm(N)$, which are the spaces of all almost Hermitian structures compatible (respectively, incompatible, i.e., with the opposite sign) with the orientation of N. Let us denote the projection by $\sigma : O(N) \to J(N)$, then for the case of N being oriented we have $\sigma_\pm : O_\pm(N) \to J^\pm(N)$.

In a soon to be available preprint [15], the authors will discuss this construction alongside some other materials for foliated Riemannian manifolds.

4 Orbifolds

The notion of V-manifolds was introduced by I. Satake [19] in 1956, however, later due to W. Thurston [20], they are known as orbifolds and have applications in several parts of mathematics and modern physics. The fact that the leaf space of a Riemannian foliation with compact leaves is an orbifold is a well-known result, and any orbifold can be realized as the leaf space of a Riemannian foliation, cf. [9]. We will reformulate some of our results for orbifolds. In this section we follow the definitions and notation of [22].

It can be shown that for any orbifold, there exists a Riemannian foliation whose leaf space is the orbifold. Now, take X and Y to be two orbifolds and denote by $L(X)$ and $L(Y)$ their linear frame bundle, respectively. The frame bundles are naturally foliated by the fibers, and the foliations are denoted by $\mathcal{F}_X$ and $\mathcal{F}_Y$. Any smooth complete orbifold diffeomorphism $f : X \to Y$ can be lifted to a foliated mapping $L(f) : L(X) \to L(Y)$.

Theorem 4 ([22]) *Let (X, g) and (Y, h) be two Riemannian orbifolds and $f : X \to Y$ be a smooth complete mapping. Then f is harmonic if and only if the induced (foliated) mapping $L(f) : (L(X), \mathcal{F}_X, g_L) \to (L(Y), \mathcal{F}_Y, h_L)$ is transversely harmonic.*

The previous theorem demonstrates that as expected some of the properties of the Riemannian orbifold correspond to the properties of the foliated Riemannian manifold. Now, we will give the counterpart of some of the results obtained earlier, for more details see [13].

Theorem 5 *Let X and Y be two almost Hermitian orbifolds, which are cosymplectic and $(1, 2)$-symplectic, respectively. Then any holomorphic complete orbifold map $\phi : X \to Y$ is harmonic.*

Let $(M^{2n+2q}, \mathcal{F})$ be a foliated Riemannian manifold with a foliation $\mathcal{F}$ and for the sake of brevity denote its normal twistor bundle by $W := Z(M, \mathcal{F})$. This W is foliated itself with foliation denoted by $\mathcal{F}_Z$. Following our earlier discussion, it is easy to see that the leaf spaces $M/\mathcal{F}$ and $W/\mathcal{F}_Z$ correspond to orbifolds, which we denote by $\mathcal{O}_M$ and $\mathcal{O}_W$, respectively. On $\mathcal{O}_W$ we also have two almost complex structures induced by $J^{\pm}$, which by abuse of notation we shall denote by $J^{\pm}$. Therefore, we have the following theorem for orbifolds.

Theorem 6 *$(\mathcal{O}_{\mathcal{W}}, J^+)$ is a complex orbifold if and only if*

- *$\mathcal{O}_M$ is conformally flat for $n \geq 3$,*
- *$\mathcal{O}_M$ is anti-self-dual for $n = 2$.*

Proposition 1 *The twistor space of a Riemannian orbifold is a manifold.*

Considering Theorem 2 and denoting the orbifolds obtained from $(N, \mathcal{F}')$ and $Z(N, \mathcal{F}')$ by $\mathcal{O}_N$ and $\mathcal{O}_{W'}$, respectively, we can state the following theorem.

Theorem 7 *Let $\mathcal{O}_M$ be a cosymplectic orbifold, $\mathcal{O}_N$ be any orbifold, and $\mathcal{O}_{W'}$ be the twistor orbifold of $\mathcal{O}_N$ with the projection $\pi : \mathcal{O}_{W'} \to \mathcal{O}_N$. Then if $\psi : \mathcal{O}_M \to \mathcal{O}_{W'}$, is J^--holomorphic, then $\pi \circ \psi$ is harmonic.*

5 Twistor Space of a Quaternion Kähler Foliated Manifold

In this section, we follow the notation of Salamon [18]. Take (M^{4n}, g) to be a quaternion Kähler manifold and let $\mathbb{H}$ be the quaternionic line bundle associated to the representation of $Sp(1)$ on $\mathbb{C}^2$. In [18], it was shown that there exists a twistor fibration $q : Z \to M$ over such a quaternion Kähler manifold, where Z is a complex manifold of dimension $2n + 1$. It can also be viewed as an S^2-bundle generated by anticommuting almost complex structures I, J, K or as $Z = \mathbb{P}(\mathbb{H})$, i.e. the projectivization of the bundle $\mathbb{H}$ and Z can be referred to as the twistor space of the quaternion Kähler manifold M. It is worth noting that $\mathbb{H}$ is not always globally well-defined but $Z = \mathbb{P}(\mathbb{H})$ is. Some of the properties of the base manifold in this case can be studied using their twistor spaces and vice versa. In particular, Salamon showed how some of the characteristic classes of the twistor space Z can be computed using cohomology groups of the manifold M.

Let $\mathcal{F}$ be a foliation of codimension $4q$ on the manifold M^{p+4q} and on $(M, \mathcal{F})$ we have a foliated quaternion Kähler structure W with local bases (I_1, I_2, I_3) on the normal bundle $N(M, \mathcal{F}) = TM/T\mathcal{F}$. We define the transversal quaternion twistor space of $\mathcal{F}$ by

$$Z\mathcal{F} = \{J \in W,\ J = \alpha_1 I_1 + \alpha_2 I_2 + \alpha_3 I_3,\ \alpha_1^2 + \alpha_2^2 + \alpha_3^2 = 1\},$$

i.e. $Z\mathcal{F}$ is the sphere bundle associated with the foliated vector bundle W and W carries a Riemannian structure such that it makes $\{I, J, K\}$ an orthonormal basis.

A quaternion Kähler foliation is defined by a cocycle modeled on a quaternion Kähler manifold N_0 and one can expect that since the twistor space $Z(N_0)$ of the manifold N_0 is a useful tool in studying its differential geometric properties, one should by using the relation between $Z(N_0)$ and $Z\mathcal{F}$ be able to study properties of the foliated manifold $(M, \mathcal{F})$.

The twistor space of a quaternion Kähler manifold plays an important role in the theory of these manifolds. It is well known that the total space of the twistor bundle is a compact Kähler manifold and that the pull-back mapping induces an injection of the cohomology of the base into the cohomology of the total space. These two facts permitted A. Fujiki to formulate and prove a generalization of Kraines's Theorem on the cohomology of compact Kähler manifolds, cf. [8, Theorem 3.22].

It has been very tempting to formulate and prove a foliated version of Fujiki's theorem. A careful reading of Salomon's theorem, cf. [18, Theorem 6.1], confirms that the canonical foliation is transversely Kähler. One should also mention that a foliated version of the Calabi conjecture has been demonstrated by Aziz El Kacimi, cf. [7]. However, there is not any version of the Leray-Hirsch theorem for foliated bundles and their basic cohomology.

Example: 3-Sasakian manifolds. On the other hand, it is possible to formulate and prove a foliated version of Fujiki's theorem for one interesting class of foliated manifolds, namely 3-Sasakian manifolds. Indeed, their canonical foliation is transversely quaternion. The main reference for the theory of 3-Sasakian manifolds is Ch. Boyer's and K. Galicki's book [5] or their paper [4]. Let us recall some useful facts about 3-Sasakian manifolds.

(i) The characteristic foliation of a 3-Sasakian manifold is transversely quaternion Kähler, cf. [5, Theorem 2.8].

(ii) The twistor space of a 3-Sasakian manifold is a trivial bundle as a foliated bundle due to the fact on the normal bundle of the characteristic foliation we have 3 linearly independent foliated almost-complex structures, cf. [5, pp. 190].

(iii) The characteristic foliation of a 3-Sasaki manifold is defined by an isometric action of the group S^3.

The above mentioned results combined with the proof provided in [8] allow us to formulate and prove the following version of [8, Theorem 3.23] in the case of a 3-Sasakian manifold.

Theorem 8 *Let M be a compact 3-Sasakian manifold of dimension $3+4q$ with positive normal scalar curvature and characteristic foliation $\mathcal{F}$. Then*

1. *for any $0 \leq k < q$ the induced linear map $L^{q-k}\colon H^{2k}(M,\mathcal{F}) \to H^{4q-2k}(M,\mathcal{F})$ is an isomorphism.*
2. *for any $k \geq 0$ we have the direct sum decomposition*

$$H^{2k}(M,\mathcal{F}) = \sum\nolimits_r L^r H_e^{2k-4r}(M,\mathcal{F}), \quad (k-q) \leq r \leq [k/2].$$

6 Suspension Construction

Let (B, g_B) and (F, g_F) be two smooth Riemannian manifolds. Let $h : \pi_1(B) \to Isom(F, g_F)$ be a representation of the fundamental group of B into the isometry group of (F, g_F). On the product Riemannian manifold $(\tilde{B} \times F, \tilde{g}_B \times g_F)$, where $(\tilde{B}, \tilde{g}_B)$ is the universal covering of B with the lifted Riemannian metric, the group $\pi_1(B)$ acts via the diagonal action δ:

$$(x, y) \in \tilde{B} \times F, \quad \alpha \in \pi_1(B), \quad \delta(\alpha)(x, y) = (x\alpha, h(\alpha)(y))$$

the action on $\tilde{B}$ being the deck transformation action.

The orbit space of the δ action $\pi_\delta : \tilde{B} \times F \to \tilde{B} \times F/\delta = M(B, F; h)$ is a smooth manifold. Since the action is isometric for $\tilde{g}_B \times g_F$, this metric defines a Riemannian metric g_h on $M(B, F; h)$. Moreover, the Riemannian foliation $\tilde{\mathcal{F}}_h$ given by the projection $p_2 : \tilde{B} \times F \to F$ onto the second factor is (δ, h)-invariant, so it defines a Riemannian foliation $\mathcal{F}_h$ on $M(B, F; h)$ for which the Riemannian metric g_h is bundle-like. The leaves of $\mathcal{F}_h$ are covering spaces of B.

Now let B and F be even-dimensional Riemannian manifolds. The bundle of normal twistors of the foliated manifold $(M(B, F; h), \mathcal{F}_h)$ will be denoted by $Z(M, \mathcal{F}_h)$. The pull-back $\pi_\delta^* Z(M, \mathcal{F}_h)$ is isomorphic to the bundle of transverse twistors of the lifted foliation, that is of $\tilde{\mathcal{F}}_h$. However, this foliation is given by a global submersion p_2, so $\pi_\delta^* Z(M, \mathcal{F}_h)$ is isomorphic to $p_2^* Z(F)$. These isomorphisms preserve the canonical Riemannian and transverse almost Hermitian structures.

Let $Z(\tilde{B} \times F)$ be the twistor space of the product Riemannian manifold $\tilde{B} \times F$, also $Z(\tilde{B})$ and $Z(F)$ be the twistor spaces of the Riemannian manifolds $\tilde{B}$ and F, respectively. Suppose that $J \in Z(\tilde{B} \times F)$ is an almost complex structure, which preserves the splitting $T(\tilde{B} \times F) = T(\tilde{B}) \oplus T(F)$, then it also decomposes as $J = J^{\tilde{B}} \oplus J^F$, where $J^{\tilde{B}}$ and J^F belong to $Z(\tilde{B})$ and $Z(F)$, respectively. On the foliated manifold $(M(B, F; h), \mathcal{F}_h)$, there exist three twistor spaces, which we will denote by $Z(M(B, F; h))$, $Z(M, \mathcal{F}_h)$ and $Z(\mathcal{F}_h)$, that are, respectively, the twistor space, the bundle of normal twistors and the bundle of tangent twistors of $M(B, F, h)$. Let $J^M \in Z(M(B, F; h))$ denote an almost complex structure that preserves the tangent bundle to the leaves, then we have the decomposition $J^M = J^L \oplus J^Q$, where J^L and J^Q belong to $Z(\mathcal{F}_h)$ and $Z(M, \mathcal{F}_h)$, respectively. Finally, we obtain a reduced twistor space $Z(M(B, F; h)) = Z(\mathcal{F}_h) \times_M Z(M, \mathcal{F}_h)$ and it is isomorphic to the quotient of the product twistors $Z(\tilde{B} \times F)$ by the induced action δ.

In order to recall some of the results from [10], we need to provide some background. Take four Riemannian manifolds (B_i, g_i), (F_i, h_i), $i = 1, 2$ and denote by G_i the fundamental group of the manifolds B_i. Let h_i be representations of the groups G_i into the groups $Isom(F_i, g_i)$, which are the isometry groups of the Riemannian manifolds (F_i, g_i). Suppose that we have a smooth map $f : B_1 \to B_2$ and let us denote by $\tilde{f} : \tilde{B}_1 \to \tilde{B}_2$ its lift to the universal coverings of these manifolds and it is (G_1, G_2)-equivariant. The map f induces a map $\pi_1(f) : G_1 \to G_2$ on the fundamental groups of the manifolds and let $h_i : G_i \to Isom(F_i, g_i)$ be the representations of the groups G_i, $i = 1, 2$, respectively. Let us choose a $(G_1, h_1; G_2, h_2)$-equivariant map $\phi : F_1 \to F_2$, i.e. $\forall\, \nu \in F_1, \gamma \in G_1, \phi(\nu.h_1(\gamma)) = \phi(\nu).h_2(\pi_1(f)(\gamma))$ – these actions are the isometry group acting from the right. Then we construct a map $\tilde{\psi} : \tilde{B}_1 \times F_1 \to \tilde{B}_2 \times F_2$ defined as $\tilde{\psi} = (\tilde{f}, \phi)$ and it is (G_1, G_2)-equivariant, therefore it induces a map $\psi : M_1(B_1, F_1; h_1) \to M_2(B_2, F_2; h_2)$. Let us recall the following.

Lemma 2 (see Lemma 3.5 in [10]) *The map $\phi : F_1 \to F_2$ is harmonic if and only if the induced map ψ is transversely harmonic.*

Using Theorem 3, it is straightforward to combine the results.

Lemma 3 *Let Riemannian manifolds $M_1(B_1, F_1; h_1)$ and $M_2(B_2, F_2; h_2)$ satisfy one of the conditions in Theorem 3. Then the map ϕ: $F_1 \to F_2$ is harmonic if and only if ψ is transversely holomorphic.*

Looking at the possible examples of suspension construction, e.g., see [10, 11], the main Riemannian manifolds candidates to use for (B, g_B) are S^1, $S^1 \times \ldots \times S^1$ or T^k. Among them, the torus T^k is more interesting for us to study its twistor space. Let J_n be the set of all linear complex structures on $\mathbb{R}^{2n}$, it can be shown that it is actually a complex manifold. Denote by $T^{2n} = \mathbb{R}^{2n}/\mathbb{Z}^{2n}$ the $2n$-dimensional real torus. Atiyah in [1] discussed how a complex structure T_J is defined on T^{2n} by each

$J \in J_n$ and showed that the union of all T_J for $J \in J_n$ is a complex manifold denoted by M_n. This manifold is a product $J_n \times T^{2n}$ with the obvious projection to the first factor J_n.

Example 1 Let (F, g, J) be a compact Kähler manifold and X a holomorphic Killing vector field. Denote by $Isom(F, g, J)$ the group of holomorphic isometries and by G_X its abelian subgroup generated by the flow of the vector field X. Take $h: \pi_1(T^k) \to G_X$ to be the representation of the group $\pi_1(T^k)$ and by suspending h we get a compact manifold $M(T^k, F; h)$, and this manifold is a fiber bundle over T^k with the fiber F. Moreover, it is a transversely Kähler foliated manifold, whose leaves are coverings of T^k. It is straightforward to see that if instead, we take the manifold F to be cosymplectic or (1,2)-symplectic, then we can produce examples for the cases in Theorem 3 and this follows from the isometry group being the group of holomorphic isometries.

Let $M_1(B_1, F_1; h_1)$ and $M_2(B_2, F_2; h_2)$ be two foliated manifolds obtained from suspension construction, that we shall denote by M_1 and M_2, respectively. Suppose that M_1 is transversely and tangentially cosymplectic and its foliation $\mathcal{F}_{h_1}$ is minimal. Moreover, let $\mathcal{F}_{h_2}$ be totally geodesic and over M_2 have the twistor spaces $\pi_1: Z(\mathcal{F}_h) \to M_2$ and $\pi_2: Z(M_2, \mathcal{F}_{h_2}) \to M_2$ constructed as earlier. Let $\psi_1: M_1 \to Z(\mathcal{F}_{h_2})$ and $\psi_2: M_2 \to Z(M_2, \mathcal{F}_{h_2})$ be two foliated maps and construct a map $\phi = (\phi_1, \phi_2)$ by $\phi_i = \pi_i \circ \psi_i, i = 1, 2$, which is a map $\phi : M_1 \to M_2$.

Proposition 2 *Let ψ_1 be tangentially J^--holomorphic and ψ_2 be transversely J^--holomorphic. Then ϕ is a harmonic map.*

Proof As an application of Theorem 2 and its tangential counterpart, one gets that ϕ_1 and ϕ_2 are leaf-wise harmonic and transversely harmonic maps, respectively. The rest is a consequence of Corollary 1. □

Acknowledgements The second author acknowledges that the research cooperation was funded by the program Excellence Initiative—Research University at the Jagiellonian University in Kraków.

References

1. Atiyah, M.F.: Some examples of complex manifolds, Bonn. Math. Schr. 6, 28 S. (1958)
2. de Bartolomeis, P.: Twistor Constructions for Vector Bundles. In: *Complex Analysis and Geometry*, eds. V. Ancona and A. Silva, Springer US, pp. 103–114 (1993)
3. Burstall, F. E.: Twistor methods for harmonic maps. In: Hansen V.L. (eds.). *Differential Geometry*. Lecture Notes in Mathematics, vol. 1263. Springer, Berlin, Heidelberg (1987)
4. Boyer, Ch.P., Galicki, K., Mann, B.M.: The geometry and topology of 3-Sasakian manifolds. J. Reine Angew. Math. 455, 183–220 (1994)
5. Boyer, C.P., Galicki, K.: Sasakian Geometry. Oxford Mathematical Monographs. Oxford University Press, Oxford (2008)

6. Duggal, K.L., Ianus, S., Pastore, A.M.: Maps interchanging f-structures and their harmonicity. Acta Appl. Math. 67, 91–115 (2001)
7. El Kacimi-Alaoui, A.: Opérateurs transversalement elliptiques sur un feuilletage riemannien et applications. Compositio Math. 73, No. 1, 57–106 (1990)
8. Fujiki, A.: On the de Rham cohomology group of a compact Kähler symplectic manifold. Advanced Studies in Pure Mathematics 10, 105–165 (1987)
9. Haefliger, A., Girbau, J., Sundararaman, D.: On deformations of transversely holomorphic foliations. J. Reine Angew. Math. 345, 122–147 (1983)
10. Konderak, J.J., Wolak, R.A.: Transversely harmonic maps between manifolds with Riemannian foliations. Q. J. Math. 54, No. 3, 335–354 (2003)
11. Konderak, J.J., Wolak, R.A.: Some remarks on transversely harmonic maps. Glasg. Math. J. 50, No. 1, 1–16 (2008)
12. Lichnerowicz, A.: Applications harmoniques et variétés kähleriennes. Seminario Mat. e. Fis. di Milano. 39, 186–195 (1969)
13. Mohseni, R., Wolak, R.A.: Twistor spaces on foliated manifolds, Int. J. Math. 32, No. 8, 2150056 (2021)
14. Mohseni, R., Wolak, R.A.: Cohomology of quaternionic foliations and orbifolds, preprint, arXiv:2202.02733
15. Mohseni, R., Wolak, R.A.: Suspension and Grassmannians, preprint
16. O'Brian, N.R., Rawnsley, J.H.: Twistor spaces. Ann. Glob. Anal. Geom. 3, 29–58 (1985)
17. Penrose, R., MacCallum, M.A.H.: Twistor theory: An approach to the quantisation of fields and space-time. Physics Reports. 6, No. 4, 241–315 (1973)
18. Salamon, S.: Harmonic and holomorphic maps. In: *Geometry Seminar "Luigi Bianchi" II: Lectures given at the Scuola Normale Superiore*, ed. E. Vesentini, Springer-Verlag Berlin Heidelberg, pp. 161–224 (1984)
19. Satake, I.: On a generalization of the notion of manifold. Proc. Natl. Acad. Sci. U.S.A. 42, No. 6, 359–363 (1956)
20. Thurston, W.: The Geometry and Topology of Three-Manifolds. Princeton University Press, (1997)
21. Wolak, R.A.: Foliated and associated geometric structures on foliated manifolds. Ann. Fac. Sci. Toulouse Math. (5). 10, No. 3, 337–360 (1989)
22. Wolak, R.A.: Orbifolds, geometric structures and foliations. Applications to harmonic maps. Rend. Semin. Mat. Univ. Politec. Torino. 73, 173–187 (2016)

Mixed 3-Sasakian Statistical Manifolds and Statistical Submersions

Crina Daniela Neacşu

Abstract Mixed 3-structures are known to be the counterpart in odd dimension of the quaternionic structures of second kind, also termed in the literature as paraquaternionic structures. The aim of this paper is to investigate the statistical structures in paraquaternionic and mixed 3-Sasakian geometry. We derive the main properties of statistical manifolds equipped with such kind of structures, focusing on the case of mixed 3-Sasakian statistical manifolds and paraquaternionic Kähler-like statistical manifolds. Moreover, we investigate the geometry of statistical submersions with total space a mixed 3-Sasakian manifold or a paraquaternionic Kähler-like statistical manifold. Moreover, we provide some illustrative examples.

Keywords Statistical submersion · Mixed 3-structure · Paraquaternionic structure · Distribution

1 Introduction

The concept of quaternionic structure of second kind, better known in the literature as paraquaternionic structure, was introduced by Libermann in [25]. The geometry of Riemannian manifolds equipped with paraquaternionic structures is an interesting research area and many remarkable results were established in the last decades, including applications in various fields like Lagrangian and Hamiltonian mechanical systems (see, e.g. [34, Chap. 6] and the references therein). On another hand, the notion of mixed 3-structure appeared in a natural way in the study of hypersurfaces of paraquaternionic manifolds performed by Ianuş, Mazzoco and Vîlcu in [18]. This concept was further refined by Caldarella and Pastore in [8], the authors introducing negative and positive mixed 3-Sasakian structures, as well as negative and positive mixed 3-cosymplectic structures. Moreover, they proved that a positive (resp. nega-

C. D. Neacşu (✉)
National University of Science and Technology Politehnica Bucharest, Splaiul Independentei 313, 060042 Bucharest, Romania
e-mail: crina.neacsu@upb.ro

© The Author(s), under exclusive license to Springer Nature Switzerland AG 2024
V. Rovenski et al. (eds.), *Differential Geometric Structures and Applications*, Springer Proceedings in Mathematics & Statistics 440,
https://doi.org/10.1007/978-3-031-50586-7_5

tive) mixed 3-Sasakian space is an Einstein manifold with negative (resp. positive) Einstein constant. Due this property, mixed 3-Sasakian spaces are of some interest in theoretical physics [20]. Several important results on the geometry of these manifolds and their submanifolds, as well as notable examples of such spaces, can be found in [5, 7, 9, 10, 19, 21, 36].

On another hand, a statistical manifold is a pseudo-Riemannian manifold equipped with an affine connection without torsion satisfying some compatibility conditions. Originally, statistical manifolds were introduced by Lauritzen [24], these being nowadays a key concept in the field of information geometry [3, 4, 6, 27, 31]. In the last period, an increased interest can be noted in the study of statistical manifolds endowed with some additional geometric structures and their statistical submersions [11–14, 22, 23, 26, 28, 30, 32, 33, 35, 40]. Motivated by all these studies, we are going to investigate the influence of the existence of quaternionic structures of second kind and mixed 3-structures on the geometry of statistical manifolds.

The paper is divided as follows. After the present Introduction, we derive in Sect. 2 some basic results regarding mixed 3-structures on statistical manifolds. In particular, we introduce and characterize mixed 3-Sasakian and mixed 3-cosymplectic statistical manifolds. Section 4 deals with paraquaternionic structures on statistical manifolds, a special attention being paid to the investigation of parahyper-Kähler-like statistical manifolds and paraquaternionic Kähler like statistical manifolds. It is proved that the tangent bundle of an almost para-Hermite-like statistical manifold can be equipped with an almost Hermite-like paraquaternionic statistical structure, this structure being parahyper-Kähler-like in some specific condition. The last two sections are devoted to the investigation of statistical submersions from statistical manifolds endowed with mixed 3-structures and paraquaternionic structures. Some examples provided along the paper illustrate the main concepts and results.

2 Mixed 3-Structures on Statistical Manifolds

Let M be a smooth manifold, g a pseudo-Riemannian metric on M and ∇ an affine connection on M. Then the pair (∇, g) is said to be *a statistical structure* on M if the torsion tensor field of ∇ vanishes and ∇g is symmetric, i.e.

$$(\nabla_X g)(Y, Z) - (\nabla_Y g)(X, Z) = 0$$

for all vector fields X, Y and Z on M. In this case, we say that (M, ∇, g) is *a statistical manifold*. On the other hand, the *dual connection* or conjugate connection ∇^* of ∇ with respect to the metric g is defined by

$$Xg(Y, Z) = g(\nabla_X Y, Z) + g(Y, \nabla^*_X Z), \tag{1}$$

for all vector fields X, Y and Z on M.

Suppose now that (M, g, ∇) is a statistical manifold and ∇^* is the dual connection of ∇ with respect to the metric g. Then it is known that the Levi-Civita connection $\bar{\nabla}$ of the metric g is given by

$$\bar{\nabla} = \frac{1}{2}(\nabla + \nabla^*). \tag{2}$$

We can define now a tensor field K on M, of type $(1, 2)$, by

$$K_V W = \nabla_V W - \bar{\nabla}_V W.$$

It follows immediately that K is symmetric, that is,

$$K_V W = K_W V,$$

and we also have

$$g(K_U V, W) = g(K_U W, V),$$

for all vector fields U, V and W on M.

Definition 1 ([37]) Let M be a smooth manifold equipped with a triple (φ, ξ, η) where φ is a field of endomorphisms of the tangent spaces, ξ is a vector field and η is a 1-form on $\bar{M}$. If we have

$$\varphi^2 = \tau(-Id + \eta \otimes \xi), \quad \eta(\xi) = 1, \tag{3}$$

where Id denotes the identity tensor field, then we say that:

(i) (φ, ξ, η) in an almost contact structure on $\bar{M}$, if $\tau = 1$.
(ii) (φ, ξ, η) in an almost paracontact structure on $\bar{M}$, is $\tau = -1$.

From the above definition, one can obtain immediately

$$\varphi\xi = 0, \quad \eta \circ \varphi = 0. \tag{4}$$

Definition 2 ([37]) A mixed 3-structure on a smooth manifold M consists in a triplet of structures $(\varphi_\alpha, \xi_\alpha, \eta_\alpha)_{\alpha=1,2,3}$ which are almost paracontact structures for $\alpha = 1, 2$ and almost contact structure for α=3, satisfying the following conditions:

$$\eta_\alpha(\xi_\beta) = 0, \tag{5}$$

$$\varphi_\alpha(\xi_\beta) = \tau_\beta \xi_\gamma, \quad \varphi_\beta(\xi_\alpha) = \tau_\alpha \xi_\gamma, \tag{6}$$

$$\eta_\alpha \circ \varphi_\beta = -\eta_\beta \circ \varphi_\alpha = \tau_\alpha \eta_\gamma, \tag{7}$$

$$\varphi_\alpha \varphi_\beta - \tau_\alpha \eta_\beta \circ \xi_\alpha = -\varphi_\beta \varphi_\alpha + \tau_\beta \eta_\alpha \circ \xi_\beta = \tau_\gamma \varphi_\gamma, \tag{8}$$

where (α, β, γ) is an even permutation of $(1, 2, 3)$ and $\tau_1 = \tau_2 = \tau_3 = -1$.

Definition 3 ([37]) Suppose M is a smooth manifold endowed with a mixed 3-structure $(\varphi_\alpha, \xi_\alpha, \eta_\alpha)_{\alpha=1,2,3}$. If there exists a pseudo-Riemannian metric g on M such that

$$g(\varphi_\alpha X, \varphi_\alpha Y) = \tau_\alpha[g(X, Y) - \varepsilon_\alpha \eta_\alpha(X)\eta_\alpha(Y)], \tag{9}$$

for any vector field X, Y on M and $\alpha = 1, 2, 3$, where $\varepsilon_\alpha = g(\xi_\alpha, \xi_\alpha) = \pm 1$, then g is called a compatible metric on M. In this case, we say that the manifold M has a metric mixed 3-structure $((\varphi_\alpha, \xi_\alpha, \eta_\alpha)_{\alpha=1,2,3}, g)$ and $(M, (\varphi_\alpha, \xi_\alpha, \eta_\alpha)_{\alpha=\overline{1,3}}, \bar{g})$ is called *a metric mixed 3-structure manifold*.

Remark 1 From (9) it follows that

$$\eta_\alpha(X) = \varepsilon_\alpha g(X, \xi_\alpha), \quad g(\varphi_\alpha X, Y) = -g(X, \varphi_\alpha Y) \tag{10}$$

for any vector fields X, Y on M and $\alpha = 1, 2, 3$. Now, it is easy to see that (10) implies

$$g(\xi_1, \xi_1) = g(\xi_2, \xi_2) = -g(\xi_3, \xi_3).$$

Consequently, we deduce that ξ_1 and ξ_2 are both either spacelike vector fields or timelike vector fields and these also determine the causality of ξ_3. Due to this, we can distinguish between positive and negative metric mixed 3-structures, according as ξ_1 and ξ_2 are both spacelike or both timelike vector fields.

We derive now immediately that the dimension of a metric mixed 3-structure manifold is $4n + 3$ and the signature of any compatible g is either $(2n + 1, 2n + 2)$, provided that structure is positive (i.e. $\varepsilon_1 = \varepsilon_2 = -\varepsilon_3 = 1$), or $(2n + 2, 2n + 1)$, provided that the structure is negative (i.e. $\varepsilon_1 = \varepsilon_2 = -\varepsilon_3 = -1$) (for detail see [8]).

Definition 4 ([8, 21]) Let $(M, (\varphi_\alpha, \xi_\alpha, \eta_\alpha)_{\alpha=\overline{1,3}}, g)$ be a metric mixed 3-structure manifold.

(i) If $(\varphi_1, \xi_1, \eta_1, g)$ and $(\varphi_2, \xi_2, \eta_2, g)$ are para-cosymplectic structures, while $(\varphi_3, \xi_3, \eta_3, g)$ is a cosymplectic structure, *i.e.* the Levi-Civita connection $\bar{\nabla}$ of g satisfies

$$\bar{\nabla}\varphi_\alpha = 0 \tag{11}$$

for all $\alpha \in \{1, 2, 3\}$, then $((\varphi_\alpha, \xi_\alpha, \eta_\alpha)_{\alpha=\overline{1,3}}, g)$ is called *a mixed 3-cosymplectic structure* on M. In this case, $(M, (\varphi_\alpha, \xi_\alpha, \eta_\alpha)_{\alpha=\overline{1,3}}, g)$ is said to be *a mixed 3-cosymplectic manifold*.

(ii) If $(\varphi_1, \xi_1, \eta_1, \bar{g})$ and $(\varphi_2, \xi_2, \eta_2, \bar{g})$ are para-Sasakian structures, while $(\varphi_3, \xi_3, \eta_3, \bar{g})$ is a Sasakian structure, *i.e.*

$$(\bar{\nabla}_X \varphi_\alpha)Y = \tau_\alpha[g(X, Y)\xi_\alpha - \varepsilon_\alpha \eta_\alpha(Y)X] \tag{12}$$

for all vector fields X, Y on M and $\alpha \in \{1, 2, 3\}$, then $((\varphi_\alpha, \xi_\alpha, \eta_\alpha)_{\alpha=\overline{1,3}}, g)$ is called *a mixed 3-Sasakian structure* on M. In this case, $(M, (\varphi_\alpha, \xi_\alpha, \eta_\alpha)_{\alpha=\overline{1,3}}, g)$ is said to be *a mixed 3-Sasakian manifold*.

We note that from (11) we derive

$$\bar{\nabla}\xi_\alpha = 0,$$

and (12) implies

$$\bar{\nabla}_X \xi_\alpha = -\varepsilon_\alpha \varphi_\alpha X,$$

for $\alpha \in \{1, 2, 3\}$ and any vector field X on M.

Definition 5 Let $(M, (\varphi_\alpha, \xi_\alpha, \eta_\alpha)_{\alpha=\overline{1,3}}, g)$ be a metric mixed 3-structure manifold. If M can be endowed with a statistical structure (∇, g), then we say that $(M, \nabla, (\varphi_\alpha, \xi_\alpha, \eta_\alpha)_{\alpha=\overline{1,3}})$ is a 3-structure statistical manifold.

Definition 6 Let $(M, (\varphi_\alpha, \xi_\alpha, \eta_\alpha)_{\alpha=\overline{1,3}}, g)$ be be a mixed 3-cosymplectic manifold (resp. a mixed 3-Sasakian manifold). If (∇, g) is a statistical structure on M such that

$$K_X \varphi_\alpha Y + \varphi_\alpha K_X Y = 0 \tag{13}$$

for all $\alpha \in \{1, 2, 3\}$ and X, Y vector fields on M, then $(M, \nabla, (\varphi_\alpha, \xi_\alpha, \eta_\alpha)_{\alpha=\overline{1,3}}, g)$ is said to be *a mixed 3-cosymplectic statistical manifold* (resp. *a mixed 3-Sasakian statistical manifold*).

Remark 2 If $(M, \nabla, (\varphi_\alpha, \xi_\alpha, \eta_\alpha)_{\alpha=\overline{1,3}}, g)$ is a mixed 3-cosymplectic or mixed 3-Sasakian statistical manifold, then so is $(M, \nabla^*, (\varphi_\alpha, \xi_\alpha, \eta_\alpha)_{\alpha=\overline{1,3}}, g)$.

Theorem 1 *If $(M, \nabla, (\varphi_\alpha, \xi_\alpha, \eta_\alpha)_{\alpha=\overline{1,3}}, g)$ is a mixed 3-structure statistical manifold, then we have for all vector fields X, Y and Z on M:*

$$(\nabla_X \omega_\alpha)(Y, Z) = g(Y, \nabla^*_X \varphi_\alpha Z - \varphi_\alpha \nabla_X Z) \tag{14}$$

and

$$\nabla_X \varphi_\alpha Y - \varphi_\alpha \nabla^*_X Y = (\bar{\nabla}_X \varphi_\alpha) Y + K_X \varphi_\alpha Y + \varphi_\alpha K_X Y, \tag{15}$$

for $\alpha = 1, 2, 3$, where $\omega_\alpha(Y, Z) = g(Y, \varphi_\alpha Z)$.

Proof Using (3), we derive easily:

$$\begin{aligned}(\nabla_U \omega_\alpha)(V, W) &= U(\omega_\alpha(V, W)) - \omega_\alpha(\nabla_U V, W) - \omega_\alpha(V, \nabla_U \omega)\\ &= Ug(V, \varphi_\alpha W) - g(\nabla_U V, \varphi_\alpha W) - g(V, \varphi_\alpha \nabla_U W),\end{aligned}$$

and in view of (1), we find

$$\begin{aligned}(\nabla_U \omega_\alpha)(V, W) &= g(\nabla_U V, \varphi_\alpha W) + g(V, \nabla^*_U \varphi_\alpha W)\\ &\quad - g(\nabla_U V, \varphi_\alpha W) - g(V, \varphi_\alpha \nabla_U W)\\ &= g(V, \nabla^*_U \varphi_\alpha W) - g(V, \nabla_U W)\\ &= g(V, \nabla^*_U \alpha W - \varphi_\alpha \nabla_U W).\end{aligned}$$

Therefore, it follows that (14) holds.

On the other hand, using the definition of K, we derive immediately

$$K_X \varphi_\alpha Y = \nabla_X \varphi_\alpha Y - \bar{\nabla}_X \varphi_\alpha Y$$

and

$$\varphi_\alpha K_X Y = \varphi_\alpha \nabla_X Y - \varphi_\alpha \bar{\nabla}_X Y.$$

By summing the last two relations, we derive

$$\nabla_X \varphi_\alpha Y + \varphi_\alpha \nabla_X Y - \bar{\nabla}_X \varphi_\alpha Y - \varphi_\alpha \bar{\nabla}_X Y = K_X \varphi_\alpha Y + \varphi_\alpha K_X Y.$$

Using now (2) in the above relation, we obtain

$$\nabla_X \varphi_\alpha Y - \varphi_\alpha \nabla^*_X Y = (\bar{\nabla}_X \varphi_\alpha) Y + K_X \varphi_\alpha Y + \varphi_\alpha K_X Y.$$

Hence, we conclude that (15) also holds. □

Theorem 2
(i) A mixed 3-structure statistical manifold $(M, \nabla, (\varphi_\alpha, \xi_\alpha, \eta_\alpha)_{\alpha=\overline{1,3}}, g)$ *is mixed 3-Sasakian if and only if:*

$$\nabla_X \varphi_\alpha Y - \varphi_\alpha \nabla^*_X Y = \tau_\alpha [g(X, Y)\xi_\alpha - \varepsilon_\alpha \eta_\alpha(Y) X] \tag{16}$$

and

$$\nabla_X \xi_\alpha = \varepsilon_\alpha [g(\nabla_X \xi_\alpha, \xi_\alpha)\xi_\alpha - \varphi_\alpha X] \tag{17}$$

or, equivalently, if and only if

$$\nabla^*_X \varphi_\alpha Y - \varphi_\alpha \nabla_X Y = \tau_\alpha [g(X, Y)\xi_\alpha - \varepsilon_\alpha \eta_\alpha(Y) X] \tag{18}$$

and

$$\nabla^*_X \xi_\alpha = \varepsilon_\alpha [g(\nabla^*_X \xi_\alpha, \xi_\alpha)\xi_\alpha - \varphi_\alpha X] \tag{19}$$

for all vector fields X and Y on M.

(ii) A mixed 3-structure statistical manifold $(M, \nabla, (\varphi_\alpha, \xi_\alpha, \eta_\alpha)_{\alpha=\overline{1,3}}, g)$ *is mixed 3-cosymplectic if and only if:*

$$\nabla_X \varphi_\alpha Y = \varphi_\alpha \nabla^*_X Y, \tag{20}$$

or, equivalently,

$$\nabla^*_X \varphi_\alpha Y = \varphi_\alpha \nabla_X Y \tag{21}$$

for all vector fields X and Y on M.

Proof (*i*) Suppose first that $(M, \nabla, (\varphi_\alpha, \xi_\alpha, \eta_\alpha)_{\alpha=\overline{1,3}}, g)$ is a mixed 3-Sasakian statistical manifold. Then we have

$$K_X\varphi_\alpha Y + \varphi_\alpha K_X Y = 0, \quad \alpha = 1, 2, 3,$$

and using the above relation in (15), we obtain

$$\nabla_X\varphi_\alpha Y - \varphi_\alpha\nabla^*_X Y = (\bar{\nabla}_X\varphi_\alpha)Y.$$

Therefore, in view of (12), the above relations imply (16). Moreover, due to Remark 2, we also obtain (18). Next, by replacing Y with ξ_α in (18), we get

$$\nabla^*_X\varphi_\alpha\xi_\alpha - \varphi_\alpha\nabla_X\xi_\alpha = \tau_\alpha[g(X, \xi_\alpha)\xi_\alpha - \varepsilon_\alpha\eta_\alpha(\xi_\alpha)X],$$

therefore, taking account of (4) and (10), we derive

$$-\varphi_\alpha\nabla_X\xi_\alpha = \tau_\alpha[g(X, \xi_\alpha)\xi_\alpha - \varepsilon_\alpha^2 g(\xi_\alpha, \xi_\alpha)X].$$

Consequently, as $g(\xi_\alpha, \xi_\alpha) = \varepsilon_\alpha = \pm 1$, the above relation reduces to

$$\varphi_\alpha\nabla_X\xi_\alpha = \tau_\alpha[\varepsilon_\alpha X - g(X, \xi_\alpha)\xi_\alpha].$$

Applying now φ_α in the last relation and using (3) and (4), one obtains

$$\tau_\alpha[-\nabla_X\xi_\alpha + \eta_\alpha(\nabla_X\xi_\alpha)\xi_\alpha] = \tau_\alpha\varepsilon_\alpha\varphi_\alpha X.$$

As $\tau_\alpha \neq 0$, the above relation can be written as

$$\nabla_X\xi_\alpha = \eta_\alpha(\nabla_X\xi_\alpha)\xi_\alpha - \varepsilon_\alpha\varphi_\alpha X$$

and using (10), we derive

$$\nabla_X\xi_\alpha = \varepsilon_\alpha[g(\nabla_X\xi_\alpha, \xi_\alpha)\xi_\alpha - \varphi_\alpha X].$$

Hence, we conclude that relation (17) holds. Similarly, we find that (19) is also satisfied. In this way, we obtained that supposing $(M, \nabla, (\varphi_\alpha, \xi_\alpha, \eta_\alpha)_{\alpha=\overline{1,3}}, g)$ is a mixed 3-Sasakian statistical manifold, then (16)–(19) hold. On the other hand, it is clear that the conditions (16) and (17) are equivalent to (18) and (19). In order to complete the proof of (*i*), it is enough to prove that assuming (16) and (17) it follows that $(M, \nabla, (\varphi_\alpha, \xi_\alpha, \eta_\alpha)_{\alpha=\overline{1,3}}, g)$ is a mixed 3-Sasakian statistical manifold. If (16) and (17) are valid, then we also have that the conditions (18) and (19) hold. Using (15), it follows that (16) is equivalent to

$$\tau_\alpha[g(X, Y)\xi_\alpha - \varepsilon_\alpha\eta_\alpha(Y)X] - (\bar{\nabla}_X\varphi_\alpha)Y = K_X\varphi_\alpha Y + \varphi_\alpha K_X Y. \tag{22}$$

Similarly, from (18), we obtain

$$-\tau_\alpha[g(X,Y)\xi_\alpha - \varepsilon_\alpha\eta_\alpha(Y)X] + (\bar{\nabla}_X\varphi_\alpha)Y = K_X\varphi_\alpha Y + \varphi_\alpha K_X Y. \tag{23}$$

By adding (22) and (23) we infer that $K_X\varphi_\alpha Y + \varphi_\alpha K_X Y = 0$ for $\alpha = 1, 2, 3$. Hence, it is clear that (22) reduces to (12) and we conclude immediately that the mixed 3-structure statistical manifold $(M, \nabla, (\varphi_\alpha, \xi_\alpha, \eta_\alpha)_{\alpha=\overline{1,3}}, g)$ is nothing more than a mixed 3-Sasakian statistical manifold.

(ii) Let the mixed 3-structure statistical manifold $(M, \nabla, (\varphi_\alpha, \xi_\alpha, \eta_\alpha)_{\alpha=\overline{1,3}}, g)$ be mixed 3-cosymplectic. Then we have (13) and using (15) we derive

$$\nabla_X\varphi_\alpha Y - \varphi_\alpha\nabla^*_X Y = (\bar{\nabla}_X\varphi_\alpha)Y,$$

which, in view of (11), implies

$$\nabla_X\varphi_\alpha Y = \varphi_\alpha\nabla^*_X Y.$$

Hence, the relation (20) holds and we also derive that (21) is valid due to Remark 2. On the other hand, it follows immediately that the conditions (20) and (21) are equivalent. Conversely, if we suppose that (20) holds, then with similar arguments as in (i) it follows that (11) and (13) hold. Hence we deduce that in this case the mixed 3-structure statistical manifold $(M, \nabla, (\varphi_\alpha, \xi_\alpha, \eta_\alpha)_{\alpha=\overline{1,3}}, g)$ is nothing more than a mixed 3-cosymplectic statistical manifold. □

3 Paraquaternionic Structures on Statistical Manifolds

Definition 7 ([37]) An *almost para-hypercomplex structure* on a differentiable manifold M consists in a triplet $H = (J_1, J_2, J_3)$ of $(1, 1)$-type tensor fields on M satisfying:

$$J_\alpha^2 = -\tau_\alpha Id, \quad J_\alpha J_\beta = -J_\beta J_\alpha = \tau_\gamma J_\gamma, \tag{24}$$

for any $\alpha \in \{1, 2, 3\}$ and for any even permutation (α, β, γ) of $(1, 2, 3)$, where $\tau_1 = \tau_2 = -1 = -\tau_3$. In this case, the pair (M, H) is called an *almost para-hypercomplex manifold*. A pseudo-Riemannian metric g on (M, H) is said to be *adapted* to $H = (J_\alpha)_{\alpha=1,2,3}$ if

$$g(J_\alpha X, J_\alpha Y) = \tau_\alpha g(X, Y)$$

for any vector fields X,Y on M and $\alpha \in \{1, 2, 3\}$. Moreover, (g, H) is said to be an *almost para-hyperhermitian structure* on M, while the triplet (M, g, H) is called an *almost para-hyperhermitian manifold*.

Definition 8 ([37]) An *almost paraquaternionic Hermitian manifold* is a triplet (M, σ, g), where M is a differentiable manifold, σ is an almost paraquaternionic structure on M, that is a subbundle of $End(TM)$ with rank 3, spanned locally by an almost para-hypercomplex structure $H = (J_\alpha)_{\alpha=1,2,3}$, and g is an adapted metric with respect to H.

Definition 9 Let (M, g) be a pseudo-Riemannian manifold equipped with an almost paraquaternionic structure σ which admits for any canonical basis $\{J_1, J_2, J_3\}$ three other tensor fields $\{J_1^*, J_2^*, J_3^*\}$ of type $(1, 1)$ on M, satisfying

$$g(J_\alpha X, Y) + g(X, J_\alpha^* Y) = 0, \tag{25}$$

for all $\alpha \in \{1, 2, 3\}$ and vector fields X and Y on M. Then, the triplet (M, ∇, g) is called an *almost Hermite-like paraquaternionic manifold.* Moreover, if we have (M, ∇, g) equipped with a torsion free linear connection ∇ such that ∇g is symmetric, then the quadruplet (M, ∇, σ, g) is called an *almost Hermite-like paraquaternionic statistical manifold.*

Remark 3 We would like to note that the tensor fields $\{J_1^*, J_2^*, J_3^*\}$ defined by (25) satisfy the paraquaternionic identities (24). Due this, one can consider the subbundle σ^* of $End(TM)$ spanned locally by $\{J_1^*, J_2^*, J_3^*\}$. We can also easily check that:

$$(J_\alpha^*)^* = J_\alpha$$

and

$$g(J_\alpha X, J_\alpha^* Y) = \tau_\alpha g(X, Y)$$

for all vector fields X, Y on M and $\alpha \in \{1, 2, 3\}$.

Definition 10 An almost Hermite-like paraquaternionic statistical manifold (M, ∇, σ, g) is called a *paraquaternionic Kähler-like statistical manifold* if for any local basis $\{J_1, J_2, J_3\}$ of σ there exist three locally defined 1-forms $\omega_1, \omega_2, \omega_3$ on M such that we have for all $\alpha \in \{1, 2, 3\}$ that

$$\nabla J_\alpha = \tau_\beta \omega_\gamma \otimes J_\beta - \tau_\gamma \omega_\beta \otimes J_\gamma. \tag{26}$$

In particular, if $\omega_1 = \omega_2 = \omega_3 = 0$ then (M, ∇, σ, g) is called a *locally parahyper-Kähler-like statistical manifold*. Moreover, if $\{J_1, J_2, J_3\}$ are defined globally on M, then $(M, \nabla, J_1, J_2, J_3, g)$ is said to be a *parahyper-Kähler-like statistical manifold.*

Remark 4 We note that, if in the above definition ∇ is the Levi-Civita connection of the metric g, then (M, ∇, σ, g) reduces to a paraquaternionic Kähler manifold, usually denoted as a triplet (M, σ, g) (see, e.g., [2, 7, 15, 17, 38, 39, 41]).

Definition 11 A paraquaternionic Kähler-like statistical manifold (M, ∇, σ, g) is said to be a *paraquaternionic-like space form* if there exists a real constant c such that the curvature tensor R with respect to ∇ is given by

$$R(X,Y)Z = \frac{c}{4}\{g(Y,Z)X - g(X,Z)Y \\ + \sum\nolimits_{\alpha=1}^{3} \tau_\alpha [g(Z, J_\alpha Y)J_\alpha X - g(Z, J_\alpha X)J_\alpha Y] \\ + \sum\nolimits_{\alpha=1}^{3} [g(X, J_\alpha Y) - g(J_\alpha X, Y)]J_\alpha Z\}$$

for any vector fields X, Y, Z on M.

Remark 5 The concept of paraquaternionic-like space form generalizes the notion of paraquaternionic space form. Indeed, if in the previous definition ∇ is the Levi-Civita connection of the metric g, then we obtain the definition of a paraquaternionic space form (see [15]). The prototypes of such spaces are the paraquaternionic projective space $\mathbb{B}P^n(c)$ with paraquaternionic sectional curvature $c > 0$, the paraquaternionic Euclidean space $\mathbb{B}^n$ having null paraquaternionic sectional curvature and the paraquaternionic hyperbolic space $\mathbb{B}H^n(c)$ with paraquaternionic sectional curvature $c < 0$.

Theorem 3 *Suppose* (M, ∇, σ, g) *is an almost Hermite-like paraquaternionic statistical manifold. Then* (M, ∇, σ, g) *is a paraquaternionic Kähler-like statistical manifold if and only if* $(M, \nabla^*, \sigma^*, g)$ *is a paraquaternionic Kähler-like statistical manifold.*

Proof Suppose that (M, ∇, σ, g) is a paraquaternionic Kähler-like statistical manifold. First, using (1) and (25), we derive

$$\begin{aligned} g((\nabla_Z J_\alpha)X, Y) &= g(\nabla_Z J_\alpha X - J_\alpha \nabla_Z X, Y) \\ &= g(\nabla_Z J_\alpha X, Y) - g(J_\alpha \nabla_Z X, Y) \\ &= Zg(J_\alpha X, Y) - g(J_\alpha X, \nabla^*_Z Y) + g(\nabla_Z X, J^*_\alpha Y) \\ &= Zg(J_\alpha X, Y) + g(X, J^*_\alpha \nabla^*_Z Y) + Zg(X, J^*_\alpha Y) \\ &\quad -g(X, \nabla^*_Z J^*_\alpha Y) \\ &= -g(X, \nabla^*_Z J^*_\alpha Y - J^*_\alpha \nabla^*_Z Y) \\ &= -g(X, (\nabla^*_Z J^*_\alpha)Y). \end{aligned} \tag{27}$$

On the other hand, making use of (26), we obtain

$$\begin{aligned} g((\nabla_Z J_\alpha)X, Y) &= g(\tau_\beta \omega_\gamma(Z) J_\beta X - \tau_\gamma \omega_\beta(Z) J_\gamma X, Y) \\ &= g(\tau_\beta \omega_\gamma(Z) J_\beta X, Y) - g(\tau_\gamma \omega_\beta(Z) J_\gamma X, Y) \\ &= g(X, -\tau_\beta \omega_\gamma(Z) J^*_\beta Y) + g(X, \tau_\alpha \omega_\beta(Z) J^*_\gamma Y) \\ &= g(X, \tau_\gamma \omega_\beta(Z) J^*_\gamma Y - \tau_\beta \omega_\gamma(Z) J^*_\beta Y). \end{aligned} \tag{28}$$

From (27) and (28) we get

$$-(\nabla^*_Z J^*_\alpha)Y = \tau_\gamma \omega_\beta(Z) J^*_\gamma Y - \tau_\beta \omega_\gamma(Z) J^*_\beta Y,$$

therefore, we deduce

$$\nabla^* J_\alpha^* = \tau_\beta \omega_\gamma \otimes J_\beta^* - \tau_\gamma \omega_\beta \otimes J_\gamma^*.$$

Thus, it follows that $(M, \nabla^*, \sigma^*, g)$ is a paraquaternionic Kähler-like statistical manifold. Conversely, if $(M, \nabla^*, \sigma^*, g)$ is a paraquaternionic Kähler-like statistical manifold, then using similar arguments, we obtain that (M, ∇, σ, g) is a paraquaternionic Kähler-like statistical manifold. □

Corollary 1 *Suppose (M, ∇, σ, g) is an almost Hermite-like paraquaternionic statistical manifold. Then (M, ∇, σ, g) is a parahyper-Kähler-like statistical manifold, if and only if so is $(M, \nabla^*, \sigma^*, g)$.*

Proof The assertion is clear from Theorem 3, because a parahyper-Kähler-like statistical manifold is a particular type of paraquaternionic Kähler-like statistical manifold with $\omega_1 = \omega_2 = \omega_3 = 0$. □

Example 1 Let (M, ∇, P, g) be am almost para-Hermitian-like manifold (for examples of such manifolds see [40, Sect. 5]). Next, we prove that the tangent bundle TM of M can be endowed with an almost Hermite-like paraquaternionic statistical structure. First, we note that TM can be equipped with a Sasaki-type metric G defined by:

$$G(A, B) = g(kA, kB) + g(\pi_* A, \pi_* B),$$

for all vector fields A, B on TM, where $\pi : TM \to M$ is the natural projection and k is the connection map associated with the Levi-Civita connection of the metric g.

We would like to remark at this stage that if $X \in \Gamma(TM)$, then there is a unique vector field on TM, denoted by X^h and called the horizontal lift of X, and a unique vector field on TM denoted by X^v and called the vertical lift of X, such that for all $U \in TM$ we have

$$\pi_* X_U^h = X_{\pi(U)}, \quad \pi_* X_U^v = 0_{\pi(U)}, \quad k X_U^h = 0_{\pi(U)}, \quad k X_U^v = X_{\pi(U)}.$$

We recall now that, according to Theorem 3.1 in [16], one can define a torsion free linear connection ∇' on TM compatible to G. Hence, it follows that (TM, ∇', G) is a statistical manifold. Using now the almost product structure P on M, we can also define three tensor fields $\{J_1, J_2, J_3\}$, on TM as follows:

$$\begin{cases} J_3 X^h = X^v, \\ J_3 X^v = -X^h, \end{cases}$$

$$\begin{cases} J_2 X^h = (\varphi X)^v, \\ J_2 X^v = (\varphi X)^h, \end{cases}$$

$$\begin{cases} J_1 X^h = -(\varphi X)^h, \\ J_1 X^v = (\varphi X)^v. \end{cases}$$

By a direct computation we obtain that $H = (J_1, J_2, J_3)$ is an almost para-hypercomplex structure on TM. Moreover, considering the 3-subbundle σ generated by H, it follows immediately that (TM, ∇', σ, G) is an almost Hermite-like paraquaternionic statistical manifold. In addition, we also obtain that the quadruplet (TM, ∇', σ, G) is a parahyper-Kähler-like statistical manifold, provided that (M, ∇, P, g) is a flat para-Kähler-like statistical manifold.

4 Statistical Submersions from Mixed 3-Sasakian Manifolds

Semi-Riemannian submersions were introduced by O'Neill, see [29]. A *semi-Riemannian submersion* between connected semi-Riemannian manifolds (M, g) and (N, g_N) is a onto smooth map $\pi : M \to N$ satisfying the next three conditions:
(i) $\pi_{*|p} : T_pM \to T\pi_(p)N$ is onto for all $p \in M$;
(ii) The fibres $\pi^{-1}(p')$, $p' \in N$, are semi-Riemannian submanifolds of M;
(iii) π_* preserves scalar products of vectors normal to fibres.

Recall that the vectors tangent to fibres are called *vertical* and those normal to fibers are called *horizontal*.
We denote by $\mathcal{V}$ the vertical distribution, by $\mathcal{H}$ the horizontal distribution, and by v and h the vertical and horizontal projections. We also recall that an horizontal vector field X on M is called *basic*, if X is π-related to a vector field X' on N. It is clear that every vector field X' on N has a unique horizontal lift X on M and X is basic.

Definition 12 ([33]) If (M, ∇, g) and (N, ∇^N, g_N) are statistical manifolds, then a semi-Riemannian submersion $\pi : M \to N$ is called a *statistical submersion* if

$$\pi_*(\nabla_X Y)_p = (\nabla^N{}_{X'} Y')_{\pi(p)}$$

for all basic vector fields X, Y on M which are π-related to X' and Y' on N, and $p \in M$.

Remark 6 If $\pi : M \to N$ is a statistical submersion, then it is known that any fiber admits a canonical statistical structure [33]. Hence any fiber is a statistical manifold. The affine connections induced on fibres by the dual connections ∇ and ∇^* from M are denoted by $\widehat{\nabla}$ and $\widehat{\nabla}^*$, respectively. We note that $\widehat{\nabla}$ and $\widehat{\nabla}^*$ can be defined by

$$\widehat{\nabla}_U V = v\nabla_U V, \ \widehat{\nabla}^*_U V = v\nabla^*_U V \tag{29}$$

for all $U, V \in \Gamma(\mathcal{V})$. Note that $\widehat{\nabla}$ and $\widehat{\nabla}^*$ are torsion free and conjugate to each other with respect to the induced metric on fibres.

For a statistical submersion $\pi : M \to N$ it is also possible to define as in the case of semi-Riemannian submersions, the O'Neill tensor fields T and A by

$$T(E, F) = T_E F = h\nabla_{vE} vF + v\nabla_{vE} hF \tag{30}$$

and

$$A(E, F) = A_E F = v\nabla_{hE} hF + h\nabla_{hE} vF, \tag{31}$$

for $E, F \in \Gamma(TM)$.

In a similar way, we can consider the O'Neill tensor fields T^* and A^* on M by putting ∇^* instead of ∇ in (30) and (31). It is easy to check now that $T^{**} = T$ and $A^{**} = A$ and also have for all vertical vector fields U and V, see [1],

$$\nabla_U V = T_U V + v\nabla_U V, \tag{32}$$

$$\nabla^*_U V = T^*_U V + v\nabla^*_U V, \tag{33}$$

$$T_U V = T_V U, \tag{34}$$

and

$$T^*_U V = T^*_V U. \tag{35}$$

We recall now that if $T_U V = 0$, for all vertical vector fields U and V on M, then π is called a statistical submersion *with isometric fibers*, see [32].

Definition 13 Let $(M, \nabla, (\varphi_\alpha, \xi_\alpha, \eta_\alpha)_{\alpha=\overline{1,3}}, g)$ be a mixed 3-structure statistical manifold and (N, ∇^N, g_N) be a statistical manifold. Then a statistical submersion $\pi : M \to N$ is called

(i) *invariant* if $\varphi_\alpha(\ker \pi_*) \subset \ker \pi_*$ for $\alpha \in \{1, 2, 3\}$,
(ii) *anti-invariant* if $\varphi_\alpha(\ker \pi_*) \subset (\ker\pi_*)^\perp$ for $\alpha \in \{1, 2, 3\}$.

Theorem 4 *Let $(M, \nabla, (\varphi_\alpha, \xi_\alpha, \eta_\alpha)_{\alpha=\overline{1,3}}, g)$ be a mixed 3-structure statistical manifold and (N, ∇^N, g_N) be a statistical manifold. If $\pi : M \to N$ is an anti-invariant statistical submersion, then the structure vector fields ξ_1, ξ_2 and ξ_3 can not be all vertical.*

Proof We assume by reduction to the absurd that ξ_1, ξ_2 and ξ_3 are all vertical. Hence we have $\xi_\alpha \in \ker \pi_*$, for $\alpha \in \{1, 2, 3\}$, and due to the fact that π is anti-invariant, we deduce that

$$\varphi_\alpha(\xi_\beta) \in (\ker \pi_*)^\perp.$$

But using (6) in the above relation, we derive

$$\varphi_\alpha(\xi_\beta) = \tau_\beta \xi_\gamma \in (\ker \pi_*)^\perp,$$

thus, we conclude that $\xi_\gamma \subset (\ker \pi_*)^\perp$. But $\xi_\gamma \subset \ker \pi_*$ and therefore we obtain a contradiction. □

Theorem 5 *Let $(M, \nabla, (\varphi_\alpha, \xi_\alpha, \eta_\alpha)_{\alpha=\overline{1,3}}, g)$ be a mixed 3-Sasakian statistical manifold and (N, ∇^N, g_N) be a statistical manifold. If $\pi : M \to N$ is a statistical submersion such that the structure vector fields ξ_1, ξ_2 and ξ_3 are all horizontal, then π is anti-invariant.*

Proof Replacing X by $\varphi_\alpha U$ and Y by V in (9), where U, V are vertical vector fields on M, we derive

$$g(\varphi_\alpha^2 U, \varphi_\alpha V) = \tau_\alpha[g(\varphi_\alpha U, V) - \varepsilon_\alpha \eta_\alpha(\varphi_\alpha U)\eta_\alpha(V)],$$

and, in view of (3) and (4), we obtain

$$\tau_\alpha g(-U + \eta_\alpha(U)\xi_\alpha, \varphi_\alpha V) = \tau_\alpha g(\varphi_\alpha U, V).$$

Because $\tau_\alpha \neq 0$, we can simplify the above relation by τ_α and taking into account (10) we have $\eta_\alpha(E) = 0$ (since U is vertical and ξ_α is horizontal). Thus, the last equation reduces to

$$-\, g(\varphi_\alpha F, E) = g(\varphi_\alpha E, F).$$

Since $(M, \nabla, (\varphi_\alpha, \xi_\alpha, \eta_\alpha)_{\alpha=\overline{1,3}}, g)$ is mixed 3-Sasakian, we can use Theorem 2. Therefore, in view of (17), we derive

$$\varphi_\alpha U = \varepsilon_\alpha \nabla_U \xi_\alpha - g(\nabla_U \xi_\alpha, \xi_\alpha)\xi_\alpha,$$

and inserting the last equation in the preceding equation, we obtain

$$g(\varphi_\alpha V, U) = -g(\varepsilon_\alpha \nabla_U \xi_\alpha, V) + g(\nabla_U \xi_\alpha, \xi_\alpha)g(\xi_\alpha, V).$$

Now, since ξ_α is horizontal and V is vertical, we have $g(\xi_\alpha, V) = 0$ and the above equation becomes

$$g(\varphi_\alpha V, U) = -\varepsilon_\alpha g(\nabla_U \xi_\alpha, V).$$

Therefore, in view of (1), since ξ_α and V are orthogonal, we get

$$g(\varphi_\alpha V, U) = \varepsilon_\alpha g(\xi_\alpha, \nabla_U^* V). \tag{36}$$

On another hand, using (17), we derive

$$\varphi_\alpha V = \varepsilon_\alpha \nabla_V \xi_\alpha - g(\nabla_V \xi_\alpha, \xi_\alpha)\xi_\alpha,$$

therefore, we have

$$g(\varphi_\alpha V, U) = \varepsilon_\alpha g(\nabla_V \xi_\alpha, U) - g(\nabla_V \xi_\alpha, \xi_\alpha)g(\xi_\alpha, U).$$

But $g(\xi_\alpha, U) = 0$ is true, thus we find from the above equation that

$$g(\varphi_\alpha V, U) = \varepsilon_\alpha g(\nabla_V \xi_\alpha, U).$$

Using now (1) and (33), as well as the symmetry of T with respect to vertical vector fields, we obtain from the last equation

$$\begin{aligned} g(\varphi_\alpha V, U) &= -\varepsilon_\alpha g(\xi_\alpha, \nabla^*_V U) \\ &= -\varepsilon_\alpha g(\xi_\alpha, T^*_V U + v\nabla^*_V U) \\ &= -\varepsilon_\alpha g(\xi_\alpha, T^*_V U) - \varepsilon_\alpha g(\xi_\alpha, v\nabla^*_V U) \\ &= -\varepsilon_\alpha g(\xi_\alpha, T^*_U V) \\ &= -\varepsilon_\alpha g(\xi_\alpha, T^*_U V + v\nabla^*_U V) \\ &= -\varepsilon_\alpha g(\xi_\alpha, \nabla^*_U V). \end{aligned}$$

Hence, we derive

$$g(\varphi_\alpha V, U) = -\varepsilon_\alpha g(\xi_\alpha, \nabla^*_U V). \tag{37}$$

Now, it is clear that (36) and (37) imply

$$g(\varphi_\alpha V, U) = 0$$

for all vertical vector fields U and V. Hence, it follows that

$$\varphi_\alpha(\ker \pi_*) \subset (\ker \pi_*)^\perp,$$

so we conclude that π is an anti-invariant statistical submersion. □

Theorem 6 *Let $(M, \nabla, (\varphi_\alpha, \xi_\alpha, \eta_\alpha)_{\alpha=\overline{1,3}}, g)$ be a mixed 3-structure statistical manifold and (N, ∇^N, g_N) be a statistical manifold. If $\pi : M \to N$ is an invariant statistical submersion, then the structure vector fields ξ_1, ξ_2 and ξ_3 are either vertical or horizontal.*

Proof Let us consider the following decompositions of the structure vector fields:

$$\xi_\alpha = \xi_\alpha^h + \xi_\alpha^v, \quad \alpha \in \{1, 2, 3\},$$

where $\xi_\alpha^v \in \ker \pi_*$ and $\xi_\alpha^h \in (\ker \pi_*)^\perp$. Since $\varphi_\alpha \xi_\alpha = 0$, applying $\varphi_\alpha \xi_\alpha$ in the above equation, we obtain

$$\varphi_\alpha \xi_\alpha^h = -\varphi_\alpha \xi_\alpha^v \in \ker \pi_*$$

since the submersion π is invariant.

If we take now a vertical vector field E, we derive from (10) that

$$g(\varphi_\alpha \xi_\alpha^h, E) = -g(\xi_\alpha^h, \varphi_\alpha E) = 0.$$

It follows that $\varphi_\alpha \xi_\alpha^h = 0$ and consequently $\varphi_\alpha \xi_\alpha^v = 0$. Thus, $\varphi_\alpha^2(\xi_\alpha^v) = 0$ holds, and, in view of (3), this implies

$$-\xi_\alpha^v + \eta_\alpha(\xi_\alpha^v)\xi_\alpha = 0,$$

or, equivalently,

$$-\xi_\alpha^v+\eta_\alpha(\xi_\alpha^v)(\xi_\alpha^v+\xi_\alpha^h)=0.$$

Hence, we derive

$$\xi_\alpha^v(-1+\eta_\alpha(\xi_\alpha^v))+\eta_\alpha(\xi_\alpha^v)\xi_\alpha^h=0$$

and this implies easily that either $\xi_\alpha^h=0$ or $\xi_\alpha^v=0$ for $\alpha\in\{1,2,3\}$. □

Theorem 7 *Let $(M,\nabla,(\varphi_\alpha,\xi_\alpha,\eta_\alpha)_{\alpha=\overline{1,3}},g)$ be a mixed 3-Sasakian statistical manifold (resp. a mixed 3-cosymplectic statistical manifold) and (N,∇^N,g_N) be a statistical manifold. If $\pi:M\to N$ is an invariant statistical submersion such that the structure vector fields ξ_1,ξ_2 and ξ_3 are vertical, then any fiber of the submersion is a mixed 3-Sasakian statistical manifold (resp. a mixed 3-cosymplectic statistical manifold).*

Proof We know that any fiber $\widehat{M}=\pi^{-1}(p'),\ p'\in N$, is a statistical manifold with the statistical structure $(\widehat{\nabla},\widehat{\nabla}^*,\hat{g})$, where the torsion free affine connections $\widehat{\nabla}$ and $\widehat{\nabla}^*$ are given by (29), these being conjugate with respect to the induced metric $\hat{g}$ on fibres.

We suppose first that $\pi:M\to N$ is an invariant statistical submersion from a mixed 3-Sasakian statistical manifold $(M,\nabla,(\varphi_\alpha,\xi_\alpha,\eta_\alpha)_{\alpha=\overline{1,3}},g)$. Using (29) and (33) we have

$$\hat{\nabla}_U(\varphi_\alpha V)-\varphi_\alpha(\hat{\nabla}_U^*V)=T_U(\varphi_\alpha V)+v\nabla_U(\varphi_\alpha V)-\varphi_\alpha(T_U^*V+v\nabla_U^*V)\tag{38}$$

for any vertical vector fields U and V. But using (30) and (31), we get

$$T_U(\varphi_\alpha V)-\varphi_\alpha(T_U^*V)=h(\nabla_U\varphi_\alpha V-\varphi_\alpha(\nabla_U^*V)),\tag{39}$$

and due to (16) we derive

$$\begin{aligned}T_U(\varphi_\alpha V)-\varphi_\alpha(T_U^*V)&=\tau_\alpha h[\hat{g}(U,V)\xi_\alpha-\varepsilon_\alpha\eta_\alpha(V)U]\\&=\tau_\alpha[\hat{g}(U,V)h\xi_\alpha-\varepsilon_\alpha\eta_\alpha(V)hU)\\&=0,\end{aligned}$$

since ξ_α and U are vertical. Therefore, (38) reduces to

$$\hat{\nabla}_U(\varphi_\alpha V)-\varphi_\alpha(\hat{\nabla}_U^*V)=v(\nabla_U\varphi_\alpha)V,\tag{40}$$

and using (12) in (40), we find

$$\hat{\nabla}_U(\varphi_\alpha V)-\varphi_\alpha(\hat{\nabla}_U^*V)=\tau_\alpha[\hat{g}(U,V)\xi_\alpha-\varepsilon_\alpha\eta_\alpha(V)U].\tag{41}$$

On the other hand, for any vertical vector field U we obtain using (29) and (17):

$$\begin{aligned}\hat{\nabla}_U\xi_\alpha&=v\nabla_U\xi_\alpha\\&=\varepsilon_\alpha v[\hat{g}(v\nabla_U\xi_\alpha,\xi_\alpha)\xi_\alpha+g(h\nabla_U\xi_\alpha,\xi_\alpha)\xi_\alpha-\varphi_\alpha U],\end{aligned}$$

thus we deduce

$$\hat{\nabla}_U \xi_\alpha = \varepsilon_\alpha [\hat{g}(\hat{\nabla}_U \xi_\alpha, \xi_\alpha)\xi_\alpha - \varphi_\alpha U]. \tag{42}$$

From (41) and (42), in view of Theorem 2(i), we conclude that the fiber $(\hat{M}, \hat{\nabla}, (\varphi_\alpha, \xi_\alpha, \eta_\alpha)_{\alpha=\overline{1,3}}, \hat{g})$ is a mixed 3-Sasakian statistical manifold.

Similarly, if we suppose that $\pi : M \to N$ is an invariant statistical submersion from a mixed 3-cosymplectic statistical manifold $(M, \nabla, (\varphi_\alpha, \xi_\alpha, \eta_\alpha)_{\alpha=\overline{1,3}}, g)$, it follows that

$$\hat{\nabla}_U \varphi_\alpha V = \varphi_\alpha \hat{\nabla}^*_U V$$

for all vertical vector fields U and V on M. Applying Theorem 2(ii) we conclude that $(\hat{M}, \hat{\nabla}, (\varphi_\alpha, \xi_\alpha, \eta_\alpha)_{\alpha=\overline{1,3}}, \hat{g})$ is a mixed 3-cosymplectic statistical manifold. □

Theorem 8 *Let $(M, \nabla, (\varphi_\alpha, \xi_\alpha, \eta_\alpha)_{\alpha=\overline{1,3}}, g)$ be a mixed 3-Sasakian statistical manifold or a mixed 3-cosymplectic statistical manifold and (N, ∇^N, g_N) be a statistical manifold. If $\pi : M \to N$ is an invariant statistical submersion such that the structure vector fields ξ_1, ξ_2 and ξ_3 are vertical, then π is a statistical submersion with isometric fibers.*

Proof If $U, V \in \ker \pi_*$, then we have proved in the previous Theorem that in the case when the total space is a mixed 3-Sasakian statistical manifold we have

$$T_U \varphi_\alpha V = \varphi_\alpha T^*_U V, \tag{43}$$

for $\alpha = 1, 2, 3$. On the other, if the total space is a mixed 3-Sasakian statistical manifold, it is also clear that equation (43) holds due to (39) and (20).

Applying repeatedly (43), we obtain taking into account (8), (34) and (35):

$$\begin{aligned} T_U \varphi_\gamma V &= \varphi_\gamma T^*_U V \\ &= \tau_\gamma \varphi_\alpha \varphi_\beta T^*_U V \\ &= \tau_\gamma \varphi_\alpha T_U \varphi_\beta V \\ &= \tau_\gamma \varphi_\alpha T_{\varphi_\beta V} U \\ &= \tau_\gamma T^*_{\varphi_\beta V} \varphi_\alpha U \\ &= \tau_\gamma T^*_{\varphi_\alpha U} \varphi_\beta V \\ &= \tau_\gamma \varphi_\beta T_{\varphi_\alpha U} V \\ &= \tau_\gamma \varphi_\beta T_V \varphi_\alpha U \\ &= \tau_\gamma \varphi_\beta \varphi_\alpha T^*_V U \\ &= \tau_\gamma \varphi_\beta \varphi_\alpha T^*_U V \\ &= -\tau_\gamma^2 \varphi_\gamma T^*_U V \\ &= -\varphi_\gamma T^*_U V. \end{aligned}$$

Hence, we derive

$$T_U \varphi_\gamma V = \varphi_\gamma T^*_U V = -\varphi_\gamma T^*_U V,$$

thus we deduce $T_U^* V = 0$. Consequently, $T_U V = 0$, and we conclude that the statistical submersion π is with isometric fibers. □

Remark 7 For the remaining of this section, we will focuss on the study of anti-invariant statistical submersions from mixed 3-structures statistical manifolds $(M, \nabla, (\varphi_\alpha, \xi_\alpha, \eta_\alpha)_{\alpha=\overline{1,3}}, g)$ with horizontal structure vector fields. If π is such a submersion, then we will denote

$$D_\alpha := \varphi_\alpha(\ker \pi_*) \subset (\ker \pi_*)^\perp, \quad \alpha = 1, 2, 3,$$

$$D = D_1 \oplus D_2 \oplus D_3 \in (\ker \pi_*)^\perp,$$

and

$$\Xi = Span\{\xi_1, \xi_2, \xi_3\}.$$

We note that D_1, D_2 and D_3 are distributions orthogonal to each other and D defines a non-trivial subbundle in $(\ker \pi_*)^\perp$. Moreover, it is clear that we have the following decompositions:

$$TM = \ker \pi_* \oplus (\ker \pi_*)^\perp$$

and

$$(\ker \pi_*)^\perp = \Xi \oplus D \oplus D^\perp,$$

where $D^\perp$ denotes the orthogonally complementary subbundle of $\Xi \oplus D$ in $(\ker \pi_*)^\perp$.

Proposition 1 *Let $(M, \nabla, (\varphi_\alpha, \xi_\alpha, \eta_\alpha)_{\alpha=\overline{1,3}}, g)$ be a mixed 3-structure statistical manifold and (N, ∇^N, g_N) be a statistical manifold. If $\pi : M \to N$ is an anti-invariant statistical submersion such that the structure vector fields ξ_1, ξ_2 and ξ_3 are horizontal, then we have:*

(i) $\varphi_\alpha D_\alpha \subset \ker \pi_*$,
(ii) $\varphi_\alpha D_\beta \subset D_\gamma$,
(iii) $\varphi_\alpha D^\perp \subset D^\perp$ *for any* $\alpha \in \{1, 2, 3\}$ *and any even permutation* (α, β, γ) *of* $(1, 2, 3)$.

Proof (*i*) Suppose $X \in D_\alpha$. Then there is a vertical vector U with $X = \varphi_\alpha U$ and using (3) we derive

$$\begin{aligned} \varphi_\alpha X &= \varphi_\alpha(\varphi_\alpha U) \\ &= \varphi_\alpha^2(U) \\ &= \tau_\alpha(-U + \eta_\alpha(U)\xi_\alpha); \end{aligned}$$

But in view of (36) we have

$$\eta_\alpha(U) = \varepsilon_\alpha g(U, \xi_\alpha) = 0,$$

because U is vertical and ξ_α is horizontal. Thus, from the last two equations, we obtain

$$\varphi_\alpha X = -\tau_\alpha U \in \ker \pi_*.$$

(ii) Suppose $X \in D_\beta$. Then we can find a vertical vector U with $X = \varphi_\beta U$ and using (8) we derive

$$\begin{aligned}\varphi_\alpha X &= \varphi_\alpha(\varphi_\beta U)\\ &= \tau_\gamma \varphi_\gamma U + \tau_\alpha \eta_\beta(U)\xi_\alpha\\ &= \tau_\gamma \varphi_\gamma U \in D_\gamma.\end{aligned}$$

(iii) If we take $X \in D^\perp$ and $Y \in \ker \pi_*$, then using (3) and (10) we derive

$$\begin{aligned}g(\varphi_\alpha X, \varphi_\alpha Y) &= -g(X, \varphi_\alpha^2 Y)\\ &= g(X, \tau + \alpha(Y))\\ &= \tau_\alpha(g(X, Y))\\ &= 0,\end{aligned}$$

and using (3) and (8) we find

$$\begin{aligned}g(\varphi_\alpha X, \varphi_\beta Y) &= -g(X, \varphi_\alpha \varphi_\beta Y)\\ &= -g(X, \tau_\gamma \varphi_\gamma Y + \tau_\alpha \eta_\beta(Y)\xi_\alpha)\\ &= -\tau_\gamma g(X, \varphi_\gamma Y)\\ &= 0.\end{aligned}$$

More than that, using (3) we find

$$\begin{aligned}g(\varphi_\alpha X, \xi_\alpha) &= -g(X, \varphi_\alpha \xi_\alpha)\\ &= 0,\end{aligned}$$

while using (3) and (6), we derive

$$\begin{aligned}g(\varphi_\alpha X, \xi_\beta) &= -g(X, \varphi_\alpha \xi_\beta)\\ &= -g(X, \tau_\beta \xi_\gamma)\\ &= -\tau_\beta g(X, \xi_\gamma)\\ &= 0,\end{aligned}$$

and the conclusion follows. □

Theorem 9 *Let $(M, \nabla, (\varphi_\alpha, \xi_\alpha, \eta_\alpha)_{\alpha=\overline{1,3}}, g)$ be a mixed 3-Sasakian statistical manifold and (N, ∇^N, g_N) be a statistical manifold. If $\pi : M \to N$ is an anti-invariant statistical submersion such that the structure vector fields ξ_1, ξ_2 and ξ_3 are horizontal, then:*

(i) the distribution Ξ is integrable,
(ii) the distribution $D \oplus D^\perp$ is not integrable.

Proof (*i*) We know that D is integrable if for all $X, Y \in D$ we have $[X, Y] \in D$. This is equivalent to show that

$$g([\xi_\alpha, \xi_\beta], Z) = 0, \quad \forall Z \in D \oplus D^\perp.$$

Now, using (17) and (6), we derive

$$\begin{aligned} g([\xi_\alpha, \xi_\beta], Z) &= g(\nabla_{\xi_\alpha}\xi_\beta - \nabla_{\xi_\beta}\xi_\alpha, Z) \\ &= g(\nabla_{\xi_\alpha}\xi_\beta, Z) - g(\nabla_{\xi_\beta}\xi_\alpha, Z) \\ &= \varepsilon_\beta[g(\nabla_{\xi_\alpha}\xi_\beta, \xi_\beta)g(\xi_\beta, Z) - g(\varphi_\beta\xi_\alpha, Z)] \\ &\quad -\varepsilon_\alpha[g(\nabla_{\xi_\beta}\xi_\alpha, \xi_\alpha)g(\xi_\alpha, Z) - g(\varphi_\alpha\xi_\beta, Z)] \\ &= -\varepsilon_\beta\tau_\alpha g(\xi_\gamma, Z) + \varepsilon_\alpha\tau_\beta g(\xi_\gamma, Z) \\ &= 0. \end{aligned}$$

Hence, the distribution Ξ is integrable.

(*ii*) We prove that there exist $X, Y \in D \oplus D^\perp$ such that $[X, Y] \notin D \oplus D^\perp$. Indeed, using (1) and (19) we obtain:

$$\begin{aligned} g([X, Y], \xi_\alpha) &= g(\nabla_X Y, \xi_\alpha) - g(\nabla_Y X, \xi_\alpha) \\ &= Xg(Y, \xi_\alpha) - g(Y, \nabla^*_X\xi_\alpha) - Yg(X, \xi_\alpha) + g(X, \nabla^*_Y\xi_\alpha) \\ &= g(X, \nabla^*_Y\xi_\alpha) - g(Y, \nabla^*_X\xi_\alpha) \\ &= \varepsilon_\alpha[g(X, g(\nabla_Y\xi_\alpha, \xi_\alpha)\xi_\alpha - \varphi_\alpha Y) - g(Y, g(\nabla_X\xi_\alpha, \xi_\alpha)\xi_\alpha - \varphi_\alpha X)] \\ &= \varepsilon_\alpha[-g(X, \varphi_\alpha Y) + g(Y, \varphi_\alpha X)], \end{aligned}$$

and in view of (10), we find

$$g([X, Y], \xi_\alpha) = 2\varepsilon_\alpha g(Y, \varphi_\alpha X),$$

and the conclusion follows. □

Example 2 Let us consider an 11-dimensional manifold $M = \mathbb{R}^{11}$. The standard coordinates on M are denoted by $(x_1, x_2, x_3, x_4, y_1, y_2, y_3, y_4, z_1, z_2, z_3)$. We consider on M the semi-Riemannian metric g defined by:

$$g = \sum_{i=1}^{4}(-dx_i^2 + dy_i^2) + \sum_{i=1}^{4} \epsilon_i dz_i^2,$$

where $\epsilon_1 = -\epsilon_2 = \epsilon_3 = 1$ or $\epsilon_1 = -\epsilon_2 = \epsilon_3 = -1$.

We also define an affine connection ∇ on M given by

$$\nabla_{\partial_{x_i}}\partial_{z_i} = \nabla_{\partial_{z_i}}\partial_{x_i} = -\partial_{y_i},$$

$$\nabla_{\partial_{x_i}}\partial_{y_i} = \nabla_{\partial_{y_i}}\partial_{x_i} = -\partial_{z_i},$$

$$\nabla_{\partial_{y_i}} \partial_{z_i} = \nabla_{\partial_{z_i}} \partial_{y_i} = -\partial_{x_i},$$

$$\nabla_{\partial_{x_i}} \partial_{x_i} = \partial_{x_i},$$

and all the remaining components being *zero*.

We also consider on M the tensor fields $\varphi_1, \varphi_2, \varphi_3$, defined by

$$\varphi_1((x_1, ..., x_4, y_1, ..., y_4, z_1, z_2, z_3) = (-y_3, y_4, -y_1, y_2, -x_3, x_4, -x_1, x_2, 0, z_3, z_2),$$

$$\varphi_2((x_1, ..., x_4, y_1, ..., y_4, z_1, z_2, z_3) = (y_4, y_3, y_2, y_1, x_4, x_3, x_2, x_1, -z_2, -z_1, 0),$$

$$\varphi_3((x_1, ..., x_4, y_1, ..., y_4, z_1, z_2, z_3)=(-x_2, x_1, -x_4, x_3, -y_2, y_1, -y_4, y_3, z_3, 0, -z_1),$$

the vector fields ξ_1, ξ_2, ξ_3 given by

$$\xi_1 = \partial_{z_1}, \xi_2 = \partial_{z_3}, \xi_3 = -\partial_{z_2}$$

as well as the 1-forms η_1, η_2, η_3 which are the dual forms to vector fields ξ_1, ξ_2, ξ_3 respectively. Then it follows that $(\mathbb{R}^{11}, \nabla, (\varphi_\alpha, \xi_\alpha, \eta_\alpha)_{\alpha\in\{1,2,3\}}, g)$ is a mixed 3-structure statistical manifold.

Next, we consider two manifolds N_1, N_2 as follows. Let $N_1 = \mathbb{R}^8$ equipped with the pseudo-Riemannian metric

$$g_{N_1} = \sum\nolimits_{i=1}^{4} (-dx_i^2 + dy_i^2)$$

and also with the statistical structure defined by

$$\nabla^{N_1}_{\partial_{x_i}} \partial_{x_i} = \partial_{x_i}$$

and all remaining components zero. Also, let $N_2 = \mathbb{R}^9$ equipped with the pseudo-Riemannian metric

$$g_{N_2} = \sum\nolimits_{i=1}^{4} (-dx_i^2 + dy_i^2) + \xi_i dz_i^2$$

and the statistical structure defined by

$$\nabla^{N_2}_{\partial_{x_i}} \partial_{x_i} = \partial_{x_i}$$

and all remaining components zero.

We define now the maps $\pi_1 : M \to N_1$ and $\pi_2 : M \to N_2$ by

$$\pi_1(x_1, ..., x_4, y_1, ..., y_4, z_1, z_2, z_3) = (x_1, x_2, x_3, x_4, y_1, y_2, y_3, y_4)$$

and

$$\pi_2(x_1, ..., x_4, y_1, ..., y_4, z_1, z_2, z_3) = (x_1 - y_1, x_2, x_4, x_3 - y_3, y_2, y_4, z_1, z_2, z_3),$$

respectively. Then a direct computation shows us that we have

$$\ker \pi_{1*} = Span\{\partial_{z_1}, \partial_{z_2}, \partial_{z_3}\}$$

and

$$(\ker \pi_{1*})^{\perp} = Span\{\partial_{x_1}, ..., \partial_{x_4}, \partial_{y_1}, ..., \partial_{y_4}\}.$$

Moreover, we derive that: $\varphi_\alpha(\ker \pi_{1*}) \subset \ker \pi_{1*}$, for all $\alpha \in \{1, 2, 3\}$ and we conclude immediately that π_1 is an invariant statistical submersion such that structure vector fields ξ_1, ξ_2, ξ_3 are all vertical. Similarly, we derive that π_2 is an anti-invariant statistical submersion such that structure vector fields ξ_1, ξ_2, ξ_3 are all horizontal.

5 Paraquaternionic Statistical Submersions

Definition 14 Suppose (M, σ, g) and (M', σ', g') are two almost Hermite-like paraquaternionic manifolds. Then:

(i) A map $f : M \to M'$ is said to be a (σ, σ')-*holomorphic map at a point* $p \in M$ if for any $J \in \sigma_p$ exists $J' \in \sigma'_{f(p)}$ such that $f_* \circ J = J' \circ f_*$,
Moreover, we say that f is a (σ, σ')-*holomorphic map* if it is a (σ, σ')-holomorphic map at each point $p \in M$.

(ii) A semi-Riemannian submersion $\pi : M \to M'$ which is a (σ, σ')-holomorphic map is said to be an *almost Hermite-like paraquaternionic submersion.*

Definition 15 Suppose (M, ∇, σ, g) and $(M', \nabla', \sigma', g')$ are two almost Hermite-like paraquaternionic statistical manifolds. Then:

(i) A statistical submersion $\pi : M \to M'$ which is a (σ, σ')-holomorphic map is said to be an *almost Hermite-like paraquaternionic statistical submersion.*

(ii) An almost Hermite-like paraquaternionic statistical submersion $\pi : M \to M'$ is called a *paraquaternionic Kähler-like statistical submersion* if (M, ∇, σ, g) is a paraquaternionic Kähler-like statistical manifold. Moreover, if (M, ∇, σ, g) is a (locally) parahyper-Kähler-like statistical manifold, then π is said to be a *(locally) parahyper-Kähler-like statistical submersion.*

Remark 8 Similarly as in quaternionic case (see [35]), if (M, σ, g) and (M', σ', g') are two almost Hermite-like paraquaternionic manifolds, then a map $f : M \to M'$ is a (σ, σ')-holomorphic map at $p \in M$ if and only if any for any canonical local basis $\{J_1, J_2, J_3\}$ of σ_p it follows that there exist a canonical local basis $\{J'_1, J'_2, J'_3\}$ of $\sigma'_{f(p)}$ such that $f_* \circ J_\alpha = J'_\alpha \circ f_*$, for $\alpha \in \{1, 2, 3\}$.

Remark 9 Suppose (M, ∇, σ, g) and $(M', \nabla', \sigma', g')$ are two almost Hermite-like paraquaternionic statistical manifolds. Then we know that $(M, \nabla^*, \sigma^*, g)$ and

$(M', \nabla'^*, \sigma'^*, g')$ are almost Hermite-like paraquaternionic statistical manifolds too. We consider a semi-Riemannian submersion $\pi : M \to M'$. It is easy to see that π is a (σ, σ')-holomorphic map if and only if π is a (σ^*, σ'^*)-holomorphic map.

Proposition 2 *Suppose (M, ∇, σ, g) and $(M', \nabla', \sigma', g')$ are two almost Hermite-like paraquaternionic statistical manifolds and $\pi : M \to M'$ is an almost Hermite-like paraquaternionic statistical submersion. Then:*

(i) $J_\alpha(\ker \pi_*) \subset \ker \pi_*$ *and* $J^*_\alpha(\ker \pi_*) \subset \ker \pi_*$,
(ii) $J_\alpha((\ker \pi_*)^\perp) \subset (\ker \pi_*)^\perp$ *and* $J^*_\alpha((\ker \pi_*)^\perp) \subset (\ker \pi_*)^\perp$,
(iii) If X is a basic vector field, with $\pi_*(X) = X'$ where X' is a vector field on M', then $J_\alpha X$ and $J^*_\alpha X$ are basic vector fields with $\pi_*(J_\alpha X) = J'_\alpha X'$ and $\pi_*(J^*_\alpha X) = J'^*_\alpha X'$, for $\alpha \in \{1, 2, 3\}$.

Proof (i) For any $U \in \ker \pi_*$, we have

$$\pi_*(J_\alpha U) = J'_\alpha(\pi_* U) = 0.$$

So, it follows that $J_\alpha U \in \ker \pi_*$, for any $U \in \ker \pi_*$ and $\alpha \in \{1, 2, 3\}$. Thus, we conclude $J_\alpha(\ker \pi_*) \subset \ker \pi_*$ and similarly we obtain $J^*_\alpha(\ker \pi_*) \subset \ker \pi_*$.

(ii) For any $X \in (\ker \pi_*)^\perp$ and $Y \in \ker \pi_*$, we obtain using (25) and (i) that

$$g(J_\alpha X, Y) = -g(X, J^*_\alpha Y) = 0$$

and the conclusion follows.

(iii) Because X is a basic vector field, it follows that X is horizontal and $\pi_*(X) = X'$. But X horizontal implies $J_\alpha X$ is also horizontal in view of (ii). Moreover, we have

$$\pi_*(J_\alpha X) = J'_\alpha(\pi_* X) = J'_\alpha X',$$

and therefore we conclude that $J_\alpha X$ is basic. Similarly, we obtain that $J^*_\alpha X$ is basic, being horizontal and π-related to $J'^*_\alpha X'$, that is $\pi_*(J^*_\alpha X) = J'^*_\alpha X'$. □

Remark 10 Suppose $\pi : M \to M'$ is an almost Hermite-like paraquaternionic statistical submersion between two almost Hermite-like paraquaternionic statistical manifolds (M, ∇, σ, g) and $(M', \nabla', \sigma', g')$. Let us denote a fiber of the submersion π by $\hat{M}$. Then it is known that $\hat{M}$ admits a statistical structure $(\hat{\nabla}, \hat{g})$, where $\hat{g}$ denotes the induced metric by g on the fiber (for details see [33]). Moreover, for any canonical basis $\{J_1, J_2, J_3\}$ of σ, one can define $\hat{J}_\alpha = J_{\alpha|\hat{M}}$, $\alpha \in \{1, 2, 3\}$, and we have a subbundle $\hat{\sigma}$ which is spanned locally by $\{\hat{J}_\alpha\}_{\alpha=1,2,3}$. Now, it can be directly checked that $(\hat{M}, \hat{\nabla}, \hat{\sigma}, \hat{g})$ is an almost Hermite-like paraquaternionic statistical manifold.

Theorem 10 *Suppose $\pi : M \to M'$ is a paraquaternionic Kähler-like statistical submersion. Then:*

(i) The fibres and the base of π are paraquaternionic Kähler-like statistical manifolds.
(ii) π is a statistical submersion with isometric fibers.
(iii) The fibres and the base of π are paraquaternionic-like space forms, provided that the total space of π is a paraquaternionic-like space form.

Proof (i) Suppose X and Y are basic vector fields on M such that $\pi_*(X) = X'$ and $\pi_*(Y) = Y'$, where X' and Y' are vector fields on the base space M' of the submersion π. Then it follows immediately that we have:

$$(\nabla'_{X} J'_{\alpha})Y' = \pi_*((\nabla_X J_\alpha)Y). \tag{44}$$

But in view of the fact that (M, ∇, σ, g) is a paraquaternionic Kähler-like statistical manifold, we know that there exist the locally 1-forms ω_1, ω_2 and ω_3 on the total space of π such that relations (26) are satisfied. Then one can consider other locally 1-forms, namely ω'_1, ω'_2 and ω'_3 on M' defined by:

$$\omega'_\alpha(X') = \omega_\alpha(X),\ \alpha \in \{1, 2, 3\}. \tag{45}$$

Using now (26), (44) and (45) we derive

$$\nabla' J'_\alpha = \tau_\beta \omega'_\gamma \otimes J'_\beta - \tau_\gamma \omega'_\beta \otimes J'_\gamma$$

for any even permutation (α, β, γ) of $(1, 2, 3)$. Thus we deduce that the base of π is a paraquaternionic Kähler-like statistical manifold.

In the following, we will show that the fibres of π are paraquaternionic Kähler-like statistical manifolds. Let us denote by $\hat{M} = \pi^{-1}(p')$, $p' \in M'$, a fiber of π. If U and V are vertical vector fields, then we have in view of (29) and (34):

$$\begin{aligned}(\nabla_U J + \alpha)V &= \nabla_U J\alpha V - J\alpha \nabla_U V \\ &= T_U J_\alpha V + \hat{\nabla}_U J_\alpha V - J_\alpha T_U V - J_\alpha \hat{\nabla}_U V \\ &= (T_U J_\alpha V - J_\alpha T_U V) + (\hat{\nabla}_U J_\alpha V - J_\alpha \hat{\nabla}_U V) \\ &= (T_U J_\alpha V - J_\alpha T_U V) + (\hat{\nabla}_U J_\alpha)V. \end{aligned} \tag{46}$$

On another hand, (26) implies

$$(\nabla_U J_\alpha)V = \tau_\beta \omega_\gamma(U) J_\beta V - \tau_\gamma \omega_\beta(U) J_\alpha V, \tag{47}$$

and from (46) and (47), by identifying the horizontal and vertical components, we obtain

$$T_U J_\alpha V - J_\alpha T_U V = 0, \alpha = 1, 2, 3 \tag{48}$$

and

$$(\hat{\nabla}_U J_\alpha)V = \tau_\beta \omega_\gamma(U) J_\beta V - \tau_\gamma \omega_\beta(U) J_\alpha V. \tag{49}$$

From (49) we deduce immediately the desired conclusion.

(ii) Using (48) we derive easily that $T_U V = 0$ and the conclusion follows.

(iii) The assertion follows using (i) and O'Neill equations for statistical submersions (see [32]). □

As an immediate consequence of the above theorem, we have the next result.

Corollary 2 *Suppose $\pi : M \to M'$ is a parahyper-Kähler-like statistical submersion. Then:*
(i) The fibres and the base of π are parahyper-Kähler-like statistical manifolds.
(ii) π is a statistical submersion with isometric fibers.

Example 3 If (M, ∇, σ, g) is an almost Hermite-like paraquaternionic statistical manifold, then it is known that the tangent bundle TM equipped with the Sasaki metric G can be endowed with a linear connection ∇' such that (TM, ∇', G) is a statistical manifold (see [16]). Moreover, we define for any canonical local basis $\{J_1, J_2, J_3\}$ of σ the next three tensor fields on TM:

$$J'_\alpha X^h = (J_\alpha X)^h, \quad J'_\alpha X^v = (J_\alpha X)^v, \quad \alpha \in \{1, 2, 3\}.$$

If $\sigma' = Span\{J'_1, J'_2, J'_3\}$, then it is easy to check that $(TM, \nabla', \sigma', g')$ is an almost Hermite-like paraquaternionic statistical submanifold and the natural projection $\pi : TM \to M$ provides us a non-trivial example of almost Hermite-like paraquaternionic statistical submersion. Moreover, π is a parahyper-Kähler-like statistical submersion with isometric fibers, provided that (M, ∇, σ, g) is flat.

References

1. Abe, N., Hasegawa, K: An affine submersion with horizontal distribution and its applications. Differ. Geom. Appl. 14, 235–250 (2001). https://doi.org/10.1016/S0926-2245(01)00034-1
2. Alekseevsky, D.V., Cortes, V.: The twistor spaces of a para-quaternionic Kähler manifold. Osaka J. Math. 45(1), 215–251 (2008). https://doi.org/10.18910/5112
3. Amari, S.: Differential-Geometrical Methods in Statistics. Springer, New-York, (1985)
4. Amari, S., Nagaoka, H.: Method of information geometry. Amer. Math. Soc., Providence, Oxford University Press, Oxford (2000)
5. Bejan, C.L., Druţă-Romaniuc, S.L.: Structures which are harmonic with respect to Walker metrics. Mediterr. J. Math. 12, 481–496 (2013). https://doi.org/10.1007/s00009-014-0409-y
6. Boyom, N.M.: Foliations-Webs-Hessian Geometry-Information Geometry-Entropy and Cohomology. Entropy 18(12), 1–92 (2016). https://doi.org/10.3390/e18120433
7. Caldarella, A.: On paraquaternionic submersions between paraquaternionic Kähler manifolds. Acta Appl. Math. 112(1), 1–14 (2010). https://doi.org/10.1007/s10440-009-9549-7
8. Caldarella, A., Pastore, A.M.: Mixed 3-Sasakian structures and curvature. Ann. Polon. Math. 96, 107–125 (2009). https://doi.org/10.4064/ap96-2-2

9. Calvaruso, G., Perrone, A.: Left-invariant hypercontact structures on three-dimensional Lie groups. Period Math Hung 69, 97–108 (2014). https://doi.org/10.1007/s10998-014-0054-z
10. Calvaruso, G., Perrone, D.: Metrics of Kaluza-Klein type on the anti-de Sitter space $\mathbb{H}_1^3$. Math. Nachr. 287(8-9), 885-902 (2014). https://doi.org/10.1002/mana.201200105
11. Chen, B.-Y., Decu, S., Vîlcu, G.-E.: Inequalities for the Casorati Curvature of Totally Real Spacelike Submanifolds in Statistical Manifolds of Type Para-Kähler Space Forms. Entropy 23, 1399 (2021). https://doi.org/10.3390/e23111399
12. Fei, T., Zhang, J.: Interaction of Codazzi couplings with (para-)Kähler geometry. Results. Math. 72(4), 2037–2056 (2017). https://doi.org/10.1007/s00025-017-0711-7
13. Furuhata, H., Hasegawa, I.: Submanifold theory in holomorphic statistical manifolds. In: Geometry of Cauchy-Riemann Submanifolds. Springer, Singapore, 179–215 (2016).
14. Furuhata, H., Hasegawa, I., Okuyama, Y., Sato, K., Shahid, M.H.: Sasakian statistical manifolds. J. Geom. Phys. 117, 179–186 (2017). https://doi.org/10.1016/j.geomphys.2017.03.010
15. García-Río, E., Matsushita, Y., Vazquez-Lorenzo, R.: Paraquaternionic Kähler manifolds. Rocky Mt. J. Math. 31(1), 237–260 (2001). https://doi.org/10.1216/rmjm/1008959679
16. Ianuş, I.: Statistical manifolds and tangent bundles. Politehn. Univ. Bucharest Sci. Bull. Ser. D Mech. Engrg. 56-57(1-4), 29–34 (1994–95).
17. Ianuş, S., Marchiafava, S., Vîlcu, G.-E.: Paraquaternionic CR-submanifolds of paraquaternionic Kähler manifolds and semi-Riemannian submersions. Cent. Eur. J. Math. 8(4), 735–753 (2010). https://doi.org/10.2478/s11533-010-0048-x
18. Ianuş, S., Mazzocco, R., Vîlcu, G.-E.: Real lightlike hypersurfaces of paraquaternionic Kähler manifolds. Mediterr. J. Math. 3 (2006), 581–592. https://doi.org/10.1007/s00009-006-0098-2
19. Ianuş, S., Ornea, L., Vîlcu, G.-E.: Submanifolds in manifolds with metric mixed 3-structures. Mediterr. J. Math. 9(1), 105–128 (2012). https://doi.org/10.1007/s00009-011-0121-0
20. Ianuş, S., Visinescu, M., Vîlcu, G.-E.: Conformal Killing-Yano tensors on manifolds with mixed 3-structures; SIGMA, Symmetry Integrability Geom. Methods Appl. 5, Paper 022, 12 pages (2009). https://doi.org/10.3842/SIGMA.2009.022
21. Ianuş, S., Vîlcu, G.-E.: Semi-Riemannian hypersurfaces in manifolds with metric mixed 3-structures. Acta Math. Hung. 127(1-2), 154–177 (2010). https://doi.org/10.1007/s10474-009-9112-z
22. A. Kazan: Conformally-projectively flat trans-Sasakian statistical manifolds. Physica A Stat. Mech. Appl. 535, 122441 (2019). https://doi.org/10.1016/j.physa.2019.122441
23. Kazan, S., Takano, K.: Anti-invariant holomorphic statistical submersions. Results Math. 78, 128 (2023). https://doi.org/10.1007/s00025-023-01904-8
24. Lauritzen, S.L.: Statistical manifolds. In: Differential Geometry in Statistical Inferences. IMS Lecture Notes Monogr. Ser., 10, Inst. Math. Statist., Hayward California, 96–163 (1987).
25. Libermann, P.: Sur les structures presque quaternioniennes de deuxième espéce. C.R. Acad. Sc. Paris 234, 1030–1032 (1952).
26. Murathan, C., Şahin, B.: A study of Wintgen like inequality for submanifolds in statistical warped product manifolds. J. Geom. 109(30), 1–18 (2018). https://doi.org/10.1007/s00022-018-0436-0
27. Murray M.K., Rice, J.W.: Differential geometry and statistics. Monographs on Statistics and Applied Probability 48, Chapman & Hall, London (1993)
28. Noda, T.: Symplectic structures on statistical manifolds. J. Aust. Math. Soc. 90(3), 371–384 (2011). https://doi.org/10.1017/S1446788711001285
29. O'Neil, B.: Semi-Riemannian geometry with applications to relativity. Academic Press, New York-London (1983)
30. Popescu, P., Rovenski, V., Stepanov, S.: On singular distributions with statistical structure. Mathematics 8, 1825 (2020). https://doi.org/10.3390/math8101825
31. Shima, H.: The geometry of Hessian structures. World Scientific (2007)
32. Takano, K.: Statistical manifolds with almost complex structures and its statistical submersions. Tensor, N.S. 65, 128–142 (2004).
33. Takano, K.: Statistical manifolds with almost contact structures and its statistical submersions. J. Geom. 85, 171–187 (2006). https://doi.org/10.1007/s00022-006-0052-2

34. M. Tekkoyun: Mechanical Systems on Manifolds. Differential Geometry - Dynamical Systems Monographs, Vol. 11, Geometry Balkan Press (2014).
35. Vîlcu, A.-D., Vîlcu, G.-E.: Statistical manifolds with almost quaternionic structures and quaternionic Kähler-like statistical submersions. Entropy 17(9), 6213–6228 (2015). https://doi.org/10.3390/e17096213
36. Vîlcu, G.-E.: On generic submanifolds of manifolds endowed with metric mixed 3-structures. Commun. Contemp. Math. 18, 1550081 (2016). https://doi.org/10.1142/S0219199715500819
37. Vîlcu, G.-E.: Paraquaternionic CR-submanifolds. in: Geometry of Cauchy-Riemann Submanifolds, Springer, Singapore, 361–390 (2016).
38. Vîlcu, G.-E.: Canonical foliations on paraquaternionic Cauchy-Riemann submanifolds. J. Math. Anal. Appl. 399, 551–558 (2013). https://doi.org/10.1016/j.jmaa.2012.10.036
39. Vîlcu, G.-E.: Mixed paraquaternionic 3-submersions. Indag. Math. 24, 474–488 (2013). https://doi.org/10.1016/j.indag.2013.01.006
40. Vîlcu, G.-E. : Almost product structures on statistical manifolds and para-Kähler-like statistical submersions. Bull. Sc. Math. 171, 103018 (2021). https://doi.org/10.1016/j.bulsci.2021.103018
41. Voicu, R., Vîlcu G.-E.: Curvature properties of pseudo-sphere bundles over paraquaternionic manifolds. Int. J. Geom. Methods Mod. Phys. 9 (2012), 1250024. https://doi.org/10.1142/S0219887812500247

Minimal Unit Vector Fields on Oscillator Groups

Alexander Yampolsky

Abstract In this paper we treat minimal left-invariant unit vector fields on oscillator group and their relations with the ones that defines a harmonic map. Particularly, if all structure constants of the oscillator group are equal to each other, then all unit left-invariant vector fields that define a harmonic map into the unit tangent bundle with Sasaki metric are minimal.

Keywords Minimal unit vector fields · Sasaki metric · Oscillator group · Harmonic maps

Mathematics Subject Classifications (2010). 53C20 · 53C25 · 53C43

1 Introduction and Preliminaries

Let (M, g) be Riemannian manifold of dimension n. Denote by $U(u^1, \ldots, u^n)$ a local chart on M, and by $\partial_i = \partial/\partial u^i$ the natural coordinate frame over U. At any $p \in U$, there is a decomposition $V = V^i \partial_i$ for each vector $V \in T_pM$. Then $(u^1, \ldots, u^n; V^1, \ldots, V^n)$ forms a local coordinate chart TU on the tangent bundle TM.

Denote by ds^2 the line element of (M, g). Denote by ∇ the Livi-Chivita connection for g. The Sasaki metric line element $d\sigma$ on the tangent bundle TM is defined by local Pithagorean theorem, namely,

$$d\sigma^2 = ds^2 + |DV|_g^2,$$

where $DV^i = du^k \nabla_k V^i = dV^i + \Gamma^i_{jk} V^j du^k$ is a covariant differential of V^i. At this point the $(V^1, \ldots, V^n)$ are independent variables. Denote by g_S the Riemannian

---

A. Yampolsky (✉)
V.N. Karazin Kharkiv National University, Kharkiv, Ukraine
e-mail: a.yampolsky@karazin.ua

© The Author(s), under exclusive license to Springer Nature Switzerland AG 2024

V. Rovenski et al. (eds.), *Differential Geometric Structures and Applications*, Springer Proceedings in Mathematics & Statistics 440,
https://doi.org/10.1007/978-3-031-50586-7_6

metric on TM with the line element $d\sigma^2$. Riemannian manifold (TM, g_S) is referred to as *tangent bundle with Sasaki metric*.

Let $V(u)$ be a smooth local tangent vector field on M. In terms of natural local coordinates $(U(u^1, \dots, u^n); V^1, \dots, V^n)$ the vector field V defines a local embedding $V : U \to (TM, g_S)$ as

$$\begin{cases} u^i = u^i, \\ V^i = V^i(u^1, \dots, u^n). \end{cases}$$

In this approach, a smooth local vector field can be understood as local explicitly given submanifold/graph $V(U) \subset (TM, g_S)$ *with induced intrinsic and extrinsic geometry.*

If V is given globally, then $V(M) \subset (TM, g_S)$ is globally given graph in (TM, g_S) diffeomorphic to the base manifold. Remark that if one do not pose any restrictions on the vector field, then $V(M)$ does not inherit intrinsic geometry of the base manifold even locally, and its extrinsic geometry might be of any kind. To see this, it is sufficient to consider the case $g_{ij} = \delta_{ij}$, when (TM, g_S) is isometric to E^{2n} with Euclidean metric, and to look on a graph $V(E^n) \subset E^{2n}$. In contrast, if $g(V, V) = 1$, then $V(M)$ gives rise to some kind of equidistant, relative to the base, submanifold in $T_1M \subset (TM, g_S)$, and the geometry of $V(M)$ becomes more predictable both, in intrinsic and extrinsic senses.

If $g(V, V) = 1$, then $V(M)$ is globally/locally given n dimensional submanifold in T_1M of codimension $(n-1)$ with the pull-back metric

$$g_V(X, Y) = g(X, Y) + g(\nabla_X V, \nabla_Y V). \tag{1}$$

One can see that this metric is a local additive deformation of the metric g on M in presence of a (unit) vector field V. At this point *one can assign* the geometry of submanifold $V(M)$ to the field V. One can talk about *sectional curvature*, *Ricci curvature*, *scalar* curvature and other properties from Riemannian geometry.

In this article, we will focus on the *extrinsic* properties of the submanifold $V(M) \subset (TM, g_S)$ related to the second fundamental form of the (local) imbedding $V(M) \subset (T_1M, g_S)$.

1.1 Second Fundamental Form of $V(M) \subset (T_1M, g_S)$

Denote by $\mathscr{X}(M)$ a module of smooth vector fields on M. At each point $p \in M$ the point-wise *Nomizu operator* $A_V : T_pM \to V^\perp \subset T_pM$ is defined by

$$A_V X = -\nabla_X V.$$

The *adjoint* Nomizu operator A_V^t is defined by

$$g(A_V X, Y) = g(X, A_V^t Y).$$

If V is smooth, then both operators form smooth $(1, 1)$-tensor fields on M. The *tangent* $V_* : \mathscr{X}(M) \to TV(M)$ and the *normal* $\tilde{n} : \mathscr{X}(M) \to T^\perp V(M)$ mappings are defined by [9]

$$V_*(X) = X^h - (A_V X)^t = X^h - (A_V X)^v, \quad \tilde{n}(X) = (A_V^t X)^h + X^t,$$

where X^t means the vertical lift of $X - g(X, V)V$ and is called by *tangentional lift* of the corresponding object.

The *rough* V*-Hessian* and the V*-harmonicity tensor* are given by

$$Hess_V(X, Y) = \tfrac{1}{2}\big((\nabla_X A_V)Y + (\nabla_Y A_V)X\big),$$
$$Hm_V(X, Y) = \tfrac{1}{2}\big(R(V, A_V X)Y + R(V, A_V Y)X\big),$$

where $(\nabla_X A_V)Y = \nabla_X(A_V Y) - A_V(\nabla_X Y)$ and R is the curvature tensor of the base manifold (M, g). The

$$trace(Hess_V) = \sum\nolimits_i (\nabla_{e_i} A_V)e_i = \bar{\Delta}\, V$$

is known as the *rough (or Bohner) Laplacian*. The unit vector field is said to be **harmonic** if

$$\bar{\Delta}\, V = |A_V|^2 V$$

with respect to some orthonormal frame $\{e_1, \dots, e_n\}$. If in addition

$$trace(Hm_V) = \sum\nolimits_i R(V, A_V e_i)e_i = 0,$$

then V defines a ***harmonic map*** $V : M \to (T_1 M, g_S)$ of the base manifold into its unit tangent bundle with the Sasaki metric. The *second fundamental form of mapping* $V : M \to (T_1 M, g_S)$ is given by

$$B_V(X, Y) = \big(Hm_V(X, Y)\big)^h + \big(Hess_V(X, Y)\big)^t.$$

It follows that the tension field of the mapping $V : M \to (T_1 M, g_S)$ is given by

$$\tau_V = trace(Hm_V)^h + (\bar{\Delta}\, V)^t.$$

The *second fundamental form of the submanifold* $(V(M), g_V) \subset (T_1 M, g_S)$ with respect to the normal vector field $\tilde{n}(Z)$ is of the form [9]

$$\tilde{\Omega}_{\tilde{n}(Z)}(V_* X, V_* Y) = g\big(Hess_V(X, Y) + A_V Hm_V(X, Y), Z^\perp\big), \tag{2}$$

where $Z^\perp = Z - g(Z, V)V$.

As a consequence, the *second fundamental form vanishes* if and only if

$$Hess_V(X,Y) + A_V Hm_V(X,Y) - g(A_V X, A_V Y)V = 0. \tag{3}$$

The equation (3) provides a condition on *totally geodesic property of the unit vector field* [11]. In spite the fact that the system (3) is overdefined system of differential equations, in some cases the system can be solved [9–12].

1.2 Minimal Unit Vector Fields

At first, minimality of a unit vector field was defined in global setting by H. Gluck and W. Ziller [2] by using the volume variation of a submanifold $V(M) \subset (T_1M, g_S)$ relative to variation of the field within the class of unit vector fields. O. Gil-Medrano and E. Llinares-Fuster [3] proved that in local setting minimality of unit vector field is equivalent to minimality of a submanifold $V(M) \subset (T_1M, g_S)$ relative to classic volume variation of a submanifold. In other words, locally given unit vector field V is locally minimal, if the submanifold $V(M) \subset (T_1M, g_S)$ has zero mean curvature vector, and minimality problem for the unit vector field reduces to trace of the shape operator for $V(M) \subset (T_1M, g_S)$.

Probably, the easiest way to rich the shape operator is a singular frame of the Nomizu operator. Since $V \in \ker A_V^t$, the orthonormal singular frames for the Nomizu operator A_V consist of

$$\{e_1, \dots, e_{n-1}, e_n\} \in \mathscr{X}(M), \quad \{f_1, \dots, f_{n-1}, f_n = V\} \in V^\perp$$

such that

$$A_V e_\alpha = \sigma_\alpha f_\alpha, \quad A_V e_n = 0, \quad A_V^t f_\alpha = \sigma_\alpha e_\alpha \quad (\alpha = 1, \dots, n-1),$$

where $\sigma_1 \geq \sigma_2 \geq \dots \geq \sigma_{n-1} \geq \sigma_n = 0$ are singular values for A_V. With respect to the singular frames the tangent and normal framing for $V(M)$ consists of [8]

$$\tilde{e}_n = e_n^h, \quad \tilde{e}_\alpha = \frac{e_\alpha^h + \sigma_\alpha f_\alpha^v}{\sqrt{1+\sigma_\alpha^2}}, \qquad \tilde{n}_\alpha = \frac{-\sigma_\alpha e_\alpha^h + f_\alpha^v}{\sqrt{1+\sigma_\alpha^2}}, \quad (\alpha = 1, \dots, n-1).$$

The mean curvature vector of local embedding $V(M) \subset T_1M$ with respect to this framing is of the form [8]

$$H_V = \frac{1}{n}\sum_{\alpha=1}^{n-1} g\Big(\sum_{i=1}^{n} \frac{(\nabla_{e_i} A_V)e_i + A_V R(V, A_V e_i)e_i}{1+\sigma_i^2}, \frac{f_\alpha}{\sqrt{1+\sigma_\alpha^2}}\Big)\tilde{n}_\alpha. \tag{4}$$

The expression (4) makes good geometric sense. Set

$$\check{H}_V = \sum\nolimits_{i=1}^{n} \frac{(\nabla_{e_i} A_V)e_i + A_V R(V, A_V e_i)e_i}{1+\sigma_i^2}.$$

It is easy to see that $g(\check{H}_V, V) = \sum_{i=1}^{n} \frac{\sigma_i^2}{1+\sigma_i^2}$. Hence, the minimality condition can be given in the form

$$\sum\nolimits_{i=1}^{n} \frac{(\nabla_{e_i} A_V)e_i + A_V R(V, A_V e_i)e_i}{1+\sigma_i^2} = \Big(\sum\nolimits_{i=1}^{n} \frac{\sigma_i^2}{1+\sigma_i^2}\Big)V.$$

The latter means that V *is minimal if and only if* $\check{H}_V$ *is collinear to* V.

1.3 The Reeb Vector Field

An *almost contact metric structure* on a smooth manifold (M^{2n+1}, g) of dimension $2n+1$ consists of a unit vector field ξ, a (1,1)-tensor field φ and a 1-form η such that

$$\varphi^2 = -I + \eta \otimes \xi, \quad \varphi\xi = 0, \quad \eta(\xi) = 1, \quad \eta \circ \varphi = 0, \tag{5}$$

$$g(\varphi X, \varphi Y) = g(X, Y) - \eta(X)\,\eta(Y), \quad \eta(X) = g(X, \xi) \tag{6}$$

for all vector fields on the manifold. It easily follows that φ is skew symmetric

$$g(\varphi X, Y) = -g(X, \varphi Y), \quad \varphi^t = -\varphi$$

and orthogonal being restricted on $\ker \eta = \xi^\perp$

$$(\varphi^t \varphi)X = (\varphi\varphi^t)X = X$$

for any $X \in \ker \eta$.

The unit vector field ξ is called *characteristic* or *Reeb vector field* of a contact metric manifold. The almost contact metric structure is denoted by (φ, ξ, η, g). An almost contact metric structure (φ, ξ, η, g) is called a (α, β)*-trans Sasakian structure* [6] if

$$(\nabla_X \varphi)Y = \alpha(g(X, Y)\xi - \eta(Y)X) + \beta(g(\varphi X, Y)\xi - \eta(Y)\varphi X) \tag{7}$$

for some smooth functions $\alpha, \beta : M^{2n+1} \to R$. The structure is called *Sasakian* if $\alpha = 1$ and $\beta = 0$ are valid; *Kenmotsu* if $\alpha = 0, \beta = 1$; and *cosymplectic* if $\alpha = 0, \beta = 0$. It worthwhile to mention that for $\dim M \geq 5$ the trans-Sasakian manifold is either α-Sasakian with $\alpha = const$ or β-Kenmotsu [7].

Theorem 1 *[13] The Reeb vector field on connected (α, β) trans-Sasakian manifold M gives rise to totally geodesic submanifold $\xi(M) \subset (T_1M, g_S)$ only in the following cases:*

- *$\beta = 0$ and $\alpha = 1$, hence M is Sasakian, or $\alpha = 0$, hence M is cosymplectic;*
- *$\alpha = 0$ and $\nabla\beta = \dfrac{\beta^2(\beta^2+1)}{1-\beta^2}\xi$. If $\beta = const$ or M is compact, then $\beta = 0$, hence M is cosymplectic.*

Corollary 1 (see also [9]) *The Hopf unit vector field on every odd-dimensional unit sphere S^{2m+1} $(m \geq 1)$ is totally geodesic.*

Theorem 2 *[13] The norm of the mean curvature vector H_ξ for the Reeb vector field ξ on (α, β) trans-Sasakian manifold of $\dim M = 2n+1$ has the form*

- *if $\dim(M) > 3$, then*

$$|H_\xi| = \tfrac{(1+(2n-1)\alpha^2)}{(2n+1)(1+\alpha^2)^{3/2}}\,|\varphi^2\nabla\alpha| = 0 \quad (\alpha = const,\ \beta = 0);$$

$$|H_\xi| = \tfrac{(1+(2n-1)\beta^2)}{(2n+1)(1+\beta^2)^{3/2}}\,|\varphi\bar{\Delta}\xi| \quad (\alpha = 0),$$

where $\bar{\Delta}$ stands for the rough Laplacian;

- *if $\dim(M) = 3$, then $|H_\xi| = \frac{|\varphi^2\nabla\alpha - \varphi\nabla\beta|}{3\sqrt{1+\alpha^2+\beta^2}}$.*

2 Minimal Unit Vector Fields on Oscillator Group

Harmonic maps are generalizations of locally minimal isometric immersion. The pulled back metric (1) on the submanifold $V(M) \subset (T_1M, g_S)$ manifests that mapping $V : M \to (T_1M, g_S)$ is isometry if and only if V is a parallel unit vector field. It means that in general one can not expect that a unit vector field, which defines a harmonic map is minimal. Surprisingly, there are examples when it happens. For example, J. C. González-Dávila and L. Vanhecke [5] proved that a left-invariant unit vector field on a three-dimensional unimodular Lie group is harmonic if and only if it is minimal; for generalized Heisenbrg group $H(n, 1)$ $(n \geq 2)$, the set of left-invariant harmonic unit vector fields is given by $\pm\xi$, $S \cap \xi^\perp$, where ξ is a distinguished unit vector field and $S \cap \xi^\perp$ is a set of unit vector fields in orthogonal complement of ξ, and they all determine harmonic maps into the unit tangent bundle. The same authors proved that left-invariant vectors of canonical frame $\xi, e_1, \dots, e_n$ on $H(n, 1)$ are minimal as well [4]. Recently, Na Xu and Ju Tan [14] found all the left-invariant harmonic unit vector fields on the oscillator groups. Besides, they determined the associated harmonic maps from the oscillator group into its unit tangent bundle equipped with the associated Sasaki metric. Our main goal is to understand if invariant unit vector fields on oscillator group which defines a harmonic map can be minimal.

The *oscillator group* $G_n(\lambda) = G(\lambda_1, \ldots, \lambda_n)$ is the connected, simply connected, solvable Lie group whose Lie algebra $\mathfrak{g}_n(\lambda)$ is the oscillator algebra $\mathfrak{g}_n(\lambda) = (\lambda_1, \ldots, \lambda_n)$ linearly spanned by $(2n+2)$-elements

$$e_1, \ldots, e_n; e_{n+1}, \ldots, e_{2n}; e_{2n+1} = \xi, e_{2n+2} = \zeta. \tag{8}$$

Denote

$$\mathcal{X} = Span(e_1, \ldots, e_n),\ \mathcal{Y} = Span(e_{n+1}, \ldots, e_{2n}),\ \mathcal{P} = Span(\xi),\ \mathcal{Q} = Span(\zeta).$$

According to (8), we have the following decomposition:

$$\mathfrak{g}_n(\lambda) = \mathcal{X} \oplus \mathcal{Y} \oplus \mathcal{P} \oplus \mathcal{Q}.$$

Let

$$V = \sum\nolimits_{i=1}^{n} (a_i e_i + a_{n+i} e_{n+i}) + a_{2n+1}\xi + a_{2n}\zeta$$

be a left-invariant unit vector fiend on the oscillator group $G_n(\lambda)$. Then

$$V = V_{\mathcal{X}} + V_{\mathcal{Y}} + a_{2n+1}\xi + a_{2n}\zeta\ .$$

Introduce the following linear operators and formes similar to ones that are used in an of almost contact geometry:

$$\varphi(V) = \sum_{i=1}^{n}(-a_{n+i}e_i + a_i e_{n+i}),\ \ \eta(V) = g(V, \xi) = a_{2n+1},\ \ \theta(V) = g(V, \zeta) = a_{2n}.$$

As a straightforward consequences of the definitions we get

$$\begin{aligned}
&\varphi^2 = -I + \eta \otimes \xi + \theta \otimes \zeta, \quad \varphi\xi = 0, \quad \varphi\zeta = 0, \quad \eta\varphi = \theta\varphi = 0,\\
&\eta(X) = g(X, \xi), \quad \theta(X) = g(\zeta, X),\\
&\eta(\xi) = 1, \quad \eta(\zeta) = 0, \quad \theta(\zeta) = 1, \quad \theta(\xi) = 0,\\
&g(\varphi X, \varphi Y) = g(X, Y) - \eta(X)\eta(Y) - \theta(X)\theta(Y).
\end{aligned} \tag{9}$$

It easily follows that φ is skew symmetric, i.e.

$$g(\varphi X, Y) = -g(X, \varphi Y), \quad \varphi^t = -\varphi$$

and orthogonal being restricted on $\mathcal{X} \oplus \mathcal{Y}$

$$(\varphi^t\varphi)X = (\varphi\varphi^t)X = X$$

for any $X \in \mathcal{X} \oplus \mathcal{Y}$. Relative to the introduced notations,

$$V = V_{\mathcal{X}} + V_{\mathcal{Y}} + \eta(V)\xi + \theta(V)\zeta.$$

Remark, also, that the canonical frame (8) is such that

$$e_{n+i} = \varphi(e_i), \quad i = 1, \dots, n.$$

Lemma 1 *The Nomizu operator for arbitrary vector field V in case of oscillator group can be given in invariant form as follows:*

$$A_V = \frac{1}{2}\big(\eta(V)\varphi + \varphi(V)\eta + \varphi^\flat(V)\xi - 2\mathbf{E}_\lambda \varphi(V)\theta\big),$$

where $\varphi^\flat = g(\cdot, \varphi(V))$ and $\mathbf{E}_\lambda = diag(\lambda_1, \dots, \lambda_n; \lambda_1, \dots, \lambda_n)$.

Proof The non-vanishing Lie brackets of the frame (8) are [1, 14]:

$$[e_i, e_{n+j}] = \delta_{ij}\xi, \quad [\zeta, e_j] = \lambda_j e_{n+j}, \quad [\zeta, e_{n+j}] = -\lambda_j e_j \quad (i, j = 1 \dots, n). \tag{10}$$

Using Koszul formula and (10), one can determine the Levi-Civita connection on $G_n(\lambda) = G(\lambda_1, \dots, \lambda_n)$ as follows [14]:

$$\begin{array}{llll}
\nabla_\xi \xi = 0, & \nabla_\xi \zeta = 0, & \nabla_\xi e_j = -\frac{1}{2}\, e_{n+j}, & \nabla_\xi e_{n+j} = \frac{1}{2}\, e_j, \\
\nabla_\zeta \xi = 0, & \nabla_\zeta \zeta = 0, & \nabla_\zeta e_j = \lambda_j e_{n+j}, & \nabla_\zeta e_{n+j} = -\lambda_j e_j, \\
\nabla_{e_j} \xi = -\frac{1}{2}\, e_{n+j}, & \nabla_{e_j} \zeta = 0, & \nabla_{e_j} e_k = 0, & \nabla_{e_j} e_{n+k} = \frac{1}{2}\, \delta_{jk}\xi, \\
\nabla_{e_{n+j}} \xi = \frac{1}{2}\, e_j, & \nabla_{e_{n+j}} \zeta = 0, & \nabla_{e_{n+j}} e_k = -\frac{1}{2}\delta_{jk}\xi, & \nabla_{e_{n+j}} e_{n+k} = 0,
\end{array} \tag{11}$$

where δ_{jk} is the Kronecker symbol and $1 \leqslant j, k \leqslant n$.

For a left-invariant vector field

$$V = \sum\nolimits_{i=1}^{n} (a_i e_i + a_{n+i} e_{n+i}) + a_{2n+1}\xi + a_{2n+2}\zeta$$

on $G_n(\lambda)$ we have [14]

$$\begin{array}{ll}
\nabla_{e_j} V = \frac{1}{2}\, a_{n+j}\xi - \frac{1}{2} a_{2n+1} e_{n+j}, & j = 1, 2, \dots, n, \\
\nabla_{e_{n+j}} V = -\frac{1}{2}\, a_j \xi + \frac{1}{2} a_{2n+1} e_j, & j = 1, 2, \dots, n, \\
\nabla_\xi V = \frac{1}{2} \sum_{i=1}^{n} (a_{n+i} e_i - a_i e_{n+i}), & \\
\nabla_\zeta V = \sum_{i=1}^{n} \lambda_i\, (a_i e_{n+i} - a_{n+i} e_i)\ . &
\end{array}$$

If $Z = \sum_{j=1}^{n} \big(z_j e_j + z_{n+j} e_{n+j}\big) + z_{2n+1}\xi + z_{2n+2}\zeta$ is another left-invariant vector field, then

$$\begin{aligned}
\nabla_Z V &= \sum\nolimits_{j=1}^{n} (z_j \nabla_{e_j} V + z_{n+j} \nabla_{e_{n+j}} V) + z_{2n+1} \nabla_\xi V + z_{2n+2} \nabla_\zeta V \\
&= \sum\nolimits_{j=1}^{n} \Big(\frac{1}{2} z_j (a_{n+j}\xi - a_{2n+1} e_{n+j}) - \frac{1}{2} z_{n+j} (a_j \xi - a_{2n+1} e_j)\Big)
\end{aligned}$$

$$+\frac{1}{2}z_{2n+1}\sum\nolimits_{j=1}^{n}\left(a_{n+j}e_j - a_je_{n+j}\right) + z_{2n+2}\sum_{j=1}^{n}\lambda_j\left(a_je_{n+j} - a_{n+j}e_j\right)$$
$$= -\frac{1}{2}a_{2n+1}\sum\nolimits_{j=1}^{n}(-z_{n+j}e_j + z_je_{n+j}) - \frac{1}{2}z_{2n+1}\sum\nolimits_{j=1}^{n}\left(-a_{n+j}e_j + a_je_{n+j}\right)$$
$$+ z_{2n+2}\sum\nolimits_{j=1}^{n}\lambda_j\left(-a_{n+j}e_j + a_je_{n+j}\right) + \frac{1}{2}\sum\nolimits_{j=1}^{n}(-z_{n+j}a_j + z_ja_{n+j})\xi$$
$$= -\frac{1}{2}\eta(V)\varphi(Z) - \frac{1}{2}\eta(Z)\varphi(V) + \theta(Z)\,\mathbf{E}_\lambda\varphi(V) + \frac{1}{2}g(V,\varphi(Z)\xi.$$

So, we have

$$A_VZ = -\nabla_ZV$$
$$= \frac{1}{2}\,\eta(V)\varphi(Z) + \frac{1}{2}\eta(Z)\varphi(V) - \theta(Z)\,\mathbf{E}_\lambda\varphi(V) + \frac{1}{2}g(\varphi(V), Z)\xi, \tag{12}$$

which completes the proof. □

Corollary 2 *As a Riemannian manifold, the oscillator group $G_n(\lambda)$ splits into metric product $H(n,1)\times I\!R$, where $H(n,1)$ is Heisenberg group and $I\!R$, and hence, $H(n,1)$ is totally geodesic submanifold in oscillator group $G_n(\lambda)$.*

Proof It is known that, as a group, the oscillator group is a semidirect product $G_\lambda = H(n,1)\rtimes I\!R$ of $(2n+1)$-dimensional Heizenberg group $H(n,1)$ and one-dimensional abelian group $I\!R$ with Lie algebra ζ. Plugging $V=\zeta$ into (12) and using (9), we get $\nabla_Z\zeta = 0$ for all Z, and $\nabla_\zeta\zeta = 0$. Thus, G_λ admits a parallel vector field ζ. As a consequence, G_λ is a Riemannian product of Heisenberg group $H(n,1)$ and 1-dimensional geodesic ζ-foliation. □

Lemma 2 *With respect to canonical frame* (8)*, matrix of the Nomizu operator for the left-invariant vector field $V = V_{\mathcal{X}} + V_{\mathcal{Y}} + \eta(V)\xi + \theta(V)\zeta$ is of the form*

$$A_V = \begin{pmatrix} 0 & -\frac{1}{2}\,\eta(V)I_n & -\frac{1}{2}V_{\mathcal{Y}} & E_\lambda V_{\mathcal{Y}} \\ \frac{1}{2}\,\eta(V)I_n & 0 & \frac{1}{2}V_{\mathcal{X}} & -E_\lambda V_{\mathcal{X}} \\ -\frac{1}{2}V_{\mathcal{Y}}^t & \frac{1}{2}V_{\mathcal{X}}^t & 0 & 0 \\ 0 & 0 & 0 & 0 \end{pmatrix},$$

where I_n is an $n\times n$ unit matrix, and $E_\lambda = diag(\lambda_1,\ldots,\lambda_n)$.

Proof Columns of A_V are formed by A_Ve_i, A_Ve_{n+i}, $A_V\xi$ and $A_V\zeta$. Using (12) and (9), we obtain

$$A_Ve_i = \tfrac{1}{2}\eta(V)e_{n+i} - \tfrac{1}{2}a_{n+i}\xi,\quad A_Ve_{n+i} = -\tfrac{1}{2}\eta(V)e_i + \tfrac{1}{2}a_i\xi,$$
$$A_V\xi = \tfrac{1}{2}\varphi(V) = \tfrac{1}{2}(-V_{\mathcal{Y}} + V_{\mathcal{X}}),\quad A_V\zeta = -\mathbf{E}_\lambda\varphi(V) = E_\lambda V_{\mathcal{Y}} - E_\lambda V_{\mathcal{X}},$$

that completes the proof. □

Lemma 3 *The invariant form of Riemannian curvature tensor of the oscillator group $G_n(\lambda)$ can be given as follows:*

$$R(X,Y)Z = \frac{1}{2}g(\varphi(X),Y)\varphi(Z) + \frac{1}{4}\big(g(\varphi(X),Z)\varphi(Y) - g(\varphi(Y),Z)\varphi(X)\big)$$
$$+\frac{1}{4}\big(\eta(Y)X - \eta(X)Y\big)\eta(Z) + \frac{1}{4}\big(\eta(X)g(Y,Z) - \eta(Y)g(X,Z)\big)\xi$$
$$+\frac{1}{4}\big(\eta(X)\theta(Y) - \eta(Y)\theta(X)\big)\big(\eta(Z)\zeta - \theta(Z)\xi\big).$$

Proof With respect to the frame (8) the components of Riemannian tensor are

$$R\left(e_i, e_{n+j}\right)e_s = -\frac{1}{2}\delta_{ij}e_{n+s} - \frac{1}{4}\delta_{js}e_{n+i}, \quad R\left(e_i, e_{n+j}\right)e_{n+s} = \frac{1}{2}\delta_{ij}e_s + \frac{1}{4}\delta_{is}e_j,$$
$$R\left(e_i, e_j\right)e_{n+s} = \frac{1}{4}\left(\delta_{js}e_{n+i} - \delta_{is}e_{n+j}\right), \quad R\left(e_{n+i}, e_{n+j}\right)e_s = \frac{1}{4}\left(\delta_{js}e_i - \delta_{is}e_j\right),$$
$$R\left(e_i, \xi\right)e_j = \frac{1}{4}\delta_{ij}\xi, \quad R\left(e_i, \xi\right)\xi = -\frac{1}{4}e_i, \quad R\left(e_{n+i}, \xi\right)e_{n+j} = \frac{1}{4}\delta_{ij}\xi,$$
$$R\left(e_{n+i}, \xi\right)\xi = -\frac{1}{4}e_{n+i},$$

where $1 \leqslant i, j, s \leqslant n$, see [14]. Letting

$$X = \textstyle\sum_{i=1}^{n}(x_i e_i + x_{n+i}e_{n+i}) + x_{2n+1}\xi + x_{2n}\zeta,$$
$$Y = \textstyle\sum_{i=1}^{n}(y_i e_i + y_{n+i}e_{n+i}) + y_{2n+1}\xi + y_{2n}\zeta,$$
$$Z = \textstyle\sum_{i=1}^{n}(z_i e_i + z_{n+i}e_{n+i}) + z_{2n+1}\xi + z_{2n}\zeta,$$

after rather long accurate computations we get what was claimed. □

Theorem 3 *The oscillator group $G_n(\lambda)$ as Riemannian manifold with the left-invariant metric is isometric to the Riemannian product $H(n,1)\times\mathbb{R}$ of $(2n+1)$-dimensional generalized Heisenberg group $H(n,1)$ and abelian subgroup $\mathbb{R}$ with Lie algebras $\mathfrak{g}_{2n+1} = Span(\xi, e_1,\dots,e_n; e_{n+1},\dots,e_{2n})$ and $Span(\zeta)$, respectively. On $G_n(\lambda)$ we get the following:*

(a) *ζ is parallel and hence minimal and totally geodesic;*
(b) *$\big(\xi,\varphi,g,\eta = g(\cdot,\xi)\big)$ defines the α-Sasakian structure on $H(n,1)$ with $\alpha = \frac{1}{2}$;*
(c) *ξ is minimal but not totally geodesic;*
(d) *the unit $V \in \xi^\perp \subset \zeta^\perp$ is minimal if and only if $\lambda_i^2 = \lambda_j^2$ for all $(i,j = 1,\dots,n)$. Moreover,*

- *in case $\lambda_1 = \ldots = \lambda_n$, each unit $V \in \xi^\perp \subset \zeta^\perp$ is minimal;*
- *in case $\lambda_1 = \ldots = \lambda_k = -\lambda_{k+1} = \ldots = -\lambda_n$, the field $V = \sum_{i=1}^{n}(a_i e_i + a_{n+i}e_{n+i})$ is minimal if*

$$\sum\nolimits_{i=1}^{k}(a_i^2 + a_{n+i}^2) - \sum\nolimits_{i=k+1}^{n}(a_i^2 + a_{n+i}^2) = 0.$$

Proof (a) If $V = \zeta$, then Lemma 2 implies $A_\zeta = 0$ and (2) implies ζ is minimal and totally geodesic.

(b) Restrict out considerations on totally geodesic $H(n,1) \subset G_n(\lambda)$. Then $\forall\, Z$ we have $\theta(Z) = 0$ and (9) imply

$$\begin{aligned}&\varphi^2 = -I + \eta \otimes \xi, \quad \varphi\xi = 0, \quad \eta\varphi = 0,\\ &\eta(X) = g(X,\xi), \quad \eta(\xi) = 1,\\ &g(\varphi X, \varphi Y) = g(X,Y) - \eta(X)\eta(Y).\end{aligned}$$

To show that $H(n,1)$ is α -Sasakian (cf. (5), (6) and (7)) we only need to prove that

$$(\nabla_Z\varphi)W = \alpha(g(Z,W)\xi - \eta(W)Z) \quad \forall\, W, Z \in \ker(\theta).$$

If $V = \xi$, then Lemma 1 implies $A_\xi = \frac{1}{2}\varphi$. In other words,

$$\varphi(Z) = -2\nabla_Z\,\xi = 2\,A_\xi Z.$$

By definition,

$$(\nabla_Z\,\varphi)W = \nabla_Z(\varphi(W)) - \varphi(\nabla_Z\,W).$$

According to (12), for all $W, Z \in \ker(\theta)$ we obtain

$$\nabla_Z W = -A_Z W = -\frac{1}{2}\,\eta(Z)\varphi(W) - \frac{1}{2}\,\eta(W)\,\varphi(Z) - \frac{1}{2}\,g(\varphi(Z),W)\,\xi.$$

Thus, we have

$$\nabla_Z(\varphi(W)) = -\frac{1}{2}\,\eta(Z)\varphi^2(W) - \frac{1}{2}\,g(\varphi(Z),\varphi(W))\xi.$$

$$\varphi(\nabla_Z W) = -\frac{1}{2}\,\eta(Z)\,\varphi^2(W) - \frac{1}{2}\,\eta(W)\,\varphi^2(Z).$$

So, we get

$$\begin{aligned}(\nabla_Z\varphi)W &= -\tfrac{1}{2}\eta(W)\varphi^2(Z) + \tfrac{1}{2}g(\varphi(Z),\varphi(W))\xi\\ &= -\tfrac{1}{2}\big(\eta(W)(-Z + \eta(Z)\xi) + (g(Z,W) - \eta(Z)\eta(W))\xi\big)\\ &= \tfrac{1}{2}\big(-g(Z,W)\xi + \eta(W)Z\big).\end{aligned}$$

Hence, the subgroup $H(n,1)$ carries the α-Sasakian structure with $\alpha = \frac{1}{2}$.

(c) Since the submanifold $H(n,1)$ carries the α-Sasakian structure with $\alpha = \frac{1}{2}$ and is totally geodesic in $G_n(\lambda)$, one can apply Theorems 1 and 2 to see that ξ, being the Reeb vector field on $H(n,1)$, is minimal but not totally geodesic vector field on $H(n,1)$ as well as on $G_n(\lambda)$.

(d) If $V \subset \xi^\perp \subset \zeta^\perp$, then $\eta(V) = \theta(V) = 0$, and (12) implies $\nabla_V V = 0$. So, V is *geodesic* vector field on $G_n(\lambda)$. The matrix of Nomizu operator takes the following form:

$$A_V = \begin{pmatrix} 0 & 0 & -\frac{1}{2}V_{\mathcal{Y}} & E_\lambda V_{\mathcal{Y}} \\ 0 & 0 & \frac{1}{2}V_{\mathcal{X}} & -E_\lambda V_{\mathcal{X}} \\ -\frac{1}{2}V^t_{\mathcal{Y}} & \frac{1}{2}V^t_{\mathcal{X}} & 0 & 0 \\ 0 & 0 & 0 & 0 \end{pmatrix} = \begin{pmatrix} \mathbf{0}_{2n} & \frac{1}{2}\varphi(V) & -\mathbf{E}_\lambda\varphi(V) \\ \frac{1}{2}\varphi(V)^t & 0 & 0 \\ 0 & 0 & 0 \end{pmatrix},$$

where $\mathbf{0}_{2n}$ is $2n \times 2n$ zero matrix. Then

$$A^t_V A_V = \begin{pmatrix} \mathbf{0}_{2n} & \frac{1}{2}\varphi(V) & 0 \\ \frac{1}{2}\varphi(V)^t & 0 & 0 \\ -\varphi(V)^t\mathbf{E}_\lambda & 0 & 0 \end{pmatrix} \begin{pmatrix} \mathbf{0}_{2n} & \frac{1}{2}\varphi(V) & -\mathbf{E}_\lambda\varphi(V) \\ \frac{1}{2}\varphi(V)^t & 0 & 0 \\ 0 & 0 & 0 \end{pmatrix}$$

$$= \begin{pmatrix} \frac{1}{4}\varphi(V)\varphi(V)^t & 0 & 0 \\ 0 & \frac{1}{4}\varphi(V)^t\varphi(V) & -\frac{1}{2}\varphi(V)^t\mathbf{E}_\lambda\varphi(V) \\ 0 & -\frac{1}{2}\varphi(V)^t\mathbf{E}_\lambda\varphi(V) & \varphi(V)^t\mathbf{E}^2_\lambda\varphi(V) \end{pmatrix}.$$

Observe, that

$$\varphi(V)^t\varphi(V) = |\varphi(V)^2| = |V|^2 = 1,$$

$$\varphi(V)^t\mathbf{E}_\lambda\varphi(V) = \sum\nolimits_{j=1}^n \lambda_j(a^2_{n+j} + a^2_j) = V^t\mathbf{E}_\lambda V,$$

$$\varphi(V)^t\mathbf{E}^2_\lambda\varphi(V) = \sum\nolimits_{j=1}^n \lambda^2_j(a^2_{n+j} + a^2_j) = V^t\mathbf{E}^2_\lambda V.$$

So finally,

$$A^t_V A_V = \begin{pmatrix} \frac{1}{4}\varphi(V)\,\varphi(V)^t & 0 & 0 \\ 0 & \frac{1}{4} & -\frac{1}{2}V^t\mathbf{E}_\lambda V \\ 0 & -\frac{1}{2}V^t\mathbf{E}_\lambda V & V^t\mathbf{E}^2_\lambda V \end{pmatrix}.$$

The matrix $A^t_V A_V$ has a block diagonal form. It is easy to see that

$$rank\left(\varphi(V)\,\varphi^t(V)\right) = 1, \quad trace\left(\varphi(V)\,\varphi(V)^t\right) = \frac{1}{4}|V|^2 = \frac{1}{4}.$$

Denote by $\sigma_1, \dots, \sigma_{2n+2}$ singular values and by bold $\mathbf{e}_1, \dots, \mathbf{e}_{2n+2}$ a corresponding singular orthonormal frame for A_V. Observe, that $\sigma^2 = 0$ is the eigenvalue of multiplicity $2n - 1$, while $\sigma^2 = \frac{1}{4}$ is the eigenvalue of multiplicity 1 with $\varphi(V)$ as the eigenvector. The rest of singular values are the roots of

$$\mu^2 - \Big(\frac{1}{4} + V^t E^2_\lambda V\Big)\mu + \frac{1}{4}\Big(V^t E^2_\lambda V - \big(V^t E_\lambda V\big)^2\Big) = 0. \tag{13}$$

In coordinate form the above equation can be expressed as

$$\mu^2 - \Big(\frac{1}{4} + \sum_{i=1}^{n} \lambda_i^2 (a_i^2 + a_{n+i}^2)\Big)\mu + \frac{1}{8} \sum_{i,j=1}^{n} (\lambda_i - \lambda_j)^2 (a_i^2 + a_{n+i}^2)(a_j^2 + a_{n+j}^2) = 0.$$

The corresponding singular vectors are of the form

$$\mathbf{e}_{2n+1} = \frac{1}{\sqrt{\alpha^2 + \beta^2}} (\alpha\xi + \beta\,\zeta), \quad \mathbf{e}_{2n+2} = \frac{1}{\sqrt{\alpha^2 + \beta^2}} (-\beta\xi + \alpha\,\zeta)\,.$$

As a result, the singular frame eigenpairs are

$$(\sigma_1 = \ldots = \sigma_{2n-1} = 0;\ \ \mathbf{e}_1 = V,\ \ \mathbf{e}_2, \ldots, \mathbf{e}_{2n-1}),$$

where $(\mathbf{e}_2, \ldots, \mathbf{e}_{2n-1})$ is arbitrary orthonormal frame in $V^\perp \subset \varphi(V)^\perp$,

$$\Big(\sigma_{2n} = \frac{1}{4},\ \ \mathbf{e}_{2n} = \varphi(V)\Big), \quad (\mu_1,\ \mathbf{e}_{2n+1}), \quad (\mu_2,\ \mathbf{e}_{2n+2}).$$

From the definition of singular frame it follows that $A_V\mathbf{e}_i = 0,\ (i = 1, \ldots 2n - 1)$. Since $\mathbf{e}_i \in \mathcal{X} \oplus \mathcal{Y} \subset \ker(\eta) \subset \ker(\theta)$ and φ is skew-symmetric on $\mathcal{X} \oplus \mathcal{Y}$, the (12) implies $\nabla_{\mathbf{e}_i}\mathbf{e}_i = 0$. Therefore,

$$(\nabla_{\mathbf{e}_i} A_V)\mathbf{e}_i = \nabla_{\mathbf{e}_i}(A_V\mathbf{e}_i) - A_V(\nabla_{\mathbf{e}_i}\mathbf{e}_i) = 0, \quad R(V, A_V\mathbf{e}_i)\mathbf{e}_i = 0,$$

where $i = 1, \ldots, 2n - 1$. Now, compute

$$(\nabla_{\mathbf{e}_{2n}} A_V)\mathbf{e}_{2n} = \nabla_{\mathbf{e}_{2n}}(A_V\mathbf{e}_{2n}) - A_V(\nabla_{\mathbf{e}_{2n}}\mathbf{e}_{2n}).$$

Since $\mathbf{e}_{2n} = \varphi(V)$ the (12) implies

$$A_V\mathbf{e}_{2n} = \tfrac{1}{2} g(\varphi(V), \varphi(V))\xi = \tfrac{1}{2}\xi,$$
$$\nabla_{\mathbf{e}_{2n}}(A_V\mathbf{e}_{2n}) = -\tfrac{1}{2}\nabla_{\varphi(V)}\xi = -\tfrac{1}{4}\varphi^2 V = \tfrac{1}{4}V.$$

Compute

$$R(V, A_V\mathbf{e}_{2n})\mathbf{e}_{2n} = \frac{1}{2} R(V, \xi)\varphi(V).$$

Using Lemma 3 restricted to $\mathcal{X} \oplus \mathcal{Y}$, we get

$$R(V, \xi)\varphi(V) = \tfrac{1}{2}\, g(\varphi(V), \xi)\varphi^2(V)$$
$$+\tfrac{1}{4}\big(g(\varphi(V), \varphi(V))\varphi(\xi) - g(\varphi(\xi), \varphi(V))\varphi(V)\big) = 0.$$

For $\mathbf{e}_{2n+1} = \frac{1}{\sqrt{\alpha^2+\beta^2}}(\alpha\xi + \beta\,\zeta)$ we observe, first, that (12) implies

$$A_V\xi = \frac{1}{2}\varphi(V), \quad A_V\zeta = -\frac{1}{2}\mathbf{E}_\lambda\varphi(V);$$

hence,

$$A_V\mathbf{e}_{2n+1} = \frac{(\alpha I_{2n} - 2\beta\mathbf{E}_\lambda)}{2\sqrt{\alpha^2+\beta^2}}\,\varphi(V), \quad \sigma_{2n+1}^2 = \left|\frac{(\alpha I_{2n} - 2\beta\mathbf{E}_\lambda)}{2\sqrt{\alpha^2+\beta^2}}\,\varphi(V)\right|^2,$$

where I_{2n} is a $2n \times 2n$ unit matrix. Further on, from (11) it follows that $\nabla_\xi\zeta = \nabla_\zeta\xi = \nabla_\xi\xi = \nabla_\zeta\zeta = 0$; hence,

$$\begin{aligned}(\nabla_{\mathbf{e}_{2n+1}}A_V)\mathbf{e}_{2n+1} &= \nabla_{\mathbf{e}_{2n+1}}(A_V\mathbf{e}_{2n+1})\\ &= \tfrac{\alpha I_{2n}-2\beta\mathbf{E}_\lambda}{2\sqrt{\alpha^2+\beta^2}}\nabla_{\mathbf{e}_{2n+1}}\varphi(V) = -\tfrac{\alpha I_{2n}-2\beta\mathbf{E}_\lambda}{2\sqrt{\alpha^2+\beta^2}}A_{\varphi(V)}\mathbf{e}_{2n+1}\,.\end{aligned}$$

Applying again (12), we calculate

$$A_{\varphi(V)}\mathbf{e}_{2n+1} = \frac{1}{\sqrt{\alpha^2+\beta^2}}(\alpha A_{\varphi(V)}\xi + \beta A_{\varphi(V)}\zeta) = -\frac{1}{2\sqrt{\alpha^2+\beta^2}}(\alpha I_{2n} - 2\beta\mathbf{E}_\lambda)V\,.$$

As a result,

$$(\nabla_{\mathbf{e}_{2n+1}}A_V)\mathbf{e}_{2n+1} = \frac{(\alpha I_{2n} - 2\beta\mathbf{E}_\lambda)^2}{4(\alpha^2+\beta^2)}V\,.$$

Since $(V, A_V\mathbf{e}_{2n+1}) \in \mathcal{X} \oplus \mathcal{Y}$ and $\mathbf{e}_{2n+1}$ is a linear combination of ξ and ζ, from Lemma 3 we conclude that

$$R\,(V, A_V\mathbf{e}_{2n+1})\,\mathbf{e}_{2n+1} = 0.$$

In a similar way, for $\mathbf{e}_{2n+2} = \frac{1}{\sqrt{\alpha^2+\beta^2}}(-\beta\xi + \alpha\zeta)$ we get

$$A_V\mathbf{e}_{2n+2} = -\frac{\beta I_{2n}+2\alpha\mathbf{E}_\lambda}{2\sqrt{\alpha^2+\beta^2}}\,\varphi(V), \quad \sigma_{2n+2}^2 = \left|\frac{\beta I_{2n} + 2\alpha\mathbf{E}_\lambda}{2\sqrt{\alpha^2+\beta^2}}\,\varphi(V)\right|^2.$$

$$(\nabla_{\mathbf{e}_{2n+2}}A_V)\mathbf{e}_{2n+2} = \frac{(\beta I_{2n} + 2\alpha\mathbf{E}_\lambda)^2\,V}{4(\alpha^2+\beta^2)}, \quad R\,(V, A_V\mathbf{e}_{2n+2})\,\mathbf{e}_{2n+2} = 0.$$

Summing up,

$$\check{H}_V = \frac{\frac{1}{4}V}{1+\frac{1}{4}} + \frac{1}{1+\sigma_{2n+1}^2}\,\frac{(\alpha I_{2n} - 2\beta\mathbf{E}_\lambda)^2 V}{4(\alpha^2+\beta^2)} + \frac{1}{1+\sigma_{2n+2}^2}\,\frac{(\beta I_{2n} + 2\alpha\mathbf{E}_\lambda)^2 V}{4(\alpha^2+\beta^2)}\,.$$

The vector field is minimal if and only if $\check{H}_V \parallel V$. It means that the following proportion relation must be fulfilled:

$$(\alpha I_{2n} - 2\beta \mathbf{E}_\lambda)^2 \sim I_{2n}, \quad (\beta I_{2n} + 2\alpha \mathbf{E}_\lambda)^2 \sim I_{2n}.$$

It follows that

$$\alpha^2 I_{2n} - 4\alpha\beta \mathbf{E}_\lambda + 4\beta^2 \mathbf{E}_\lambda^2 \sim \mathrm{I}_{2n}, \quad \beta^2 \mathrm{I}_{2n} + 4\alpha\beta \mathbf{E}_\lambda + 4\alpha^2 \mathbf{E}_\lambda^2 \sim I_{2n}. \tag{14}$$

Adding, we get

$$(\alpha^2 + \beta^2) I_{2n} + 4(\alpha^2 + \beta^2)\mathbf{E}_\lambda^2 \sim I_{2n} \quad \Rightarrow \quad \mathbf{E}_\lambda^2 \sim I_{2n}.$$

In this case, from (14) we conclude

$$4\alpha\beta \mathbf{E}_\lambda \sim I_{2n}.$$

In case $\alpha = 0$ and $\beta \neq 0$ we have only $E_\lambda^2 \sim I_{2n}$, which implies

$$\lambda_1^2 = \dots \lambda_n^2 = \lambda^2, \quad \sigma_{2n+1}^2 = \lambda^2, \quad \sigma_{2n+2}^2 = \frac{1}{4}.$$

The Vieta's theorem for (13) implies that this is possible if and only if $V^t \mathbf{E}_\lambda V = 0$. Then

$$\check{H}_V = \frac{\frac{1}{4}V}{1+\frac{1}{4}} + \frac{\lambda^2 V}{1+\lambda^2} + \frac{\frac{1}{4}V}{1+\frac{1}{4}};$$

hence, V is minimal under assumption that $V^t \mathbf{E}_\lambda V = 0$.

If $\beta = 0$ and $\alpha \neq 0$, then

$$\sigma_{2n+1}^2 = \frac{1}{4}, \quad \sigma_{2n+2}^2 = \lambda^2,$$

and again the Vieta's theorem for (13) implies that it is possible if and only if $V^t \mathbf{E}_\lambda V = 0$. Then

$$\check{H}_V = \frac{\frac{1}{4}V}{1+\frac{1}{4}} + \frac{\frac{1}{4}V}{1+\frac{1}{4}} + \frac{\lambda^2 V}{1+\lambda^2};$$

hence, V is minimal. The condition $V^t \mathbf{E}_\lambda V = 0$ makes sense if $\mathbf{E}_\lambda \nsim I_{2n}$, which means that among the structure constants there are ones of differen sign. Suppose that $\lambda_1 = \dots = \lambda_k = \lambda > 0$, while $\lambda_{k+1} = \dots = \lambda_n = -\lambda < 0$. Then the vector field $V = \sum_{i=1}^n (a_i e_i + a_{n+i} e_{n+i})$ is minimal if

$$V^t \mathbf{E}_\lambda V = 0 \sim \sum\nolimits_{i=1}^k (a_i^2 + a_{n+i}^2) - \sum\nolimits_{i=k+1}^n (a_i^2 + a_{n+i}^2) = 0\,.$$

If $\alpha\,\beta \neq 0$, then

$$\mathbf{E}_\lambda \sim I_{2n};$$

hence, $\lambda_i = \lambda$. In this case, $\sigma_{2n+1} = \frac{1}{4} + \lambda^2$, $\sigma_{2n+2} = 0$ and $\alpha = \lambda$, $\beta = -2\lambda^2$, $\alpha^2 + \beta^2 = \lambda^2(1 + 4\lambda^2)$. Therefore,

$$\check{H}_V = \frac{\frac{1}{4}V}{1 + \frac{1}{4}} + \frac{\left(1 + \frac{\lambda^2}{4}\right)V}{1 + 1 + \frac{\lambda^2}{4}} = \left(\frac{\sigma_1^2}{1 + \sigma_1^2} + \frac{\sigma_{2n+1}^2}{1 + \sigma_{2n+1}^2}\right)V;$$

hence, V is **minimal** without any restrictions. □

Now we can compare our result with the result of N. Xu and J. Tan:

Theorem 4 *[14] The set of left-invariant unit vector fields on the oscillator group $G_n(\lambda) = G\,(\lambda_1, \ldots, \lambda_n)$, such that the corresponding maps into the unit tangent bundles are harmonic, is given by*

$$\{\pm\xi\}\bigcup\{\pm\zeta\}\bigcup\left(\mathcal{S}\bigcap\left\{\sum\nolimits_{j=1}^{n}\left(a_j e_j + a_{n+j} e_{n+j}\right)\right\}\right),$$

where $\left(a_i^2 + a_{n+i}^2\right)\left(a_j^2 + a_{n+j}^2\right) = 0$ for $\lambda_i^2 \neq \lambda_j^2$.

Corollary 3 *On the oscillator group with structure constants $\lambda_i = \lambda_j$ $(\forall i, j = 1, \ldots, n)$ each left-invariant unit vector field that defines a **harmonic map** into the unit tangent bundle with Sasaki metric **is minimal**.*

J.C. González-Dávila and L. Vanhecke [4] studied minimality of left-invariant unit vector fields on generalized Heisenberg group and proved that elements $e_1, \ldots, e_n$; $e_{n+1}, \ldots, e_{2n}, \xi$ of the frame (8) are minimal. Theorem 3 presents a more general result.

References

1. Biggs, R. and Remsing, C.: Some remarks on the oscillator group. Diff. Geom. and its Appl. 35, 199–209, (2014). https://doi.org/10.1016/j.difgeo.2014.03.003.
2. Gluck, H., Ziller, W.: On the volume of a unit vector field on the three sphere. Com. Math. Helv. 61, 177–192 (1986). eudml.org/doc/140047.
3. Gil-Medrano, O. and Llinares-Fuster, E.: Minimal unit vector fields. Tohôku Math. J. 54(1), 71–84 (2002). https://doi.org/10.2748/tmj/1113247180.
4. González-Dávila, J.C. and Vanhecke, L.: Examples of minimal unit vector fields. Ann. Global Anal. Geom. 18, 385–404 (2000). https://doi.org/10.1023/A:1006788819180
5. González-Dávila, J.C. and Vanhecke, L.: Invariant harmonic unit vector fields on Lie groups. Bollettino dell'Unione Matematica Italiana, Serie 8, Vol. 5-B, No. 2, 377–403 (2002). www.bdim.eu/item?id=BUMI_2002_8_5B_2_377_0
6. Obiña, J.A.: New Classes of almost Contact metric structures, Publ. Math. Debrecen 32: 187–193 (1985). www.publi.math.unideb.hu/load_doc.php?p=3718&t=pap

7. Marrero, J.C.: The local structure of trans-Sasakian manifolds. Ann. di Mat. Pura ed Appl. 162 (4), 77–86 (1992). https://doi.org/10.1007/BF01760000
8. Yampolsky, A.: On the mean curvature of a unit vector field, Publ. Math. Debrecen 60, 2/3, 131–155 (2002). https://doi.org/10.5486/PMD.2002.2562
9. Yampolsky, A.: A totally geodesic property of Hopf vector fields, Acta Math. Hung. 101, 93–112 (2003). https://doi.org/10.1023/B:AMHU.0000003895.90745.38
10. Yampolsky, A.: Full description of totally geodesic unit vector fields on Riemannian 2-manifold. Mat. Fiz. Anal. Geom., 11:3, 355–365 (2004). www.mathnet.ru/eng/jmag213
11. Yampolsky, A.: On special types of minimal and totally geodesic unit vector fields. Proc. 7-th Intl. Conf. in Geometry, Integrability and Quantization June 2–10, Varna, Bulgaria, 292–306 (2005). www.emis.de/proceedings/Varna/vol7/P07Yampolsky.pdf
12. Yampolsky, A.: Invariant totally geodesic unit vector fields on three-dimensional Lie groups. J. Math. Phys., Anal., Geom. 3 (2), 253–276 (2007). www.jmag.ilt.kharkiv.ua/index.php/jmag/article/view/jm03-0253e/521
13. Yampolsky, A.: On properties of the Reeb vector field of (α, β) trans-Sasakian structure, Turkish J. Math, V. 46: No. 6, Art. 19 (2022). https://doi.org/10.55730/1300-0098.3271
14. Xu, N. and Tan, J.: Invariant harmonic unit vector fields on the oscillator groups, Czech. Math. J., 69, 4, 907–924 (2019). https://doi.org/10.21136/CMJ.2019.0538-17

Growth and Structure of Equicontinuous Foliated Spaces

Manuel F. Moreira Galicia

Abstract Molino's description of Riemannian foliations on compact manifolds extends to compact equicontinuous foliated spaces as developed by Àlvarez Lòpez and Manuel Moreira. This extension is particularly considered when leaves are densely packed, and the pseudogroup exhibits strong quasi-analytic behavior. Notably, this extension leads to the establishment of an association with a structural local group within such a foliated space. Application of this framework results in a partial generalization of Carrière and Breuillard-Gelander's results, creating a connection between the structural local group and the growth of leaves. Additionally, we present an illustrative examples, as well as a scenario with zero topological codimension. In this context, instances of weak solenoids embodying this characteristic have been previously established by Dyer, Hurder, and Lukina.

Keywords Molino theory · Minimal action · Strong quasi-analytic action · Local groups action · Equicontinuous foliated space · Equicontinuous pseudogroup · Growth of leaves · Coarse quasi-isometry

1 Introduction

Riemannian foliations were introduced by Reinhart [52] using requiring isometric transverse dynamics. It was pointed out by Ghys in [44, Appendix E] (see also Kellum's paper [37]) that equicontinuous foliated spaces should be considered as the "topological Riemannian foliations", and therefore many of the results about Riemannian foliations should have versions for equicontinuous foliated spaces. Some

M. F. Moreira Galicia (✉)
Department of Physic and Chemistry, Faculty of Pharmacy, University of San Carlos, Guatemala City, Guatemala
e-mail: morgal2002@profesor.usac.edu.gt; morgal2002@gmail.com
URL: https://portal.ccqqfar.usac.edu.gt/

Geometría y Topología Facultad de Matemáticas, Universidad de Santiago de Compostela, 15782 Santiago de Compostela, Spain

© The Author(s), under exclusive license to Springer Nature Switzerland AG 2024
V. Rovenski et al. (eds.), *Differential Geometric Structures and Applications*, Springer Proceedings in Mathematics & Statistics 440,
https://doi.org/10.1007/978-3-031-50586-7_7

steps in this direction were given by Álvarez and Candel [4, 5], showing that, under reasonable conditions, their leaf closures form minimal foliated spaces, and their generic leaves are quasi-isometric to each other, like in the case of Riemannian foliations. In the same direction, Matsumoto [39] proved that any minimal equicontinuous foliated space has a non-trivial transverse invariant measure, which is unique up to scaling if the space is compact—observe that this unicity implies ergodicity. The magnitude of the generalization from Riemannian foliations to equicontinuous foliated spaces was made precise by Álvarez and Candel [5] (see also Tarquini's paper [55]), giving a topological description of Riemannian foliations within the class of equicontinuous foliated spaces.

Most of the known properties of Riemannian foliations follow from a description due to Molino [43, 44]. However, so far, there was no version of Molino's description for equicontinuous foliated spaces—the indicated properties of equicontinuous foliated spaces were obtained by other means. The goal of our work in Álvarez and Moreira [1] is to develop such a version of Molino's theory, and use it to study the growth of their leaves, following the study of the growth of Riemannian foliations by Carrière [10] and Breuillard-Gelander [13]. In more recent results, Dyer, Hurder, and Lukina established Molino's spaces in [21] for equicontinuous matched-box manifolds. Notably, the requirement of strong quasi-analyticity of the action is not required. Additionally, in [20], they demonstrated that if the group H is finite, then the manifold X is virtually homogeneous. To understand our results better, let us briefly recall Molino's theory.

2 Molino's Theory for Riemannian Foliations

The needed basic concepts from foliation theory can be seen in [14, 33, 34].

Let $\mathcal{F}$ be a (smooth) foliation of codimension q on a manifold M. Let $T\mathcal{F} \subset TM$ denote the vector subbundle of vectors tangent to the leaves, and $N\mathcal{F} = TM/T\mathcal{F}$ its normal bundle. Recall that there is a natural flat leafwise partial connection on $N\mathcal{F}$ so that any local normal vector field is leafwise parallel if and only if it is locally projectable by the distinguished submersions; terms like "leafwise flat", "leafwise parallel" and "leafwise horizontal" will refer to this partial connection. It is said that $\mathcal{F}$ is:

- *Riemannian* if $N\mathcal{F}$ has a leafwise parallel Riemannian structure;
- *transitive* if the group of its foliated diffeomorphisms acts transitively on M;
- *transversely parallelizable (TP)* if there is a leafwise parallel global frame of $N\mathcal{F}$, called *transverse parallelism*; and a
- *Lie foliation* if moreover, the transverse parallelism is a basis of a Lie algebra with the operation induced by the vector fields bracket.

These conditions are successively stronger. Molino's theory describes Riemannian foliations on compact manifolds in terms of minimal Lie foliations, and using TP foliations as an intermediate step:

1st step. If $\mathcal{F}$ is Riemannian and M compact, then there is an $\mathrm{O}(q)$-principal bundle, $\hat{\pi} : \widehat{M} \to M$, with an $\mathrm{O}(q)$-invariant TP foliation, $\widehat{\mathcal{F}}$, such that $\widehat{\pi}$ is a foliated map whose restrictions to the leaves are the holonomy covers of the leaves of $\mathcal{F}$.

2nd step. If $\mathcal{F}$ is TP and M compact, then there is a fiber bundle $\pi : M \to W$ whose fibers are the leaf closures of $\mathcal{F}$, and the restriction of $\mathcal{F}$ to each fiber is a Lie foliation.

Since the structure of a Lie foliation is unique in the minimal case, we end up with a Lie algebra associated to $\mathcal{F}$, called *structural Lie algebra*. The proofs of the above statements strongly use the differential structure of $\mathcal{F}$.

In the first step, $\hat{\pi} : \widehat{M} \to M$ is the $\mathrm{O}(q)$-principal bundle of orthonormal frames for some leafwise parallel metric on $N\mathcal{F}$, and $\widehat{\mathcal{F}}$ is given by the corresponding flat leafwise horizontal distribution. Then $\widehat{\mathcal{F}}$ is TP by a standard argument.

In the second step, foliated flows are used to produce the fiber bundle trivializations whose fibers are the leaf closures; this works because there are foliated flows in any transverse direction since $\mathcal{F}$ is TP.

When $\mathcal{F}$ is minimal (the leaves are dense), any leaf closure $\widehat{M}_0$ of $\widehat{\mathcal{F}}$ is a principal subbundle of $\hat{\pi} : \widehat{M} \to M$, yielding the following:

Minimal case. If $\mathcal{F}$ is minimal and Riemannian, and M is compact, then, for some closed subgroup $H \subset \mathrm{O}(q)$, there is an H-principal bundle, $\hat{\pi}_0 : \widehat{M}_0 \to M$, with an H-invariant minimal Lie foliation, $\widehat{\mathcal{F}}_0$, such that $\widehat{\pi}_0$ is a foliated map, whose restrictions to the leaves are the holonomy covers of the leaves of $\mathcal{F}$.

A useful description of Lie foliations was also given by Fedida [22, 23], but it will not be considered here.

The differential structure cannot be used in our generalization; instead, we use the holonomy pseudogroup. Thus, let us briefly indicate the holonomy properties of Riemannian foliations, which will play an important role.

2.1 Holonomy of Riemannian Foliations

Recall that a *pseudogroup* is a maximal collection of local transformations of a space, which contains the identity map, and is closed under the operations of composition, inversion, restriction and combination. It can be considered as a generalized dynamical system, and all basic dynamical concepts have pseudogroup versions. They are relevant in foliation theory because the holonomy pseudogroup of a foliation $\mathcal{F}$ describes the transverse dynamics of $\mathcal{F}$. Such a pseudogroup is well determined up to certain *equivalence* of pseudogroups introduced by Haefliger [30, 31]. We may say that $\mathcal{F}$ is *transversely modeled* by a class of local transformations of some space if its holonomy pseudogroup can be generated by that type of local transformations. Riemannian, TP and Lie foliations can be respectively characterized by being transversely modeled by local isometries of some Riemannian manifold, local parallelism preserving diffeomorphisms of some parallelizable manifold, and local left translations of a Lie group. In this sense, Riemannian foliations are the transversely rigid ones, and TP foliations have a stronger type of transverse rigidity.

When the ambient manifold M is compact, Haefliger [32] has observed that the holonomy pseudogroup $\mathcal{H}$ of $\mathcal{F}$ satisfies a property called compact generation. If moreover $\mathcal{F}$ is Riemannian, then Haefliger [31, 32] has also strongly used the following properties of $\mathcal{H}$: completeness, quasi-analyticity, and existence of a closure $\overline{\mathcal{H}}$, which is also complete and quasi-analytic. Here, $\overline{\mathcal{H}}$ is defined by taking the closure of the set of 1-jets of maps in $\mathcal{H}$ in the space of 1-jets.

For a compactly generated pseudogroup $\mathcal{H}$ of local isometries of a Riemannian manifold T, Salem has given a version of Molino's theory [54], [44, Appendix D]. In particular, in the minimal case, it turns out that there is a Lie group G, a compact subgroup $K \subset G$ and a dense finitely generated subgroup $\Gamma \subset G$ such that $\mathcal{H}$ is equivalent to the pseudogroup generated by the action of Γ on the homogeneous space G/K (this was also observed by Haefliger [31]).

2.2 Growth of Riemannian Foliations

Molino's theory has many consequences for a Riemannian foliation $\mathcal{F}$ on a compact manifold M: classification in particular cases, growth, cohomology, tautness, tenseness and global analysis. In all of them, Molino's theory is used to reduce the study to the case of Lie foliations with dense leaves, where it usually becomes a problem of Lie theory. We concentrate on the consequences about growth of the leaves and their holonomy covers. This study was begun by Carrière [10], and recently continued by Breuillard-Gelander, as a consequence of their study of a topological Tits alternative [13]. Their results state the following, where $\mathfrak{g}$ is the structural Lie algebra of $\mathcal{F}$:

Carrière's Theorem. The holonomy covers of the leaves are Følner if and only if $\mathfrak{g}$ is solvable, and of polynomial growth if and only if $\mathfrak{g}$ is nilpotent. In the second case, the degree of their polynomial growth is bounded by the nilpotence degree of $\mathfrak{g}$.

Breuillard-Gelander's Theorem. The growth of the holonomy covers of the leaves is either polynomial or exponential.

2.3 Equicontinuous Foliated Spaces

A *foliated space* $X \equiv (X, \mathcal{F})$ is a topological space X equipped with a partition $\mathcal{F}$ into connected manifolds (*leaves*), which can be locally described as the fibers of topological submersions. It will be assumed that X is locally compact and Polish. A foliated space should be considered as a "topological foliation". In this sense, all topological notions of foliations have obvious versions for foliated spaces.

In particular, the *holonomy pseudogroup* $\mathcal{H}$ of X is defined on a locally compact Polish space T. Many basic results about foliations also have straightforward generalizations; for example, $\mathcal{H}$ is compactly generated if X is compact. Even leafwise differential concepts are easy to extend. However this task may be difficult or impossible for transverse differential concepts. For instance, the normal bundle of a foliated space does not make any sense in general: it would be the tangent bundle of a topological space in the case of a space foliated by points. Thus, the concept of Riemannian foliation cannot be extended by using the normal bundle; instead, this can be done via the holonomy pseudogroup as follows.

The transverse rigidity of a Riemannian foliation can be translated to the foliated space X by requiring equicontinuity of $\mathcal{H}$; in fact, the equicontinuity condition is not compatible with combinations of maps, and therefore it is indeed required for some generating subset $S \subset \mathcal{H}$ which is closed by the operations of composition and inversion; such an S is called a pseudo$*$group with the terminology of Matsumoto [39]. This gives rise to the concept of equicontinuous foliated space.

In the topological setting, the quasi-analyticity of $\mathcal{H}$ does not follow from the equicontinuity assumption. Thus it will be required as an additional assumption when needed. Indeed, it does not work well enough when T is not locally connected. So we use a property called strong quasi-analyticity, which is stronger than quasi-analyticity only when T is not locally connected.

Álvarez and Candel [4] have proved that, if $\mathcal{H}$ is compactly generated, equicontinuous, and strongly quasi-analytic, then it is complete and has a closure $\overline{\mathcal{H}}$. Here, $\overline{\mathcal{H}}$ is the pseudogroup generated by the homeomorphisms on small enough open subsets O of T that is limited in the compact-open topology of maps in $\mathcal{H}$ defined on those sets O.

Transitive and Lie foliations have the following topological versions. It is said that the foliated space X is:

- *homogeneous* if its group of foliated transformations acts transitively on X.
- *G-foliated space* if it is transversely modeled by local left translations in some locally compact Polish local group G (if X is minimal).

2.4 Topological Molino's Theory

Our first main result is the following topological version of the minimal case in Molino's theory.

Theorem 1 (Álvarez-Moreira [1]) *Let $X \equiv (X, \mathcal{F})$ be a compact Polish foliated space, and $\mathcal{H}$ its holonomy pseudogroup. Suppose that X is minimal and equicontinuous, and $\overline{\mathcal{H}}$ is strongly quasi-analytic. Then there is a compact Polish minimal foliated space $\widehat{X}_0 \equiv (\widehat{X}_0, \widehat{\mathcal{F}}_0)$, an open continuous foliated map $\hat{\pi}_0 : \widehat{X}_0 \to X$, and a locally compact Polish local group G such that $\widehat{X}_0$ is a G-foliated space, the fibers* *of $\hat{\pi}_0$ are homeomorphic to each other, and the restrictions of $\hat{\pi}_0$ to the leaves of $\widehat{\mathcal{F}}_0$ are the holonomy covers of the leaves of $\mathcal{F}$.*

The proof of Theorem 1 is different from Molino's proof in the Riemannian foliation case because there may not be the normal bundle of $\mathcal{F}$. To define $\widehat{X}_0$, we first construct what should be its holonomy pseudogroup, $\widehat{\mathcal{H}}_0$ on a space $\widehat{T}_0$. To some extent, this was achieved by Álvarez and Candel [5], proving that, with the assumptions of Theorem 1, there is a locally compact Polish local group G, a compact subgroup $K \subset G$ and a dense finitely generated sub-local group $\Gamma \subset G$ such that $\mathcal{H}$ is equivalent to the pseudogroup generated by the local action of Γ on G/K, like in the Riemannian foliation case. Hence $\widehat{\mathcal{H}}_0$ should be the pseudogroup generated by the local action of Γ on G. This may look like a big step towards the proof, but the realization of compactly generated pseudogroups as holonomy pseudogroups of compact foliated spaces is impossible in general, as shown by Meigniez [42]. This difficulty is overcome as follows.

Take a "good" cover of X by distinguished open sets, $\{U_i\}$, with corresponding distinguished submersions $p_i : U_i \to T_i$, and elementary holonomy transformations $h_{ij} : T_{ij} \to T_{ji}$, where $T_{ij} = p_i(U_i \cap U_j)$. Let $\mathcal{H}$ denote the corresponding representative of the holonomy pseudogroup on $T = \bigsqcup_i T_i$, generated by the maps h_{ij}. Then the construction of $\widehat{\mathcal{H}}_0$ must be associated to $\mathcal{H}$ in a natural way, so that it becomes induced by some "good" cover by distinguished open sets of a compact foliated space. In the Riemannian foliation case, the good choices of $\widehat{T}_0$ and $\widehat{\mathcal{H}}_0$ are the following ones:

- Let P be the bundle of orthonormal frames for any $\mathcal{H}$-invariant metric on T. Fix $x_0 \in T$ and $\hat{x}_0 \in P_{x_0}$. Then, as subspace of P,

$$\begin{aligned}\widehat{T}_0 &= \overline{\{\, h_*(\hat{x}_0) \mid h \in \mathcal{H},\ x_0 \in dom\, h \,\}} \\ &= \{\, g_*(\hat{x}_0) \mid g \in \overline{\mathcal{H}},\ x_0 \in dom\, g \,\}\,. \end{aligned} \tag{1}$$

- $\widehat{\mathcal{H}}_0$ is generated by the differentials of the maps in $\mathcal{H}$.

These differential concepts can be modified in the following way. In (1), each $g_*(\hat{x}_0)$ determines the germ of g at x_0, $\boldsymbol{\gamma}(g, x_0)$, by the strong quasi-analyticity of $\overline{\mathcal{H}}$. Therefore it also determines $\boldsymbol{\gamma}(f, x)$, where $f = g^{-1}$ and $x = g(x_0)$—this little change, using $\boldsymbol{\gamma}(f, x)$ instead of $\boldsymbol{\gamma}(g, x_0)$, is not really necessary, but it helps to simplify the notation in some involved arguments. So

$$\widehat{T}_0 \equiv \{\, \boldsymbol{\gamma}(f, x) \mid f \in \overline{\mathcal{H}},\ x \in dom\, f,\ f(x) = x_0 \,\}. \tag{2}$$

The projection $\hat{\pi}_0 : \widehat{T}_0 \to T$ corresponds via (2) to the source map $\boldsymbol{\gamma}(f, x) \mapsto x$. The differentials of maps $h \in \mathcal{H}$, acting on orthonormal references, correspond via (2) to the maps $\hat{h}$ defined by

$$\hat{h}(\boldsymbol{\gamma}(f, x)) = \boldsymbol{\gamma}(f h^{-1}, h(x))\,.$$

Let us describe the topology of $\widehat{T_0}$ using (2). Let $\overline{S}$ be a pseudo$*$group generating $\overline{\mathcal{H}}$ and satisfying the equicontinuity and strong quasi-analyticity conditions. Endow $\overline{S}$ with the compact-open topology on partial maps with open domains, as defined by Abd-Allah-Brown [7], and consider the subspace

$$\overline{S} * T = \{ (f, x) \in \overline{S} | x \in dom\ f \} \subset \overline{S} \times T.$$

Then the topology of $\widehat{T_0}$ corresponds via (2) to the quotient topology by the germ map $\boldsymbol{\gamma} : \overline{S} * T \to \boldsymbol{\gamma}(\overline{S} * T) \equiv \widehat{T_0}$, which is of course different from the sheaf topology on germs. This point of view, replacing orthonormal frames by germs, can be readily translated to the foliated space setting, obtaining good choices of $\widehat{T_0}$ and $\widehat{\mathcal{H}_0}$ under the conditions of Theorem 1.

Now, consider triples (x, γ, i), where $x \in U_i$, $\gamma \in \widehat{T}_{i,0} := \hat{\pi}_0^{-1}(T_i)$ and $p_i(x) = \hat{\pi}_0(\gamma)$. Declare (x, γ, i) equivalent to (y, δ, j) if $x = y$ and $\delta = \widehat{h_{ij}}(\gamma)$. Then $\widehat{X}_0$ is defined as the corresponding quotient space. Let $[x, \gamma, i]$ denote the equivalence class of each triple (x, γ, i). The foliated structure $\widehat{\mathcal{F}}_0$ on $\widehat{X}_0$ is determined by requiring that, for each fixed index i, the elements of the type $[x, \gamma, i]$ form a distinguished open set $\widehat{U}_{i,0}$, with distinguished submersion $\hat{p}_{i,0} : \widehat{U}_{i,0} \to \widehat{T}_{i,0}$ given by $\hat{p}_{i,0}([x, \gamma, i]) = \gamma$. The projection $\hat{\pi}_0 : \widehat{X}_0 \to X$ is defined by $\hat{\pi}_0([x, \gamma, i]) = x$. The properties stated in Theorem 1 are satisfied with these definitions.

It is also proved that, up to foliated homeomorphisms (respectively, local isomorphisms), $\widehat{X}_0$ (respectively, G) is independent of the choices involved. Hence G can be called the *structural local group* of $\mathcal{F}$.

2.5 Growth of Equicontinuous Foliated Spaces

Our second main result, is the following weak topological version of the above theorems of Carrière and Breuillard-Gelander.

Theorem 2 *Let X be a foliated space satisfying the conditions of Theorem 1, and let G be its structural local group. Then one of the following properties holds:*

- *G can be approximated by nilpotent local Lie groups; or*
- *the holonomy covers of all leaves of X have exponential growth.*

(The precise definition of *approximation* of a local group is given in Definition 19.) Like in the case of Riemannian foliations, Theorem 1 reduces the proof of Theorem 2 to the case of minimal G-foliated spaces, where it becomes a problem about local groups. Then, since any locally compact Polish local group can be approximated by local Lie groups in the above sense, the result follows by applying the same arguments as Breuillard-Gelander.

The paper is completed indicating some examples where Theorems 1 and 2 may have interesting applications, and propose some open problems.

3 Equicontinuous Pseudogroups

Definition 1 *(Sacksteder* [53], *Haefliger* [32]*)* A *pseudogroup* on a space T is a collection $\mathcal{H} \subset \mathrm{Loct}(T)$ (of local transformations) such that

- the identity map of T belongs to $\mathcal{H}$ ($\mathrm{id}_T \in \mathcal{H}$);
- if $h, h' \in \mathcal{H}$, then the composite $h'h$ is in $\mathcal{H}$ ($\mathcal{H}^2 \subset \mathcal{H}$);
- $h \in \mathcal{H}$ implies that $h^{-1} \in \mathcal{H}$ ($\mathcal{H}^{-1} \subset \mathcal{H}$);
- if $h \in \mathcal{H}$ and U is open in $dom\, h$, then the restriction $h : U \to h(U)$ is in $\mathcal{H}$;
- if a combination (union) of maps in $\mathcal{H}$ is defined and is a homeomorphism, then it is in $\mathcal{H}$.

Remark 1 The following properties hold:

- $\mathrm{id}_U \in \mathcal{H}$ for every open subset $U \subset T$.
- A local transformation $h \in \mathrm{Loct}(T)$ belongs to $\mathcal{H}$ if and only if it locally belongs to $\mathcal{H}$ (any point $x \in dom\, h$ has a neighborhood $V_x \subset dom\, h$ such that $h|_{V_x} \in \mathcal{H}$).
- Any intersection of pseudogroups on T is a pseudogroup on T.

Example 1 $\mathrm{Loct}(T)$ is the pseudogroup that contains any other pseudogroup on T.

Definition 2 A *sub-pseudogroup* of a pseudogroup $\mathcal{H}$ on T is a pseudogroup on T contained in $\mathcal{H}$. The *restriction* of $\mathcal{H}$ to an open subset $U \subset T$ is the pseudogroup

$$\mathcal{H}|_U = \{\, h \in \mathcal{H} \mid dom\, h \cup \mathrm{im}\, h \subset U \,\}\ .$$

The pseudogroup *generated* by a set $S \subset \mathrm{Loct}(T)$ is the intersection of all pseudogroups that contain S (the smallest pseudogroup on T containing S).

Definition 3 Let $\mathcal{H}$ be a pseudogroup on T. The *orbit* of each $x \in T$ is the set

$$\mathcal{H}(x) = \{\, h(x) \mid h \in \mathcal{H},\ x \in dom\, h \,\}\ .$$

The orbits form a partition of T. The space of orbits, equipped with the quotient topology, is denoted by $T/\mathcal{H}$. It is said that $\mathcal{H}$ is

- *(topologically) transitive* if some orbit is dense; and
- *minimal* when all orbits are dense.

The following notion, less restrictive than the concept of pseudogroup, is useful to study some properties of pseudogroups.

Definition 4 *(Matsumoto* [39]*)* A *pseudo∗group* on a space T is a family $S \subset \mathrm{Loct}(T)$ that is closed by the operations of composition and inversion.

Remark 2 Any intersection of pseudo∗groups on T is a pseudo∗group.

Definition 5 Any pseudo∗group contained in another pseudo∗group is called a *sub-pseudo∗group*. The pseudo∗group *generated* by a set $S_0 \subset \operatorname{Loct}(T)$ is the intersection of all pseudo∗groups containing S_0 (the smallest pseudo∗group containing S_0).

Remark 3 Let S be a pseudo∗group on T, and let S_1 be the collection of restrictions of all maps in S to all open subsets of their domains. Then S_1 is also a pseudo∗group on T, and S is a sub-pseudo∗group of S_1.

Definition 6 In Remark 3, it will be said that S_1 is the *localization* of S. If $S = S_1$, then the pseudo∗group S is called *local*.

Remark 4 Let $S_0 \subset \operatorname{Loct}(T)$. The pseudo∗group S generated by S_0 consists of all compositions of maps in S_0 and their inverses. The pseudogroup $\mathcal{H}$ generated by S_0 consists of all $h \in \operatorname{Loct}(T)$ that locally belong to the localization of S.

Remark 5 If two local pseudo∗groups, S_1 and S_2, generate the same pseudogroup $\mathcal{H}$, then $S_1 \cap S_2$ is also a local pseudo∗group that generates $\mathcal{H}$

Let $\mathcal{H}$ and $\mathcal{H}'$ be pseudogroups on respective spaces T and T'.

Definition 7 *(Haefliger* [30, 31]*)* A *morphism*[1] $\Phi : \mathcal{H} \to \mathcal{H}'$ is a maximal collection of homeomorphisms of open sets of T to open sets of T' such that

- if $\phi \in \Phi$, $h \in \mathcal{H}$ and $h' \in \mathcal{H}'$, then $h'\phi h \in \Phi$ ($\mathcal{H}'\Phi\mathcal{H} \subset \Phi$);
- the family of the domains of maps in Φ cover T; and
- if $\phi, \phi' \in \Phi$, then $\phi'\phi^{-1} \in \mathcal{H}'$ ($\Phi\Phi^{-1} \subset \mathcal{H}'$).

A morphism Φ is called an *equivalence* if the family $\Phi^{-1} = \{\phi^{-1} \mid \phi \in \Phi\}$ is also a morphism.

Remark 6 An equivalence $\Phi : \mathcal{H} \to \mathcal{H}'$ can be characterized as a maximal family of homeomorphisms of open sets of T to open sets of T' such that $\mathcal{H}'\Phi\mathcal{H} \subset \Phi$, and $\Phi^{-1}\Phi$ and $\Phi\Phi^{-1}$ generate $\mathcal{H}'$ and $\mathcal{H}'$, respectively.

Remark 7 Any morphism $\Phi : \mathcal{H} \to \mathcal{H}'$ induces a map between the corresponding orbit spaces, $T/\mathcal{H} \to T/\mathcal{H}'$. This map is a homeomorphism if Φ is an equivalence.

Definition 8 Let Φ_0 be a family of homeomorphisms of open subsets of T to open subsets of T' such that

- the union of domains of maps in Φ_0 meets all $\mathcal{H}$-orbits; and
- $\Phi_0\mathcal{H}\Phi_0^{-1} \subset \mathcal{H}'$.

Then there is a unique morphism $\Phi : \mathcal{H} \to \mathcal{H}'$ containing Φ_0, which is said to be *generated* by Φ_0. If moreover:

[1] This is usually called *étalé morphism*. We simply call it morphism because no other type of morphism will be considered here.

- the union of images of maps in Φ_0 meet all $\mathcal{H}'$-orbits; and
- $\Phi_0^{-1}\mathcal{H}\Phi_0 \subset \mathcal{H}$;

then Φ is an equivalence.

Definition 9 *(Haefliger* [32]*)* A pseudogroup $\mathcal{H}$ on a locally compact space T is said to be *compactly generated* if

- there is a relatively compact open subset $U \subset T$ meeting all $\mathcal{H}$-orbits;
- there is a finite set $S = \{h_1, \dots, h_n\} \subset \mathcal{H}|_U$ that generates $\mathcal{H}|_U$; and
- each h_i is the restriction of some $\tilde{h}_i \in \mathcal{H}$ with $\overline{dom\, h_i} \subset dom\, \tilde{h}_i$.

Remark 8 Compact generation is very subtle [24, 41]. Haefliger asked when compact generation implies realizability as a holonomy pseudogroup of a compact foliated space. The answer is not always affirmative [42].

Definition 10 *(Haefliger* [30]*)* A pseudogroup $\mathcal{H}$ is called *quasi-analytic* if every $h \in \mathcal{H}$ is the identity around some $x \in dom\, h$, whenever h is the identity on some open set whose closure contains x.

If a pseudogroup $\mathcal{H}$ on a space T is quasi-analytic, then every $h \in \mathcal{H}$ with connected domain is the identity on $dom\, h$ if it is the identity on some non-empty open set. Because of this, quasi-analyticity is interesting when T is locally connected, but local connectivity is too restrictive in our setting. Then, instead of requiring local connectivity, the following stronger version of quasi-analyticity will be used.

Definition 11 *(Álvarez-Candel* [4]*)* A pseudogroup $\mathcal{H}$ on a space T is said to be *strongly quasi-analytic* if it is generated by some sub-pseudo$*$group $S \subset \mathcal{H}$ such that any transformation in S is the identity on its domain if it is the identity on some non-empty open subset of its domain.

Remark 9 In [4], the term used for the above property is "quasi-effective". However the term "strongly quasi-analytic" seems to be more appropriate.

Remark 10 If the condition on $\mathcal{H}$ to be strongly quasi-analytic is satisfied with a sub-pseudo$*$group S, it is also satisfied with the localization of S. It follows that this property is hereditary by taking sub-pseudogroups and restrictions to open subsets.

Definition 12 *(Haefliger* [30]*)* A pseudogroup $\mathcal{H}$ on a space T is said to be *complete* if, for all $x, y \in T$, there are relatively compact open neighborhoods, U_x of x and V_y of y, such that, for all $h \in \mathcal{H}$ and $z \in U_x \cap dom\, h$ with $h(z) \in V_y$, there is some $g \in \mathcal{H}$ such that $dom\, g = U_x$ and with the same germ as h at z.

The pseudogroups considered from now on will be assumed to act on locally compact Polish[2] spaces; i.e., locally compact, Hausdorff and second countable spaces [36, Theorem 5.3].

[2] Recall that a space is called *Polish* if it is separable and completely metrizable.

4 Molino's Theory for Equicontinuous Foliated Spaces

4.1 Preliminaries on Equicontinuous Foliated Spaces

The needed basic concepts from equicontinuous foliated spaces can be seen in Moore and Schocher [46], Candel and Conlon [14, Chap. 11] and Ghys [24–26].

Let X and Z be locally compact Polish spaces. A *foliated chart* in X of *leaf dimension n*, *transversely modeled* on Z, is a pair (U, ϕ), where $U \subseteq X$ is open and $\phi : U \to B \times T$ is a homeomorphism for some open $T \subset Z$ and some open ball B in $\mathbb{R}^n$. It is said that U is a *distinguished open set*. The sets $P_y = \phi^{-1}(B \times \{y\})$ $(y \in T)$ are called *plaques* of this foliated chart. For each $x \in B$, the set $S_x = \phi^{-1}(\{x\} \times T)$ is called a *transversal* of the foliated chart. This local product structure defines a local projection $p : U \to T$, called *distinguished submersion*, given as composition of ϕ with the second factor projection $\pi_2 : B \times T \to T$.

Let $\mathcal{U} = \{U_i, \phi_i\}$ be a family of foliated charts in X of leaf dimension n modeled transversally on Z and covering X. Assume further that the foliated charts are *coherently foliated* in the sense that, if P and Q are plaques in different charts of $\mathcal{U}$, then $P \cap Q$ is open both in P and Q. Then $\mathcal{U}$ is called a *foliated atlas* on X of *leaf dimension n* and *transversely modeled* on Z. A maximal foliated atlas $\mathcal{F}$ of leaf dimension n and transversely modeled on Z is called a *foliated structure* on X of *leaf dimension n* and *transversely modeled* on Z. Any foliated atlas $\mathcal{U}$ of this type is contained in a unique foliated structure $\mathcal{F}$; then it is said that $\mathcal{U}$ *defines* (or is an atlas of) $\mathcal{F}$. If $Z = \mathbb{R}^m$, then X is a manifold of dimension $n + m$, and $\mathcal{F}$ is traditionally called a *foliation* of *dimension n* and *codimension m*. The reference to Z will be omitted.

For a foliated structure $\mathcal{F}$ on X of dimension n, the plaques form a basis of a topology on X called the *leaf topology*. With the leaf topology, X becomes an n-manifold whose connected components are called *leaves* of $\mathcal{F}$. $\mathcal{F}$ is determined by its leaves.

A foliated atlas $\mathcal{U} = \{U_i, \phi_i\}$ of $\mathcal{F}$ is called *regular* if

- each $\overline{U_i}$ is a compact subset of a foliated chart (W_i, ψ_i) and $\phi_i = \psi_i|_{U_i}$;
- the cover $\{U_i\}$ is locally finite; and,
- if (U_i, ϕ_i) and (U_j, ϕ_j) are elements of $\mathcal{U}$, then each plaque P of (U_i, ϕ_i) meets at most one plaque of (U_j, ϕ_j).

In this case, there are homeomorphisms $h_{ij} : T_{ij} \to T_{ji}$ such that $h_{ij} p_i = p_j$ on $U_i \cap U_j$, where $p_i : U_i \to T_i$ is the distinguished submersion defined by (U_i, ϕ_i) and $T_{ij} = p_i(U_i \cap U_j)$. Observe that the cocycle condition $h_{ik} = h_{jk} h_{ij}$ is satisfied on $T_{ijk} = p_i(U_i \cap U_j \cap U_k)$. For this reason, $\{U_i, p_i, h_{ij}\}$ is called a *defining cocycle* of $\mathcal{F}$ with values in Z—we only consider defining cocycles induced by regular foliated atlases. The equivalence class of the pseudogroup $\mathcal{H}$ generated by the maps h_{ij} on $T = \bigsqcup_{i \in I} T_i$ is called the *holonomy pseudogroup* of the foliated space $(X, \mathcal{F})$; $\mathcal{H}$ is the representative of the holonomy pseudogroup of $(X, \mathcal{F})$ induced by the defining cocycle $\{U_i, p_i, h_{ij}\}$. This T can be identified with a *total* (or *complete*) *transversal* to

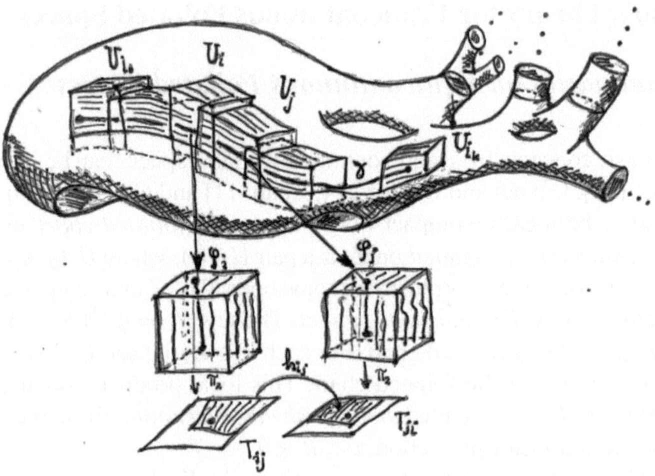

Fig. 1 The *transverse dynamics* of X modeled by the local Lie Groups G is described by its *holonomy pseudogroup*; the pseudogroup $\mathcal{H}$ generated by the local homeomorphisms $h_{ij} : T_{ij} \to T_{ji}$ on the total transverse space $T := \bigsqcup_i T_i$. The constituent elements of this pseudogroup are identified as holonomy transformations, and every T_i is an open subspace of G, and every h_{ij} is the restriction of some local left translation of G. **Arte Ciencia 1. Edison Moreira 2023**

the leaves in the sense that it meets all leaves and is locally given by the transversals defined by foliated charts. All compositions of maps h_{ij} form a pseudo∗group S that generates $\mathcal{H}$, called the *holonomy pseudo∗group* of $\mathcal{F}$ induced by $\{U_i, p_i, h_{ij}\}$. There is a canonical identity between the space of leaves and the space of $\mathcal{H}$-orbits, $X/\mathcal{F} \equiv T/\mathcal{H}$.

A foliated atlas (respectively, defining cocycle) contained in another one is called *sub-foliated atlas* (respectively, *sub-foliated cocycle*).

The *holonomy group* of each leaf L is defined as the germ group of the corresponding orbit. It can be considered as a quotient of $\pi_1(L)$ by taking "chains" of sets U_i along loops in L; this representation of $\pi_1(L)$ is called the *holonomy representation*. The kernel of the holonomy representation is equal to $q_*\pi_1(\widetilde{L})$ for a regular covering space $q : \widetilde{L} \to L$, which is called the *holonomy cover* of L. If $\mathcal{F}$ admits a countable defining cocycle, then the leaves in some dense G_δ subset of M have trivial holonomy groups [14, 33, 34], and therefore they can be identified with their holonomy covers (Fig. 1).

It is said that a foliated space is (*topologically*) *transitive* or *minimal* if any representative of its holonomy pseudogroup is such. Transitiviness (respectively, minimality) of a foliated space means that some leaf is dense (respectively, all leaves are dense).

Haefliger [32] has observed that, if X is compact, then $\mathcal{H}$ is compactly generated, which can be seen as follows. There is some defining cocycle $\{U_i', p_i', h_{ij}'\}$, with $p_i' : U_i' \to T_i'$, such that $\overline{U_i} \subset U_i'$, $T_i \subset T_i'$ and p_i' extends p_i. Therefore each h_{ij}' is an extension of h_{ij} so that $\overline{dom\, h_{ij}} \subset dom\, h'_{ij}$. Moreover $\mathcal{H}$ is the restriction to T of the pseudogroup $\mathcal{H}'$ on $T' = \bigsqcup_i T'_i$ generated by the maps h'_{ij}, and T is a relatively compact open subset of T' that meets all $\mathcal{H}'$-orbits.

Definition 13 It is said that a foliated space is *equicontinuous* if any representative of its holonomy pseudogroup is equicontinuous.

Remark 11 The above definition makes sense by Lemma 1.

Definition 14 Let G be a locally compact Polish local group. A minimal foliated space is called a *G-foliated space* if its holonomy pseudogroup can be represented by a pseudogroup given by Example 3 on a local group locally isomorphic to G.

4.2 Groupoid of Germs of a Pseudogroup

Definition 15 A *groupoid* $\mathfrak{G}$ is a small category where every morphism is an isomorphism. This means that $\mathfrak{G}$ is a set (of *morphisms*) equipped with the structure defined by an additional set T (of *objects*), and the following *structural* maps:

- the *source* and *target* maps $s, t : \mathfrak{G} \to T$;
- the *unit* map $T \to \mathfrak{G}$, $x \mapsto 1_x$;
- the *operation* (or *multiplication*) map $\mathfrak{G} \times_T \mathfrak{G} \to \mathfrak{G}$, $(\delta, \gamma) \mapsto \delta\gamma$, where

$$\mathfrak{G} \times_T \mathfrak{G} = \{\, (\delta, \gamma) \in \mathfrak{G} \times \mathfrak{G} \mid t(\gamma) = s(\delta) \,\} \subset \mathfrak{G} \times \mathfrak{G} \;;$$

- and the *inversion* map $\mathfrak{G} \to \mathfrak{G}$, $\gamma \mapsto \gamma^{-1}$;

such that the following conditions are satisfied:

- $s(\delta\gamma) = s(\gamma)$ and $t(\delta\gamma) = t(\delta)$ for all $(\delta, \gamma) \in \mathfrak{G} \times_T \mathfrak{G}$;
- for all $\gamma, \delta, \epsilon \in \mathfrak{G}$ with $t(\gamma) = s(\delta)$ and $t(\delta) = s(\epsilon)$, we have $\epsilon(\delta\gamma) = (\epsilon\delta)\gamma$ (associativity);
- $1_{t(\gamma)}\gamma = \gamma 1_{s(\gamma)} = \gamma$ (units or identity elements); and
- $s(\gamma) = t(\gamma^{-1})$, $t(\gamma) = s(\gamma^{-1})$, $\gamma^{-1}\gamma = 1_{s(\gamma)}$ and $\gamma\gamma^{-1} = 1_{t(\gamma)}$ for all $\gamma \in \mathfrak{G}$ (inverse elements).

Moreover, if $\mathfrak{G}$ and T are equipped with topologies so that all of the above structural maps are continuous, then $\mathfrak{G}$ is called a *topological groupoid*.

Remark 12 For a groupoid $\mathfrak{G}$, observe that $s(1_x) = t(1_x) = x$ for all $x \in T$, and therefore the source and target maps $s, t : \mathfrak{G} \to T$ are surjective, and the unit map $T \to \mathfrak{G}$ is injective. If moreover $\mathfrak{G}$ is a topological groupoid, then the unit map $T \to \mathfrak{G}$ is a topological embedding, and therefore the topology of T is determined by the topology of $\mathfrak{G}$; indeed, we can consider T as a subspace of $\mathfrak{G}$ if desired.

Definition 16 A topological groupoid is called *étalé* if the source and target maps are local homeomorphisms.

Let $\mathcal{H}$ be a pseudogroup on a space T. Note that the domain of the evaluation partial map $\mathrm{ev} : \mathcal{H} \times T \rightarrowtail T$ is

$$\mathcal{H} * T = \{ (h, x) \in \mathcal{H} \times T \mid x \in dom\, h \} \subset \mathcal{H} \times T \ .$$

Define an equivalence relation on $\mathcal{H} * T$ by setting $(h, x) \sim (h', x')$ if $x = x'$ and $h = h'$ on some neighborhood of x in $dom\ h\ \cap\ dom\ h'$. The equivalence class of each $(h, x) \in \mathcal{H} * T$ is called the *germ* of h at x, which will be denoted by $\boldsymbol{\gamma}(h, x)$. The corresponding quotient set is denoted by $\mathfrak{G}$, and the quotient map, $\boldsymbol{\gamma} : \mathcal{H} * T \to \mathfrak{G}$, is called the *germ map*. It is well known that $\mathfrak{G}$ is a groupoid with set of units T, where the source and target maps $s, t : \mathfrak{G} \to T$ are given by $s(\boldsymbol{\gamma}(h, x)) = x$ and $t(\boldsymbol{\gamma}(h, x)) = h(x)$, the unit map $T \to \mathfrak{G}$ is defined by $1_x = \boldsymbol{\gamma}(\mathrm{id}_T, x)$, the operation map $\mathfrak{G} \times_T \mathfrak{G} \to \mathfrak{G}$ is given by $\boldsymbol{\gamma}(g, h(x))\,\boldsymbol{\gamma}(h, x) = \boldsymbol{\gamma}(gh, x)$, and the inversion map is defined by $\boldsymbol{\gamma}(h, x)^{-1} = \boldsymbol{\gamma}(h^{-1}, h(x))$.

For $x, y \in T$, let us use the notation $\mathfrak{G}_x = s^{-1}(x)$, $\mathfrak{G}^y = t^{-1}(y)$ and $\mathfrak{G}_x^y = \mathfrak{G}_x \cap \mathfrak{G}^y$; in particular, the group $\mathfrak{G}_x^x$ will be called the *germ group* of $\mathcal{H}$ at x. Points in the same $\mathcal{H}$-orbit have isomorphic germ groups (if $y \in \mathcal{H}(x)$, an isomorphism $\mathfrak{G}_y^y \to \mathfrak{G}_x^x$ is given by conjugation with any element in $\mathfrak{G}_x^y$); hence the germ groups of the orbits make sense up to isomorphism. Under pseudogroup equivalences, corresponding orbits have isomorphic germ groups. The set $\mathfrak{G}_x$ will be called the *germ cover* of the orbit $\mathcal{H}(x)$ with basis point x. The target map restricts to a surjective map $\mathfrak{G}_x \to \mathcal{H}(x)$ whose fibers are bijective to $\mathfrak{G}_x^x$ (if $y \in \mathcal{H}(x)$, a bijection $\mathfrak{G}_x^x \to \mathfrak{G}_x^y$ is given by left product with any element in $\mathfrak{G}_x^y$); thus $\mathfrak{G}_x$ is finite if and only if both $\mathfrak{G}_x^x$ and $\mathcal{H}(x)$ are finite. Moreover germ covers basisd on points in the same orbit are also bijective (if $y \in \mathcal{H}(x)$, a bijection $\mathfrak{G}_y \to \mathfrak{G}_x$ is given by right product with any element in $\mathfrak{G}_x^y$); therefore the germ covers of the orbits make sense up to bijections.

Definition 17 It is said that $\mathcal{H}$ is:

- *locally free* if all of its germ groups are trivial; and
- *strongly locally free* if $\mathcal{H}$ is generated by a sub-pseudo$*$group $S \subset \mathcal{H}$ such that, for all $h \in S$ and $x \in dom\, h$, if $h(x) = x$ then $h = \mathrm{id}_{dom\, h}$.

Remark 13 The condition of being (strongly) locally free is stronger than the condition of being (strongly) quasi-analytic. If $\mathcal{H}$ is locally free and satisfies the condition of strong quasi-analyticity with a sub-pseudo$*$group $S \subset \mathcal{H}$, generating $\mathcal{H}$, then $\mathcal{H}$ also satisfies the condition of being strongly locally free with S.

Remark 14 If the condition on $\mathcal{H}$ to be strongly locally free is satisfied with a sub-pseudo$*$group S, then it is also satisfied with the localization of S. It follows that this property is hereditary by taking sub-pseudogroups and restrictions to open subsets

The *sheaf topology* on $\mathfrak{G}$ has a basis consisting of the sets $\{ \boldsymbol{\gamma}(h, x) \mid x \in dom\, h \}$ for $h \in \mathcal{H}$. Equipped with the sheaf topology, $\mathfrak{G}$ is an étalé groupoid.

Àlvares and Moreira introduce in [1] an additional topology on $\mathfrak{G}$, called the *bi-compact-open topology*.

4.3 Local Groups and Local Actions

The needed basic concepts from local groups and actions can be seen in [35].

Definition 18 A *local group* is a quintuple $G \equiv (G, e, \cdot, ', \mathfrak{D})$ satisfying the following conditions:

(1) $(G, \mathfrak{D})$ is a topological space.
(2) $\cdot$ is a function from a subset of $G \times G$ to G.
(3) $'$ is a function from a subset of G to G.
(4) There is a subset O of G such that

- O is an open neighborhood of e in G;
- $O \times O$ is a subset of the domain of $\cdot$;
- O is a subset of the domain of $'$;
- for all $a, b, c \in O$, if $a \cdot b, b \cdot c \in O$, then $(a \cdot b) \cdot c = (a \cdot b) \cdot c$;
- for all $a \in O, a' \in O, a \cdot e = e \cdot a = a$ and $a' \cdot a = a \cdot a' = e$;
- the map $\cdot : O \times O \to G$ is continuous; and
- the map $' : O \to G$ is continuous.

(5) The set $\{e\}$ is closed in G.

Asserting that a local group satisfies some topological property usually means that the property is satisfied on some open neighborhood of e.

A *local homomorphism* of a local group G to a local group H is a continuous partial map $\phi : G \rightarrowtail H$, whose domain is a neighborhood of e in G, which is compatible in the usual sense with the identity elements, the operations and inversions. If moreover ϕ restricts to a homeomorphism between some neighborhoods of the identities in G and H, then it is called a *local isomorphism*, and G and H are said to be *locally isomorphic*. A local group locally isomorphic to a Lie group is called a *local Lie group*.

The collection of all sets O satisfying (4) is denoted by ΨG. This is a neighborhood basis of e in G; all of these neighborhoods are symmetric with respect to the inverse operation (3). Let $\Phi(G, n)$ denote the collection of subsets A of G such that the product of any collection of at most n elements of A is defined, and the set A^n of such products is contained in some $O \in \Psi G$.

Let $H \subset G$. It is said that H is a *subgroup* of G if $H \in \Phi(G, 2)$, $e \in H$, $H' = H$ and $H^2 = H$; and H is a *sub-local group* of G if H is itself a local group with respect to the induced operations and topology.

Let ΥG denote the set of all pairs (H, V) of subsets of G so that $e \in H$, $V \in \Psi G$, $a \cdot b \in H$ for all $a, b \in V \cap H$, and $c' \in H$ for all $c \in V \cap H$. Then a subset $H \subset G$ is a sub-local group if and only if there exists some V such that $(H, V) \in \Upsilon G$ [35, Theorem 26].

Let ΠG denote the family of pairs (H, V) of subsets of G so that $e \in H$, $V \in \Psi G \cap \Phi(G, 6)$, $a \cdot b \in H$ for all $a, b \in V^6 \cap H$, $c' \in H$ for all $c \in V^6 \cap H$, and

$V^2 \setminus H$ is open. Given $(H, V) \in \Pi G$, there is a (completely regular, Hausdorff) space $G/(V, H)$ and a continuous open surjection $T : V^2 \to G/(V, H)$ such that $T(a) = T(b)$ if and only if $a' \cdot b \in H$ (cf. [35, Theorem 29]). For another pair in ΠG of the form (H, W), the spaces $G/(H, V)$ and $G/(H, W)$ are locally homeomorphic at the identity class. Thus the concept of coset space of H is well defined in this sense, as "a germ of a topological space". The notation G/H may be used in this sense. It will be said that G/H has a certain topological property when some $G/(H, V)$ has that property around $T(e)$.

Let ΔG be the set of pairs (H, U) such that $(H, U) \in \Pi G$ and, for all $a \in H \cap U^4$ and $b \in U^2$, $b' \cdot (a \cdot b) \in H$. A subset $H \subset G$ is called a normal sub-local group of G if there exists U such that $(H, U) \in \Delta G$. If $(H, U) \in \Delta G$ then the quotient space $G/(H, U)$ admits the structure of a local group (see [35, Theorem 35] for the pertinent details) and the natural projection $T : U^2 \to G/(H, U)$ is a local homomorphism. As before, another such pair (H, V) produces a locally isomorphic quotient local group.

As usual, $a \cdot b$ and a' will be denoted by ab and a^{-1}.

Local groups were first studied by Jacoby [35], giving local versions of important theorems for topological groups. For instance, Jacoby characterized local Lie groups as the locally compact local groups without small subgroups[3] [35, Theorem 96]. Also, any finite dimensional metrizable locally compact local group is locally isomorphic to the direct product of a Lie group and a compact zero-dimensional topological group [35, Theorem 107]. In particular, this property shows that any locally Euclidean local group is a local Lie group, which is an affirmative answer to a local version of Hilbert's 5th problem. However the proof of Jacoby is incorrect because he did not realize that, in local groups, associativity for three elements does not imply associativity for any finite sequence of elements [48, 51]. Fortunately, a completely new proof of the local Hilbert's 5th problem was given by Goldbring [27]. Moreover van den Dries and Goldbring [18, 19] proved that any locally compact local group is locally isomorphic to a topological group, and therefore all other theorems for local groups of Jacoby hold as well because they are known for locally compact topological groups as in the book of Montgomery and Zipping [45].

Definition 19 It is said that a local group G can be *approximated* by a class $\mathbb{C}$ of local groups if, for all $W \in \Psi G \cap \Phi(G, 2)$, there is some $V \in \Psi G$ and a sequence of compact normal subgroups $F_n \subset V$ such that $V \subset W$, $F_{n+1} \subset F_n$, $\bigcap_n F_n = \{e\}$, $(F_n, V) \in \Delta G$ and $G/(F_n, V) \in \mathbb{C}$.

Theorem 3 (Jacoby [35]) *Any locally compact second countable local group G can be approximated by local Lie groups.*

Definition 20 A *local action* of a local group G on a space X is a paro map $G \times X \rightarrowtail X$, $(g, x) \mapsto gx$, defined on some open neighborhood of $\{e\} \times X$, such that $ex = x$ for all $x \in X$, and $g_1(g_2x) = (g_1g_2)x$, provided both sides are defined.

[3] A local group is said to have no small subgroups when some neighborhood of the identity element contains no nontrivial subgroup.

Remark 15 The local transformations given by any local action of a local group on a space generate a pseudogroup.

A local action of a local group G on a space X is called *locally transitive* at some point $x \in X$ if there is a neighborhood W of e in G such that the local action is defined on $W \times \{x\}$, and $Wx := \{ gx \mid g \in W \}$ is a neighborhood of x in X. Given another local action of G on a space Y, a paro map $\phi : X \rightarrowtail Y$ is called *equivariant* is $\phi(gx) = g\phi(x)$ for all $x \in X$ and $g \in G$, provided both sides are defined.

Example 2 Let H be a sub-local group of G. If $(H, V) \in \Pi G$ and $T : V^2 \to G/(H, V)$ is the natural projection, then the map $V \times G/(H, V) \to G/(H, V)$, $(v, T(g)) \mapsto T(vg)$, defines a local action of G on $G/(H, V)$.

Remark 16 If G is a local group locally acting on X and the local action is locally transitive at $x \in X$, then there is a sub-local group H of G such that $(H, V) \in \Pi G$ for some V and the orbit paro map $G \rightarrowtail X$, $g \mapsto gx$, induces an equivariant paro map $G/(H, V) \rightarrowtail X$, which restricts to a homeomorphism between neighborhoods of $T(e)$ and x.

5 Equicontinuous Pseudogroups

Álvarez and Candel introduced the following structure to define equicontinuity for pseudogroups [4]. Let[4] $\{T_i, d_i\}$ be a family of metric spaces such that $\{T_i\}$ is a covering of a set T, each intersection $T_i \cap T_j$ is open in (T_i, d_i) and (T_j, d_j), and, for all $\epsilon > 0$, there is some $\delta(\epsilon) > 0$ so that the following property holds: for all i, j and $z \in T_i \cap T_j$, there is some open neighborhood $U_{i,j,z}$ of z in $T_i \cap T_j$ (with respect to the topology induced by d_i and d_j) such that

$$d_i(x, y) < \delta(\epsilon) \Longrightarrow d_j(x, y) < \epsilon$$

for all $\epsilon > 0$ and all $x, y \in U_{i,j,z}$. Such a family is called a *cover of T by quasi-locally equal metric spaces*. Two such families are called *quasi-locally equal* when their union is also a cover of T by quasi-locally equal metric spaces. This is an equivalence relation whose equivalence classes are called *quasi-local metrics* on T. For each quasi-local metric $\mathfrak{Q}$ on T, the pair $(T, \mathfrak{Q})$ is called a *quasi-local metric space*. Such a $\mathfrak{Q}$ induces a topology[5] on T so that, for each $\{T_i, d_i\}_{i \in I} \in \mathfrak{Q}$, the family of open balls of all metric spaces (T_i, d_i) form a basis of open sets. Any topological concept or property of $(T, \mathfrak{Q})$ refers to this underlying topology. $(T, \mathfrak{Q})$ is locally compact and Polish if and only if it is Hausdorff, paracompact and separable [4].

[4] The notation will be simplified by using, for instance, $\{T_i, d_i\}$ instead of $\{(T_i, d_i)\}$.

[5] In fact, it induces a uniformity. We could even use any uniformity to define equicontinuity, but such generality will not be used here.

Definition 21 *(Álvarez-Candel* [4]*)* Let $\mathcal{H}$ be a pseudogroup on a quasi-local metric space $(T, \mathfrak{Q})$. Then $\mathcal{H}$ is said to be (*strongly*[6]) *equicontinuous* if there exists some $\{T_i, d_i\}_{i\in I} \in \mathfrak{Q}$ and some sub-pseudo$*$group $S \subset \mathcal{H}$, generating $\mathcal{H}$, such that, for every $\epsilon > 0$, there is some $\delta(\epsilon) > 0$ so that

$$d_i(x, y) < \delta(\epsilon) \Longrightarrow d_j(h(x), h(y)) < \epsilon$$

for all $h \in S$, $i, j \in I$ and $x, y \in T_i \cap h^{-1}(T_j \cap \operatorname{im} h)$.

A pseudogroup $\mathcal{H}$ acting on a space T will be called (*strongly*) *equicontinuous* when it is equicontinuous with respect to some quasi-local metric inducing the topology of T.

Remark 17 If the condition on $\mathcal{H}$ to be equicontinuous is satisfied with a sub-pseudo$*$group S, then it is also satisfied with the localization of S. It follows that equicontinuity is hereditary by taking sub-pseudogroups and restrictions to open subsets.

Lemma 1 (Álvarez-Candel [4, Lemma 8.8]) *Let $\mathcal{H}$ and $\mathcal{H}'$ be equivalent pseudogroups on locally compact Polish spaces. Then $\mathcal{H}$ is equicontinuous if and only if $\mathcal{H}'$ is equicontinuous.*

Proposition 1 (Àlvarez-Candel [4, Proposition 8.9]) *Let $\mathcal{H}$ be a compactly generated and equicontinuous pseudogroup on a locally compact Polish quasi-local metric space $(T, \mathfrak{Q})$, and let U be any relatively compact open subset of $(T, \mathfrak{Q})$ that meets every $\mathcal{H}$-orbit. Suppose that $\{T_i, d_i\}_{i\in I} \in \mathfrak{Q}$ satisfies the condition of equicontinuity. Let E be any system of compact generation of $\mathcal{H}$ on U, and let $\bar{g}$ be an extension of each $g \in E$ with $\overline{\operatorname{dom} g} \subset \operatorname{dom} \bar{g}$. Also, let $\{T_i'\}_{i\in I}$ be any shrinking*[7] *of $\{T_i\}_{i\in I}$. Then there is a finite family $\mathcal{V}$ of open subsets of $(T, \mathfrak{Q})$ whose union contains U and such that, for any $V \in \mathcal{V}$, $x \in U \cap V$, and $h \in \mathcal{H}$ with $x \in \operatorname{dom} h$ and $h(x) \in U$, the domain of $\tilde{h} = \bar{g}_n \cdots \bar{g}_1$ contains V for any composite $h = g_n \cdots g_1$ defined around x with $g_1, \dots, g_n \in E$, and moreover $V \subset T_{i_0}'$ and $\tilde{h}(V) \subset T_{i_1}'$ for some $i_0, i_1 \in I$.*

Remark 18 The statement of Proposition 1 is stronger than the completeness of $\mathcal{H}|_U$. Since we can choose U large enough to contain two arbitrarily given points of T, it follows $\mathcal{H}$ is complete.

Proposition 2 (Álvarez-Candel [4, Proposition 9.9]) *Let $\mathcal{H}$ be a compactly generated, equicontinuous and strongly quasi-analytic pseudogroup on a locally compact Polish space T. Suppose that the conditions of equicontinuity and strong quasi-analyticity are satisfied with a sub-pseudo$*$group $S \subset \mathcal{H}$, generating $\mathcal{H}$. Let A, B*

---

[6] This adverb, used in [4], will be omitted for the sake of simplicity.

[7] A *shrinking* of an open cover $\{U_i\}$ of a space X is an open cover $\{U_i'\}$ of X, with the same index set, such that $\overline{U_i'} \subset U_i$ for all i. Similarly, if $\{U_i\}$ is a cover of a subset $A \subset X$ by open subsets of X, a *shrinking* of $\{U_i\}$, as a cover of A by open subsets of X, is a cover $\{U_i'\}$ of A by open subsets of X, with the same index set, so that $\overline{U_i'} \subset U_i$ for all i.

be open subsets of T such that $\overline{A}$ is compact and contained in B. If x and y are close enough points in T, then

$$f(x) \in A \Rightarrow f(y) \in B$$

for all $f \in S$ whose domain contains x and y.

Theorem 4 (Álvarez-Candel [4, Theorem 11.11]) *Let $\mathcal{H}$ be a compactly generated and equicontinuous pseudogroup on a locally compact Polish space T. If $\mathcal{H}$ is transitive, then $\mathcal{H}$ is minimal.*

Theorem 4 can be restated by saying that the orbit closures form a partition of the space. The following result states that indeed the orbit closures are orbits of a pseudogroup if strong quasi-analyticity is also assumed.

Theorem 5 (Álvarez-Candel [4, Theorem 12.1]) *Let $\mathcal{H}$ be a strongly quasi-analytic, compactly generated and equicontinuous pseudogroup on a locally compact Polish space T. Let $S \subset \mathcal{H}$ be a sub-pseudo$*$group generating $\mathcal{H}$ so that $\mathcal{H}$ satisfies the conditions of equicontinuity and strong quasi-analyticity with S. Let $\widetilde{\mathcal{H}}$ be the set of maps h between open subsets of T that satisfy the following property: for every $x \in \operatorname{dom} h$, there exists a neighborhood O_x of x in $\operatorname{dom} h$ so that the restriction $h|_{O_x}$ is in the closure of $C(O_x, T) \cap S$ in $C_{c\text{-}o}(O_x, T)$. Then:*

(i) *$\widetilde{\mathcal{H}}$ is closed by composition, combination and restriction to open sets;*
(ii) *any map in $\widetilde{\mathcal{H}}$ is a homeomorphism around every point of its domain;*
(iii) *$\overline{\mathcal{H}} = \widetilde{\mathcal{H}} \cap \operatorname{Loct}(T)$ is a pseudogroup that contains $\mathcal{H}$;*
(iv) *$\overline{\mathcal{H}}$ is equicontinuous;*
(v) *the orbits of $\overline{\mathcal{H}}$ are equal to the closures of the orbits of $\mathcal{H}$; and*
(vi) *$\widetilde{\mathcal{H}}$ and $\overline{\mathcal{H}}$ are independent of the choice of S.*

Remark 19 In Theorem 5, let $\overline{S}$ be the set of local transformations that are in the union of the closures of $C(O, T) \cap S$ in $C_{\text{c-o}}(O, T)$ with O running on the open sets of T. According to the proof of [4, Theorem 12.1], $\overline{S}$ is a pseudo$*$group that generates $\overline{\mathcal{H}}$. Moreover, if $\mathcal{H}$ satisfies the equicontinuity condition with S and some representative $\{T_i, d_i\}$ of a quasi-local metric, then $\overline{\mathcal{H}}$ satisfies the equicontinuity condition with $\overline{S}$ and $\{T_i, d_i\}$.

Remark 20 From the proof of [4, Theorem 12.1], it also follows that, with the notation of Remark 19, any $x \in \overline{U}$ has a neighborhood O in T such that the closure of

$$\{ h \in C(O, T) \cap S \mid h(O) \cap \overline{U} \neq \emptyset \}$$

in $C_{\text{c-o}}(O, T)$ is contained in $\operatorname{Loct}(T)$, and therefore in $\overline{S}$.

Example 3 Let G be a locally compact Polish local group with a left invariant metric, let $\Gamma \subset G$ be a dense sub-local group, and let $\mathcal{H}$ be the minimal pseudogroup generated by the local action of Γ by local left translations on G. The local left and right translations in G by each $g \in G$ will be denoted by L_g and R_g. The restrictions of the local left translations L_γ ($\gamma \in \Gamma$) to open subsets of their domains form a sub-pseudo∗group $S \subset \mathcal{H}$ that generates $\mathcal{H}$. Obviously, $\mathcal{H}$ satisfies with S the condition of being strongly locally free, and therefore strongly quasi-analytic. Moreover $\mathcal{H}$ satisfies with S the condition of being equicontinuous (indeed isometric) by considering any left invariant metric on G. Observe that any local right translation R_g ($g \in G$) generates an equivalence $\mathcal{H} \to \mathcal{H}$.

Now, suppose that $\mathcal{H}$ is compactly generated. Then $\overline{\mathcal{H}}$ is generated by the local action of G on itself by local left translations. The sub-pseudo∗group $\overline{S} \subset \overline{\mathcal{H}}$ consists of the restrictions of the local left translations L_g ($g \in G$) to open subsets of their domains. Observe that $\overline{\mathcal{H}}$ satisfies the condition of being strongly locally free, and therefore strongly quasi-analytic, with $\overline{S}$.

Lemma 2 (Álvarez-Moreira [1]) *Let G and G' be locally compact Polish local groups with left invariant metrics, let $\Gamma \subset G$ and $\Gamma' \subset G'$ be dense sub-local groups, and let $\mathcal{H}$ and $\mathcal{H}'$ be the pseudogroups generated by the local actions of Γ and Γ' by local left translations on G and G'. Suppose that $\mathcal{H}$ and $\mathcal{H}'$ are compactly generated. Then $\mathcal{H}$ and $\mathcal{H}'$ are equivalent if and only if G is locally isomorphic to G'.*

Proof Consider the notation and observations of Example 3 for both G and G'; in particular, $S \subset \mathcal{H}$ and $S' \subset \mathcal{H}'$ denote the sub-pseudo∗groups of restrictions of local translations L_γ and $L_{\gamma'}$ ($\gamma \in \Gamma$ and $\gamma' \in \Gamma'$) to open subsets of their domains. Let e and e' denote the identity elements of G and G'. Let $\Phi : \mathcal{H} \to \mathcal{H}'$ be en equivalence. Since $\mathcal{H}'$ is minimal, after composing Φ with the equivalence generated by some local right translation in G if necessary, we can assume that $\phi(e) = e'$ for some $\phi \in \Phi$ with $e \in dom\,\phi$.

Let U be a relatively compact open symmetric neighborhood of e in G with $\overline{U} \subset dom\,\phi$. Let $\{f_1, \dots, f_n\}$ be a symmetric system of compact generation of $\mathcal{H}$ on U. Thus each f_i has an extension $\tilde{f}_i \in \mathcal{H}$ so that $\overline{dom\,f_i} \subset dom\,\tilde{f}_i \subset dom\phi$.

Claim 1 We can assume that $\tilde{f}_i \in S$ and $\phi\tilde{f}_i\phi^{-1} \in S'$ for all i.

Each point in $dom\,\tilde{f}_i \cap dom\,\phi$ has an open neighborhood O such that $O \subset dom\,\tilde{f}_i$, $\tilde{f}_i|_O \in S$ and $\phi\tilde{f}_i\phi^{-1}|_{\phi(O)} \in S'$. Take a finite covering $\{O_{ij}\}$ ($j \in \{1, \dots, k_i\}$) of the compact set $\overline{dom\,f_i}$ by sets of this type. Let $\{P_{ij}\}$ be a shrinking of $\{O_{ij}\}$, as a cover of $\overline{dom\,f_i}$ by open subsets of $dom\,\tilde{f}_i$. Then the restrictions $g_{ij} = f_i|_{P_{ij} \cap U}$ ($i \in \{1, \dots, n\}$ and $j \in \{1, \dots, k_i\}$) generate $\mathcal{H}|_U$, each $\tilde{g}_{ij} = \tilde{f}_i|_{O_{ij}}$ is in S and extends g_{ij}, $\overline{dom\,g_{ij}} \subset dom\,\tilde{g}_{ij}$, and $\phi\tilde{g}_{ij}\phi^{-1} \in S'$, showing Claim 1.

According to Claim 1, the maps $f'_i = \phi f_i \phi^{-1}$ form a symmetric system of compact generation of $\mathcal{H}'$ on $U' = \phi(U)$, which can be checked with the extensions $\tilde{f}'_i = \phi\tilde{f}_i\phi^{-1}$. Let $S_0 \subset S$ and $S'_0 \subset S'$ be the sub-pseudo∗groups consisting of the restrictions of compositions of maps f_i and f'_i to open subsets of their domains, respectively. They generate $\mathcal{H}$ and $\mathcal{H}'$. It follows from Claim 1 that $\phi f \phi^{-1} \in S'$ for all $f \in S_0$. On the other hand, by Proposition 1, there is a smaller open neighborhood of the identity, $V \subset U$, such that, for all $h \in \mathcal{H}$ and $x \in V \cap dom\,h$ with $h(x) \in U$, there is some $f \in S_0$ such that $dom\,f = V$ and $\boldsymbol{\gamma}(f, x) = \boldsymbol{\gamma}(h, x)$.

Let W be another symmetric open neighborhood of the identity such that $W^2 \subset V$. Let us show that $\phi : W \to \phi(W)$ is a local isomorphism. Let $\gamma \in W \cap \Gamma$. The restriction $L_\gamma : W \to \gamma W$ is well defined and belongs to S. Hence there is some $f \in S_0$ so that $dom\ f = V$ and $\boldsymbol{\gamma}(f, e) = \boldsymbol{\gamma}(L_\gamma, e)$. Since f is also a restriction of a local left translation in G, it follows that $f = L_\gamma$ on W. So $\phi L_\gamma \phi^{-1}|_{\phi(W)} \in S'$; i.e., there is some $\gamma' \in \Gamma'$ such that $\phi L_\gamma \phi^{-1} = L_{\gamma'}$ on $\phi(W)$. In fact,

$$\phi(\gamma) = \phi L_\gamma(e) = \phi L_\gamma \phi^{-1}(e') = L_{\gamma'}(e') = \gamma' .$$

Hence, for all $\gamma, \delta \in \Gamma$,

$$\begin{aligned} \phi(\gamma\delta) &= \phi L_\gamma(\delta) = L_{\phi(\gamma)}\phi(\delta) = \phi(\gamma)\phi(\delta) , \\ \phi(\gamma)^{-1} &= L^{-1}_{\phi(\gamma)}(e') = (\phi L_\gamma \phi^{-1})^{-1}(e') \\ &= \phi L_{\gamma^{-1}}\phi^{-1}(e') = L_{\phi(\gamma^{-1})}(e') = \phi(\gamma^{-1}) . \end{aligned}$$

Since ϕ, and the product and inversion maps are continuous, it follows that $\phi(gh) = \phi(g)\phi(h)$ and $\phi(g^{-1}) = \phi(g)^{-1}$ for all $g, h \in W$. □

Example 4 This generalizes Example 3. Let G be a locally compact Polish local group with a left-invariant metric, $K \subset G$ a compact subgroup, and $\Gamma \subset G$ a dense sub-local group. Take some V so that $(H, V) \in \Pi(G)$. The left invariant metric on G can be assumed to be also K-right invariant by the compactness of K, and therefore it defines a metric on $G/(K, V)$. Then the canonical local action of Γ on some neighborhood of the identity class in $G/(K, V)$ induces a transitive equicontinuous pseudogroup $\mathcal{H}$ on a locally compact Polish space; in fact, this is a pseudogroup of local isometries.

Assume that $\mathcal{H}$ is compactly generated. Then $\overline{\mathcal{H}}$ is generated by the canonical local action of G on some neighborhood of the identity class in $G/(K, V)$. Moreover the sub-pseudo∗group $\overline{S} \subset \overline{\mathcal{H}}$ consists of the local translations of the local action of G on $G/(K, V)$.

Examples 3 and 4 are particular cases of pseudogroups induced by local actions (Remark 15). The following result indicates their relevance.

Theorem 6 (Álvarez-Candel [5, Theorem 5.2]) *Let $\mathcal{H}$ be a transitive, compactly generated and equicontinuous pseudogroup on a locally compact Polish space, and suppose that $\overline{\mathcal{H}}$ is strongly quasi-analytic. Then $\mathcal{H}$ is equivalent to a pseudogroup of the type described in Example 4.*

Remark 21 From the proof of [5, Theorems 3.3 and 5.2], it also follows that, in Theorem 6, if moreover $\overline{\mathcal{H}}$ is strongly locally free, then $\mathcal{H}$ is equivalent to a pseudogroup of the type described in Example 3.

6 Growth of Equicontinuous Pseudogroups and Foliated Spaces

6.1 Coarse Quasi-isometries and Growth of Metric Spaces

A *net* in a metric space M, with metric d, is a subset $A \subset M$ that satisfies $d(x, A) \leq C$ for some $C > 0$ and all $x \in M$; the term C-*net* is also used. A *coarse quasi-isometry* between M and another metric space M' is a bi-Lipschitz bijection between nets of M and M'; in this case, M and M' are said to be *coarsely quasi-isometric* (in the sense of Gromov) [29]. If such a bi-Lipschitz bijection, as well as its inverse, has dilation $\leq \lambda$, and it is defined between C-nets, then it will be said that the coarse quasi-isometry has *distortion* (C, λ). A family of coarse quasi-isometries with a common distortion will be called a family of *equi-coarse quasi-isometries*, and the corresponding metric spaces are called *equi-coarsely quasi-isometric*.

The version of growth for metric spaces given here is taken from [6].

Recall that, given non-decreasing functions[8] $u, v : [0, \infty) \to [0, \infty)$, it is said that u is *dominated* by v, written $u \preccurlyeq v$, when there are $a, b \geq 1$ and $c \geq 0$ such that $u(r) \leq a\, v(br)$ for all $r \geq c$. If $u \preccurlyeq v \preccurlyeq u$, then it is said that u and v represent the same *growth type* or have *equivalent growth*; this is an equivalence relation, and "$\preccurlyeq$" defines a partial order relation between growth types called *domination*. For a family of pairs of non-decreasing functions $[0, \infty) \to [0, \infty)$, *equi-domination* means that those pairs satisfy the above condition of domination with the same constants a, b, c. A family of functions $[0, \infty) \to [0, \infty)$ will be said to have *equi-equivalent growth* if they equi-dominate one another.

For a complete connected Riemannian manifold L, the growth type of each mapping $r \mapsto \mathrm{vol}B(x, r)$ is independent of x and is called the *growth type* of L. Another definition of *growth type* can be similarly given for metric spaces whose bounded sets are finite, where the number of points is used instead of the volume.

Let M be a metric space with metric d. A *quasi-lattice* Γ of M is a C-net of M for some $C \geq 0$ such that, for every $r \geq 0$, there is some $K_r \geq 0$ such that $card(\Gamma \cap B(x, r)) \leq K_r$ for every $x \in M$. It is said that M is of *coarse bounded geometry* if it has a quasi-lattice. In this case, the *growth type* of M can be defined as the growth type of any quasi-lattice Γ of M; i.e., it is the growth type of the *growth function* $r \mapsto v_\Gamma(x, r) = card(B(x, r) \cap \Gamma)$ for any $x \in \Gamma$. This definition is independent of Γ.

For a family of metric spaces, if they satisfy the above condition of coarse bounded geometry with the same constants C and K_r, then they are said to have *equi-coarse bounded geometry*. If moreover the lattices involved in this condition have growth functions with equi-equivalent growth, then these metric spaces are said to have *equi-equivalent growth*.

---

[8] Usually, growth types are defined by using non-decreasing functions $\mathbb{Z}^+ \to [0, \infty)$, but non-decreasing functions $[0, \infty) \to [0, \infty)$ give rise to an equivalent concept.

The condition of coarse bounded geometry is satisfied by complete connected Riemannian manifolds of bounded geometry, and also by discrete metric spaces with a uniform upper bound on the number of points in all balls of each given radius, J. Block and S. Weinberger [12]. In those cases, the two given definitions of growth type are equal.

Lemma 3 (Álvarez-Candel [4]) *Two coarsely quasi-isometric metric spaces of coarse bounded geometry have the same growth type. Moreover, if a family of metric spaces are equi-coarsely quasi-isometric to each other, then they have equi-equivalent growth.*

6.2 Quasi-isometry and Growth Types of Orbits

Let $\mathcal{H}$ be a pseudogroup on a space T, and E a symmetric set of generators of $\mathcal{H}$. Let $\mathfrak{G}$ be the groupoid of germs of maps in $\mathcal{H}$.

For each $h \in \mathcal{H}$ and $x \in dom\, h$, let $|h|_{E,x}$ be the length of the shortest expression of $\boldsymbol{\gamma}(h, x)$ as a product of germs of maps in E (being 0 if $\boldsymbol{\gamma}(h, x) = \boldsymbol{\gamma}(\mathrm{id}_T, x)$). For each $x \in T$, define metrics d_E on $\mathcal{H}(x)$ and $\mathfrak{G}_x$ by

$$d_E(y, z) = \min\{\, |h|_{E,y} \mid h \in \mathcal{H},\ y \in dom\, h,\ h(y) = z \,\}\,,$$
$$d_E(\boldsymbol{\gamma}(f, x), \boldsymbol{\gamma}(g, x)) = |fg^{-1}|_{E,g(x)}\ .$$

Notice that

$$d_E(f(x), g(x)) \le d_E(\boldsymbol{\gamma}(f, x), \boldsymbol{\gamma}(g, x)).$$

Moreover, on the germ covers, d_E is right invariant in the sense that, if $y \in \mathcal{H}(x)$, the bijection $\mathfrak{G}_y \to \mathfrak{G}_x$, given by right multiplication with any element in $\mathfrak{G}_x^y$, is isometric; so the isometry types of the germ covers of the orbits make sense without any reference to basis points. In fact, the definition of d_E on $\mathfrak{G}_x$ is analogous to the definition of the right invariant metric d_S on a group Γ induced by a symmetric set of generators S: $d_S(\gamma, \delta) = |\gamma\delta^{-1}|$ for $\gamma, \delta \in \Gamma$, where $|\gamma|$ is the length of the shortest expression of γ as a product of elements of S (being 0 if $\gamma = e$).

Assume that $\mathcal{H}$ is compactly generated and T locally compact. Let $U \subset T$ be a relatively compact open subset that meets all $\mathcal{H}$-orbits, let $\mathcal{G} = \mathcal{H}|_U$, and let E be a symmetric system of compact generation of $\mathcal{H}$ on U. With these conditions, the quasi-isometry type of the $\mathcal{G}$-orbits with d_E may depend on E [4, Sect. 6]. So the following additional condition on E is considered.

Definition 22 *(Álvarez-Candel* [4, Definition 4.2]*)* With the above notation, it is said that E is *recurrent* if, for any relatively compact open subset $V \subset U$ that meets all $\mathcal{G}$-orbits, there exists some $R > 0$ such that $\mathcal{G}(x) \cap V$ is an R-net in $\mathcal{G}(x)$ with d_E for all $x \in U$.

The role played by V in Definition 22 can be played by any relatively compact open subset meeting all orbits [4, Lemma 4.3]. Furthermore there exists a recurrent system of compact generation on U [4, Corollary 4.5].

An obvious modification of the arguments of the proof of [4, Theorem 4.6] gives the following.

Theorem 7 (Álvarez-Moreira [1, Theorem 5.4]) *Let $\mathcal{H}$ and $\mathcal{H}'$ be compactly generated pseudogroups on locally compact spaces T and T', let U and U' be relatively compact open subsets of T and T' that meet all orbits of $\mathcal{H}$ and $\mathcal{H}'$, let $\mathcal{G}$ and $\mathcal{G}'$ denote the restrictions of $\mathcal{H}$ and $\mathcal{H}'$ to U and U', and let E and E' be recurrent symmetric systems of compact generation of $\mathcal{H}$ and $\mathcal{H}'$ on U and U', respectively. Suppose that there exists an equivalence $\mathcal{H} \to \mathcal{H}'$, and consider the induced equivalence $\mathcal{G} \to \mathcal{G}'$ and homeomorphism $U/\mathcal{G} \to U'/\mathcal{G}'$. Then the germ covers of the $\mathcal{G}$-orbits with d_E are equicoarsely quasi-isometric to the germ covers of the corresponding $\mathcal{G}'$-orbits with $d_{E'}$.*

Corollary 1 *With the notation and conditions of Theorem 7, the corresponding orbits of $\mathcal{G}$ and $\mathcal{G}'$, as well as their germ covers, have equi-equivalent growth.*

Proof This follows from Theorem 7. □

Example 5 Let G be a locally compact Polish local group with a left-invariant metric, let $\Gamma \subset G$ be a dense finitely generated sub-local group, and let $\mathcal{H}$ denote the pseudogroup generated by the local action of Γ on G by local left translations. Suppose that $\mathcal{H}$ is compactly generated, and let $\mathcal{G} = \mathcal{H}|_U$ for some relatively compact open neighborhood U of the identity element e in G, which meets all $\mathcal{H}$-orbits because Γ is dense. For every $\gamma \in \Gamma$ with $\gamma U \cap U \neq \emptyset$, let h_γ denote the restriction $U \cap \gamma^{-1}U \to \gamma U \cap U$ of the local left translation by γ. There is a finite symmetric set $S = \{s_1, \dots, s_k\} \subset \Gamma$ such that $E = \{h_{s_1}, \dots, h_{s_k}\}$ is a recurrent system of compact generation of $\mathcal{H}$ on U; in fact, by reducing Γ if necessary, we can assume that S generates Γ. The recurrence of E means that there is some $N \in \mathbb{N}$ such that

$$U = \bigcup_{h \in E^N} h^{-1}(V \cap \operatorname{im} h) , \tag{3}$$

where E^N is the family of compositions of at most N elements of E.

For each $x \in U$, let

$$\Gamma_{U,x} = \{\, \gamma \in \Gamma \mid \gamma x \in U \,\} .$$

Let $\mathfrak{G}$ denote the topological groupoid of germs of $\mathcal{G}$. The map $\Gamma_{U,x} \to \mathfrak{G}_x$, $\gamma \mapsto \boldsymbol{\gamma}(h_\gamma, x)$, is bijective. For $\gamma \in \Gamma_{U,x}$, let $|\gamma|_{S,U,x} := |h_\gamma|_{E,x}$. Thus $|e|_{S,U,x} = 0$, and, if $\gamma \neq e$, then $|\gamma|_{S,U,x}$ equals the minimum $n \in \mathbb{N}$ such that there are $i_1, \dots, i_n \in \{1, \dots, k\}$ with $\gamma = s_{i_n} \cdots s_{i_1}$ and $s_{i_m} \cdots s_{i_1} \cdot x \in U$ for all $1 \le m \le n$. Moreover d_E on $\mathfrak{G}_x$ corresponds to the metric $d_{S,U,x}$ on $\Gamma_{U,x}$ given by

$$d_{S,U,x}(\gamma, \delta) = |\delta\gamma^{-1}|_{S,U,\gamma(x)} .$$

Observe that, for all $\gamma \in \Gamma_{U,x}$ and $\delta \in \Gamma_{U,\gamma\cdot x}$,

$$\delta\gamma \in \Gamma_{U,x}\,, \qquad |\delta\gamma|_{S,U,x} \le |\gamma|_{S,U,x} + |\delta|_{S,U,\gamma\cdot x}\,, \tag{4}$$

$$\gamma^{-1} \in \Gamma_{U,\gamma\cdot x}\,, \qquad |\gamma|_{S,U,x} = |\gamma^{-1}|_{S,U,\gamma\cdot x}\,. \tag{5}$$

In this example, we will be interested on the growth type of the orbits of $\mathcal{G}$ with d_E, or, equivalently, the growth type of the metric spaces $(\Gamma_{U,x}, d_{S,U,x})$. The following result was used by Breuillard-Gelander to study this growth type when G is a Lie group.

Proposition 3 (Breuillard-Gelander [13, Proposition 10.5]) *Let G be a non-nilpotent connected real Lie group and Γ a finitely generated dense subgroup. For any finite set $S = \{s_1, \dots, s_k\}$ of generators of Γ, and any neighborhood B of e in G, there are elements $t_i \in \Gamma \cap s_i B$ ($i \in \{1, \dots, k\}$) which freely generate a free semi-group. If G is not solvable, then we can choose the elements t_i so that they generate a free group.*

6.3 Growth of Equicontinuous Pseudogroups

Let G be a locally compact Polish local group with a left-invariant metric, let $\Gamma \subset G$ be a dense finitely generated sub-local group, and let $\mathcal{H}$ denote the pseudogroup generated by the local action of Γ on G by local left translations. Suppose that $\mathcal{H}$ is compactly generated. Let $\mathcal{G} = \mathcal{H}|_U$ for some relatively compact open neighborhood U of the identity element e in G, which meets all $\mathcal{H}$-orbits because Γ is dense. Let E be a recurrent symmetric system of compact generation of $\mathcal{H}$ on U. Let $\mathfrak{G}$ be the groupoid of germs of maps in $\mathcal{G}$.

Theorem 8 (Álvarez-Moreira [1]) *With the above notation and conditions, one of the following properties holds:*

- *G can be approximated by nilpotent local Lie groups; or*
- *the germ covers of all $\mathcal{G}$-orbits have exponential growth with d_E.*

Proof By Theorem 3, there is some $U_0 \in \Psi G$, contained in any given element of $\Psi G \cap \Phi(G, 2)$, and there exists a sequence of compact normal subgroups $F_n \subset U_0$ such that $F_{n+1} \subset F_n$, $\bigcap_n F_n = \{e\}$, $(F_n, U_0) \in \Delta G$, and $G/(F_n, U_0)$ is a local Lie group. Let $T_n : U_0^2 \to G/(F_n, U_0)$ denote the canonical projection. Take an open neighborhood U_1 of e such that $\overline{U_1} \subset U_0$. Then $F_n\overline{U_1} \subset U_0$ for n large enough by the properties of the sequence F_n. Let $U_2 = F_n U_1$ for such an n. Thus U_2 is saturated by the fibers of T_n, and $\overline{U_2} \subset U_0$. Then $U' := T_n(U_2)$ is a relatively compact open neighborhood of the identity in the local Lie group $G' := G/(F_n, U_0)$. Let $\Gamma' = T_n(\Gamma \cap U_0^2)$, which is a dense sub-local group of G', and let $\mathcal{H}'$ denote the pseudogroup on G' generated by the local action of Γ' by local left translations.

For every $\gamma \in \Gamma \cap U_0$ with $\gamma U_2 \cap U_2 \neq \emptyset$, let h_γ denote the restriction $U_2 \cap \gamma^{-1}U_2 \to \gamma U_2 \cap U_2$ of the local left translation by γ. There is a finite symmetric set $S = \{s_1, \dots, s_k\} \subset \Gamma$ such that $E_2 = \{h_{s_1}, \dots, h_{s_k}\}$ is a recurrent system of compact generation of $\mathcal{H}$ on U_2. By reducing Γ if necessary, we can suppose that S generates Γ. For every $\delta \in \Gamma'$ with $\delta U' \cap U' \neq \emptyset$, let h'_δ denote the restriction $U' \cap \delta^{-1}U' \to \delta U' \cap U'$ of the local left translation by δ. We can assume that $s_1, \dots, s_k$ are in U_2, and therefore we can consider their images $s'_1, \dots, s'_k$ by T_n. Moreover each h_{s_i} induces via T_n the map h'_{s_i}, and $E' = \{h'_{s_1}, \dots, h'_{s_k}\}$ is a system of compact generation of $\mathcal{H}'$ on U'. By increasing E_2 if necessary, we can assume that E' is also recurrent. Fix any open set V' in G' with $\overline{V'} \subset U'$. Then $V = T_n^{-1}(V')$ satisfies $\overline{V} \subset U_2$.

Claim 2 For each finite subset $F \subset \Gamma \cap U_2$, we have $U_2 \subset \bigcup_{\gamma \in \Gamma \setminus F} \gamma V$.

Since U_2 and V are saturated by the fibers of T_n, Claim 2 follows by showing that $U' \subset \bigcup_{\gamma \in \Gamma' \setminus F'} \gamma V'$, where $F' = T_n(F)$. Suppose that this inclusion is false. Then there is some finite symmetric subset $F \subset \Gamma \cap U_2$ and some $x \in U'$ such that $((\Gamma' \setminus F')x) \cap V' = \emptyset$. By the recurrence of E', there is some $N \in \mathbb{N}$ satisfying (3) with U' and E'. Since $\Gamma'_{U',x}$ is infinite because Γ' is dense in G', it follows that there is some $\gamma \in \Gamma'_{U',x} \setminus F'$ such that

$$|\gamma|_{S',U',x} > N + \max\{\, |\epsilon|_{S',U',x} | \epsilon \in F' \cap \Gamma'_{U',x} \,\} . \tag{6}$$

By (3), there is some $h \in E'^N$ such that $\gamma x \in h^{-1}(V' \cap \operatorname{im} h')$. We have $h = h'_\delta$ for some $\delta \in \Gamma'$. Note that $\delta \in \Gamma'_{U',\gamma' x}$ and $|\delta|_{S',U',\gamma x} \leq N$. Hence

$$\begin{aligned} |\gamma|_{S',U',x} \leq |\delta\gamma|_{S',U',x} &+ |\delta^{-1}|_{S',U',\delta\gamma x} \\ &= |\delta\gamma|_{S',U',x} + |\delta|_{S',U',\gamma' x} \leq |\delta\gamma|_{S',U',x} + N \end{aligned}$$

by (4) and (5), obtaining that $\delta\gamma \notin F'$ by (6). However, $\delta\gamma x \in V'$, obtaining a contradiction, which completes the proof of Claim 2.

Claim 3 For each finite subset $F \subset \Gamma \cap U_2$, we have $\overline{U_2} \subset \bigcup_{\gamma \in \Gamma \setminus F} \gamma V$.

Take a relatively compact open subset $O_1 \subset G$ such that $\overline{U_1} \subset O_1$ and $F_n \overline{O_1} \subset U_0$. Let $O_2 = F_n O_1$ and $\mathcal{K} = \mathcal{H}|_{O_2}$. Then Claim 3 follows by applying Claim 2 to O_2.

According to Claim 3, by increasing S if necessary, we can suppose that

$$\overline{U_2} \subset \bigcup_{i<j} (s_i \cdot V \cap s_j \cdot V) = \bigcup_{i<j} (s_i^{-1} \cdot V \cap s_j^{-1} \cdot V) . \tag{7}$$

Suppose that G cannot be approximated by nilpotent local Lie groups. Then we can assume that the local Lie group G' is not nilpotent. Moreover we can suppose that G' is a sub-local Lie group of a simply connected Lie group L. Let Δ be the dense subgroup of L whose intersection with G' is Γ'. Then, by Proposition 3, there are elements $t'_1, \dots, t'_k$ in Δ, as close as desired to $s'_1, \dots, s'_k$, which are free generators of a free semi-group. If the elements t_i' are close enough to s_i', then they are in U'. So there are elements $t_i \in U_2$ such that $T_n(t_i) = t_i'$. By the compactness of $\overline{U_2}$, and because U_2 and V are saturated by the fibers of T_n, if $t_1', \dots, t_k'$ are close enough to $s_1', \dots, s_k'$, then (7) gives

$$\overline{U_2} \subset \bigcup_{i<j} (t_i^{-1} V \cap t_j^{-1} V) . \tag{8}$$

We now adapt the argumentation of [13, Lemma 10.6] proof. Let $\widehat{\Gamma} \subset \Gamma$ be the sub-local group generated by $t_1, \dots, t_k$; thus $\widehat{S} = \{t_1^{\pm 1}, \dots, t_k^{\pm 1}\}$ is a symmetric set of generators of $\widehat{\Gamma}$, and $S \cup \widehat{S}$ is a symmetric set of generators of Γ. With $\widehat{E} = \{h_{t_1}^{\pm 1}, \dots, h_{t_k}^{\pm 1}\}$, observe that $E_2 \cup \widehat{E}$ is a recurrent system of compact generation of $\mathcal{H}$ on U_2. Given $x \in U_2$, let $\mathsf{S}(n)$ be the sphere with center e and radius $n \in \mathbb{N}$ in $\widehat{\Gamma}_{U_2,x}$ with $d_{\widehat{S},U,x}$. By (8), for each $\gamma \in \mathsf{S}(n)$, there are indices $i < j$ such that $\gamma x \in t_i^{-1} V \cap t_j^{-1} V$. So the points $t_i \gamma x$ and $t_j \gamma x$ are in V, obtaining that $t_i \gamma, t_j \gamma \in \mathsf{S}(n+1)$. Moreover all elements obtained in this way from elements of $\mathsf{S}(n)$ are pairwise distinct because $t_1', \dots, t_k'$ freely generate a free semigroup. Hence $card(\mathsf{S}(n+1)) \geq 2card(\mathsf{S}(n))$, giving $card(\mathsf{S}(n)) \geq 2^n$. So $(\widehat{\Gamma}_{U_2,x}, d_{\widehat{S},U_2,x})$ has exponential growth. Since $\widehat{\Gamma}_{U_2,x} \subset \Gamma_{U_2,x}$ and $d_{S\cup\widehat{S},U_2,x} \leq d_{\widehat{S},U_2,x}$ on $\widehat{\Gamma}_{U_2,x}$, it follows that $(\Gamma_{U_2,x}, d_{S\cup\widehat{S},U_2,x})$ also has exponential growth. So $(\mathfrak{G}_x, d_{E_2\cup\widehat{E}})$ has exponential growth, obtaining that $(\mathfrak{G}_x, d_E)$ has exponential growth by Corollary 1. □

6.4 Growth of Equicontinuous Foliated Spaces

Let $X \equiv (X, \mathcal{F})$ be a compact Polish foliated space. Let $\{U_i, p_i, h_{ij}\}$ be a defining cocycle of $\mathcal{F}$, where $p_i : U_i \to T_i$ and $h_{ij} : T_{ij} \to T_{ji}$, and let $\mathcal{H}$ be the induced representative of the holonomy pseudogroup. As we saw in Sect. 5, $\mathcal{H}$ can be considered as the restriction of some compactly generated pseudogroup $\mathcal{H}'$ to some relatively compact open subset, and $E = \{h_{ij}\}$ is a system of compact generation on T. Moreover Álvarez and Candel [4] observed that E is recurrent. According to Theorems 7, it follows that the quasi-isometry type of the $\mathcal{H}$-orbits and their germ covers with d_E are independent of the choice of $\{U_i, p_i, h_{ij}\}$ under the above conditions; thus they can be considered as quasi-isometry types of the corresponding leaves and their holonomy covers.

This has the following interpretation when X is a smooth manifold. In this case, given any Riemannian metric g on X, for each leaf L, the differentiable (and coarse) quasi-isometry type of $g|_L$ is independent of the choice of g; they depend only on $\mathcal{F}$ and L; in fact, it is coarsely quasi-isometric to the corresponding $\mathcal{H}$-orbit, and therefore they have the same growth type [10] (this is an easy consequence of the existence of a uniform bound of the diameter of the plaques). Similarly, the germ covers of the $\mathcal{H}$-orbits are also quasi-isometric to the holonomy covers of the corresponding leaves.

Theorem 2 follows from these observations and Theorem 8.

7 Examples and Open Problems

Theorems 1, 2 and 8 may be relevant in the following examples; most of them are taken from [14, Chap. 11].

Example 6 (*Foliated Polish Space*) Any locally free action of a connected Lie group on a locally compact Polish space, $\phi : H \times X \to X$, defines a foliated structure $\mathcal{F}$ on X whose leaves are the orbits [14, Theorem 11.3.14], [49]. Moreover, $\mathcal{F}$ is equicontinuous if ϕ is equicontinuous.

Example 7 (*Matchboxes manifold*) A *matchbox manifold* is a foliated continuum[9] $X \equiv (X, \mathcal{F})$ transversely modeled on a totally disconnected space. The case of a single leaf is discarded, and it is assumed that X is C^1 in the sense that the changes of foliated coordinates are C^1 along the leaves, with transversely continuous leafwise derivatives. An example of matchbox manifold is given by any inverse limit of smooth proper covering maps of compact n-manifolds, called an n-dimensional *solenoid*; if moreover, any composite of a finite number of bounding maps is a normal covering, then it is called a *McCord solenoid* [40]. A matchbox manifold X is equicontinuous if and only if it is a solenoid [17, Theorem 7.9]; and X is homogeneous if and only if it is a McCord solenoid [17, Theorem 1.1]; this is the case where it is a G-foliated space. See [8, 15] for a generalization using inverse limits of compact branched manifolds.

Example 8 (*Bebutov Foliated Spaces*) There exists a construction that yields a foliated space from a topological flow, known as the *Bebutov Foliated Space*, as derived from Nemytsky [47] (see also Auslander [9]. I present this construction within the framework of rewriting, which is the context adopted in this study. Let $C_b(\mathbb{R})$ be the space of continuous bounded functions $\mathbb{R} \to \mathbb{R}$, with the topology of uniform convergence. For a function $f \in C_b(\mathbb{R})$ and $t \in \mathbb{R}$, let $f_t \in C_b(\mathbb{R})$ be defined by $f_t(r) = f(r+t)$. It is said that f is *almost periodic* if $\{ f_t \mid t \in \mathbb{R} \}$ is equicontinuous [11, 28], which means that $\mathfrak{M}(f) := \overline{\{ f_t \mid t \in \mathbb{R} \}}$ is compact in $C_b(\mathbb{R})$. An equicontinuous flow $\Phi : \mathbb{R} \times \mathfrak{M}(f) \to \mathfrak{M}(f)$ is defined by $\Phi_t(g) = g_t$. If f is non-constant, then Φ is nonsingular, defining an equicontinuous foliated structure $\mathcal{F}$ on $\mathfrak{M}(f)$. If f is non-periodic, then $\mathcal{F}$ does not reduce to a single leaf. More generality, we can consider an almost-periodic non-periodic continuous function f on any connected Lie group with values in a Hilbert space.

This construction is universal in the following sense. For any compact Polish space X and any oriented equicontinuous minimal foliated structure $\mathcal{F}$ on X of dimension one, there is a non-singular equicontinuous flow $\phi : \mathbb{R} \times X \to X$ whose orbits are the leaves of $\mathcal{F}$. Take a sequence of real-valued continuous functions h_n on X that separate points. Let (e_n) be a complete orthonormal system of a separable Hilbert space $\mathfrak{H}$. Then the function $h : X \to \mathfrak{H}$, given by

[9] Recall that a *continuum* is a non-empty compact connected metrizable space.

$$h(x) = \sum_n \frac{h_n(x)}{2^n \cdot \max|h_n|} e_n ,$$

is continuous and separates points, and therefore it is a topological embedding because X is compact. Given any point $x_0 \in X$, let $f \in C_b(\mathbb{R}, \mathfrak{H})$ be defined by $f(r) = h\phi_r(x_0)$. From the equicontinuity of ϕ and the compactness of X, it follows that f is almost periodic. Thus the compact minimal equicontinuous foliated space $\mathfrak{M}_f$ is defined. The mapping $\phi_t(x_0) \mapsto f_t$ defines a continuous map of the ϕ-orbit of x_0 onto the Φ-orbit of f, which extends to an equivariant continuous map $\psi : X \to \mathfrak{M}_f$. This map is injective because h separates points, and it is surjective because Φ is minimal. Hence ψ is a foliated homeomorphism since X is compact and $\mathfrak{M}_f$ Hausdorff.

More generally, we can define in the same way a compact minimal equicontinuous foliated space $\mathfrak{M}_f$ for any bounded continuous function f on M with values in a Hilbert space, which is almost-periodic in the same sense.

Example 9 (*Space of Manifolds*) For each $n \in \mathbb{Z}_+$, let $\mathcal{M}_*(n)$ denote the set[10] of isometry classes $[M, x]$ of pointed complete connected Riemannian n-manifolds (M, x). The C^∞ convergence [50] defines a Polish topology on $\mathcal{M}_*(n)$ [3, Theorem 1.1]. The corresponding space is denoted by $\mathcal{M}_*^\infty(n)$, and its closure operator by C_∞. For any complete connected Riemannian manifold $M \equiv (M, g)$, a canonical continuous map $\iota : M \to \mathcal{M}_*^\infty(n)$ is defined by $\iota(x) = [M, x]$. A concept of *weak aperiodicity* of M was introduced in [3]. On the other hand, M is called *almost periodic* if, for all $m \in \mathbb{N}$, $\epsilon > 0$ and $x \in M$, there is a set H of diffeomorphisms of M so that $\sup|\nabla^k h^* g| < \epsilon$ for all $h \in H$ and $k \leq m$, and $\{ h(x) \mid h \in H \}$ is a net in M. If M is weakly aperiodic and almost periodic, then $C_\infty(\iota(M))$ canonically becomes a compact minimal equicontinuous foliated space of dimension n, as follows from [3, Theorem 1.2 and Lemma 12.5 (ii)] and [38, Lemma 7.2 and Theorem 4.1] (see also [50, Chapter 10, Theorem 3.3], and [16]).

Example 10 (*Inverse Limit of Minimal Lie Foliations*) Investigating the Inverse Limit of Minimal Lie Foliations: A Homogeneous Foliated Space Perspective.

This example, originally introduced by S. Hurder and documented in Àlvares and Barral [2], has spurred open questions regarding the size of the class of inverse limits of minimal Lie foliations.

In the work of Jacoby [35], a theorem asserts that any metrizable locally compact local group with finite topological dimension is locally isomorphic to the direct product of a Lie group and a compact zero-dimensional topological group. In light of this theorem, a question posed by S. Hurder emerges: Can any compact minimal foliated homogeneous space with finite "topological codimension" be realized as an inverse limit of minimal Lie foliations, as in Àlvarez Barral [2, Example 8.4].

[10] The logical problems of this definition can be avoided because any complete connected Riemannian manifold is equipotent to some subset of $\mathbb{R}$.

This inquiry aims to extend the findings of [17] (also see [8]). Interestingly, an affirmative answer to this question has been established for homogeneous matchbox manifolds in the scenario of codimension zero.

Should this conjecture prove accurate, harnessing Molino's description could facilitate the proof that any equicontinuous foliated space meeting the conditions of Theorem A manifests as an inverse limit of Riemannian foliations. This exploration carries significant implications for the understanding of minimal Lie foliations and their potential generalizations.

Problem 1 In the Examples 6–9, understand the specific application of Theorems 1 and 2.

Problem 2 Use Theorem 1 to classify particular classes of equicontinuous foliated spaces.

Question 1 Is it possible to improve Theorem 2 for special types of structural local groups?

Question 2 Is there any consequence of Theorems 1 and 2 in usual foliation theory, assuming that the minimal sets are equicontinuous?

The following questions refer to extensions of known properties of Riemannian foliations, where Theorem 1 could play an important role.

Question 3 For compact minimal equicontinuous foliated spaces, does the leafwise heat flow of leafwise differential forms preserve transverse continuity at infinite time?

Question 4 Is it possible to give useful extensions of tautness and tenseness to equicontinuous foliated spaces, and relate them to some kind of basic cohomology?

References

1. Álvarez-López, J.A. and Moreira Galicia, M.: *Topological Molino's theory*, Pacific. J. Math., 280:257–314, 2016
2. Álvarez-López, J.A. and Barral Lijó, R.: Molino's description and foliated homogeneity, Topology Appl. 260 (2019), 148–177
3. Álvarez-López, J.A., Barral Lijó, R. and Candel, A.: A universal Riemannian foliated space, Topology Appl. 198 (2016), 47–85
4. Álvarez-López, J.A. and Candel, A.: Equicontinuous foliated spaces, Math. Z. 263 (2009), 725–774
5. Álvarez-López, J.A. and Candel, A.: Topological description of Riemannian foliations with dense leaves, Pacific J. Math. 248 (2010), 257–276
6. Álvarez-López, J.A. and Candel, A.: *Generic coarse geometry of leaves*, Vol. 2223, Springer, 2018
7. Abd-Allah, A. and Brown, R.: A compact-open topology on partial maps with open domain, J. London Math. Soc. (2) 21 (1980), 480–486
8. Alcalde-Cuesta, F., Lozano-Rojo, A. and Macho-Stadler, M.: Transversely Cantor laminations as inverse limits, Proc. Amer. Math. Soc. 139 (2011), 2615–2630

9. Auslander, J.: *Minimal flows and their extensions*, North-Holland Mathematics Studies, Vol. 153, North-Holland Publishing Co., Amsterdam, 1988
10. Carrière, Y.: Feuilletages riemanniens à croissance polynômiale, Comment. Math. Helv., 63 : 1–20, 1988
11. Besicovich, A.: *Almost periodic functions*, Dover Publications, New York, 1954
12. Block, J. and Weinberger, S.: Large scale homology theories and geometry, Geometric Topology (Athens, GA, 1993) (RI), AMS/IP Stud. Adv. Math., no. 2.1, Amer. Math. Soc., 1997, pp. 522–569
13. Breuillard, E. and Gelander, T.: A topological Tits alternative, Ann. of Math. 166 (2007), 427–474
14. Candel, A. and Conlon, L.: *Foliations, I*, Graduate Studies in Mathematics, Amer. Math. Soc., Providence, RI, 2000
15. Clark, A., Fokkink, R. and Lukina, O.: The Schreier continuum and ends, Houston J. Math., 40(2):569–599, 2014
16. Cheeger, J.: Finiteness theorems for Riemannian manifolds, Amer. J. Math. 92 (1970), 61–75
17. Clark, A. and Hurder, S.: Homogeneous matchbox manifolds, Trans. Amer. Math. Soc. 365 (2013), 3151–3191
18. van den Dries, L. and Goldbring, I.: Globalizing locally compact local groups, J. Lie Theory 20 (2010), 519–524
19. van den Dries, L. and Goldbring, I.: Erratum to: Globalizing locally compact local groups, J. Lie Theory 22 (2012), 489–490
20. Dyer, J., Hurder, S. and Lukina, O.: The discriminant invariant of Cantor group actions, Topology Appl. 208 (2016), 64–92
21. Dyer, J., Hurder, S. and Lukina, O.: Molino theory for matchbox manifolds, Pacific J. Math. 289 (2017), no. 1, 91–151
22. Fedida, E.: Sur les feuilletages de Lie, C.R. Ac. Sc. Paris 272 (1971), 999–1002
23. Fedida, E.: *Feuilletages de Lie, feuilletages du plan*, Lect. Notes Math., vol. 352, Springer-Verlag, 1973, pp. 183–195
24. Ghys, É.: Groupes d'holonomie des feuilletages de Lie, Nederl. Akad. Wetensch. Indag. Math. 47 (1985), 173–182
25. Ghys, É.: Laminations par surfaces de Riemann, Panoramas & Syntheses 8 (2000), 49–95
26. Ghys, É.: Laminations par surfaces de Riemann, In: *Dynamique et Géométrie Complexes*, Panoramas & Synthèses, 8:49–95, 1999
27. Goldbring, I.: Hilbert's fifth problem for local groups, Ann. of Math. 172 (2010), 1269–1314
28. Gottschalk, W.H.: Almost periodicity, equi-continuity and total boundedness, Bull. Amer. Math. Soc. 52 (1946), 633–636
29. Gromov, M.: *Asymptotic invariants of infinite groups*, Geometric Group Theory (G.A. Niblo and M.A. Roller, eds.), vol. 2, Cambridge University Press, 1993
30. Haefliger, A.: Pseudogroups of local isometries, In: *Differential Geometry*, Santiago de Compostela, 1984 (L.A. Cordero, ed.), Pitman, 1985, pp. 174–197
31. Haefliger, A.: Leaf closures in Riemannian foliations, A Fête on Topology (New York), Academic Press, 1988, pp. 3–32
32. Haefliger, A.: Foliations and compactly generated pseudogroups, In: *Foliations: Geometry and Dynamics*, Warsaw 2000, World Sci. Publi. 2002, pp. 275–296
33. Hector, G. and Hirsch, U.: *Introduction to the geometry of foliations, part A*, Aspects of Mathematics, vol. E1, Friedr. Vieweg and Sohn, Braunschweig, 1981
34. Hector, G. and Hirsch, U.: *Introduction to the geometry of foliations, part B*, Aspects of Mathematics, vol. E3, Friedr. Vieweg and Sohn, Braunschweig, 1983
35. Jacoby, R.: Some theorems on the structure of locally compact local groups, Ann. of Math. 66 (1957), 36–69
36. Kechris, A.: *Classical descriptive set theory*, Springer-Verlag, Berlin, 1991.
37. Kellum, M.: Uniformly quasi-isometric foliations, Ergodic Theory Dynam. Systems 13 (1993), 101–122

38. Lessa, P.: Reeb stability and the Gromov-Hausdorff limits of leaves in compact foliations, arXiv:1304.6204, 2013
39. Matsumoto, S.: The unique ergodicity of equicontinuous laminations, Hokkaido Math. J. 39 (2010), 389–403
40. McCord, C.: Inverse limit sequences with covering maps, Trans. A.M.S, 114:197–209, 1965
41. Meigniez, G.: Feuilletages de Lie résolubles, Ann. Fac. Sci. Toulouse Math. 4 (1995), 801–817
42. Meigniez, G.: A compactly generated pseudogroup which is not realizable, J. Math. Soc. Japan 62 (2010), 1205–1218
43. Molino, P.: Geométrie globale des feuilletags riemanniens, Ned. Akad. von Wet., Indag. Math. 85 (1982), 45–76
44. Molino, P.: *Riemannian foliations*, Progress in Mathematics, no. 73, Birkhäuser Boston, Inc., Boston, MA, 1988
45. Montgomery, D. and Zippin, L.:*Topological transformation groups*, Interscience Publishers, New York-London, 1955
46. Moore, C.C. and Schochet, C.: *Global analysis on foliated spaces*, Mathematical Sciences Research Institute Publications, vol. 9, Springer-Verlag, New York, 1988, With appendices by S. Hurder, Moore, Schochet and Robert J. Zimmer.
47. Nemytsky, V.V.: Topological problems in the theory of dynamical systems, Stability and Dynamical Systems, Transl. Series 1, Vol. 6, AMS, (1961), 414–497
48. Olver, P.: Non-associative local Lie groups, J. Lie Theory 6 (1996), 23–51
49. Palais, R.: On the existence of slices for actions of non-compact Lie groups, Ann. of Math. 73 (1961), 295–323
50. Petersen, P.: *Riemannian geometry*, Graduate Texts in Mathematics, vol. 171, 3rd edition, Springer-Verlag, New York, Berlin, Heidelberg, 2016
51. Plaut, C.: *Associativity and the local version of Hilbert's fifth problem*, Notes, University of Tennessee, 1993
52. Reinhart, B.: Foliated manifolds with bundle-like metrics, Ann. of Math. 69 (1959), 119–132
53. Sacksteder, R.: Foliations and pseudogroups, Amer. J. Math. 87 (1965), 79–102
54. Salem, E.: Une généralisation du théoréme de Myers-Steenrod aux pseudogrupes d'isométries locales, Ann. Inst. Fourier (Grenoble) 38 (1988), 185–200
55. Tarquini, C.: Feuilletages conformes, Ann. Inst. Fourier 54, (2004), 453–480

Lichnerowicz-Type Laplacians in the Bochner Technique

Vladimir Rovenski, Sergey Stepanov, and Irina Tsyganok

A theory can be judged to be successful only if its consequences help us understand the basic structure or the beauty of geometry.

S.-T. Yau

Abstract In the well-known monograph of A. Besse the following is written: the Bochner technique is a method of proving vanishing theorems for null space of a Laplace operator admitting a Weitzenböck decomposition and further of estimating its lowest nonzero eigenvalue. In this article, we consider a generalized form of the well-known Lichnerowicz Laplacian, show how the Bochner technique works for this operator and provide some its important examples.

Keywords Riemannian manifold · Lichnerowicz-type Laplacian · Bochner technique · Weitzenböck decomposition formula · Singular distribution

Mathematical Subject Classifications (2010) 53C21 · 58J05

V. Rovenski (✉)
Department of Mathematics, University of Haifa, 3498838 Haifa, Israel
e-mail: vrovenski@univ.haifa.ac.il

S. Stepanov · I. Tsyganok
Department of Mathematics, Financial University, 125993 Moscow, Russia
e-mail: s.e.stepanov@mail.ru

I. Tsyganok
e-mail: i.i.tsyganok@mail.ru

© The Author(s), under exclusive license to Springer Nature Switzerland AG 2024
V. Rovenski et al. (eds.), *Differential Geometric Structures and Applications*,
Springer Proceedings in Mathematics & Statistics 440,
https://doi.org/10.1007/978-3-031-50586-7_8

1 Preliminaries on the Bochner Technique

1.1 The Classical Bochner Technique

The classical Bochner technique is one of the main analytical methods of global differential geometry for more than seventy years. The countdown starts from the first monograph [70] devoted to this method. It is a general method for proving the so-called *vanishing theorems* (see details in [6, 27, 60]). Historically, the first vanishing theorem states that various topological and geometric invariants (such as the Betti numbers and the dimension of the vector space of Killing differential forms) of an n-dimensional compact and oriented Riemannian manifold (M, g) vanish under certain restrictions on the curvature of (M, g). In our opinion, one of the important recent goals of the classical Bochner technique are the results on the geometry in the large of Ricci solitons and almost Ricci solitons (e.g., [16, pp. 153–176]; [42, Chap. 3]; [4, 5, 19]), which are the important components of the Ricci flow theory. This theory, besides being known after G. Perelman's proof of the H. Poincaré conjecture (for details see [42, Introduction]), has a wide range of applications in differential geometry and theoretical physics.

The first application of the Bochner technique was realized by K. Kodaira in the 1950s (see [68, pp. 351; 361–363]). But the Bochner method as a theory was founded by S. Bochner, K. Yano, A. Lichnerowicz and many others in the 1950s and 1960s to study the relationship between curvature and topology of compact Riemannian and Kählerian manifolds (see, for example, [68]). At the same time, it was based on the idea of K. Kodaira. Metrhods and types of theorems developed during this period still form the basis for many developments today.

The founders of the method used Hopf's lemma and Green's theorem (see also [70]). Hopf's lemma states that if a smooth scalar function f defined on a compact boundaryless Riemannian manifold (M, g) satisfies the conditions $\Delta f \geq 0$ or $\Delta f \leq 0$ for the *Beltrami Laplacian* $\Delta f := \operatorname{div}(\operatorname{grad} f)$, then f is a constant function. Green's theorem states that $\int_M \operatorname{div} X\, \Omega_g = 0$ for a smooth vector field X on a compact boundaryless Riemannian manifold (M, g) and the Riemannian volume form Ω_g. As a result, we have a number of theorems based on the classical Bochner technique, which show that some geometrically interesting tensor fields and mappings vanish under the assumption of positive or negative curvature of Riemannian manifolds. In our opinion, the most famous results of the classical Bochner technique are the theorem of D. Meyer and S. Gallot (see [44, p. 351]) on vanishing of the Betti numbers of compact Riemannian manifolds and the theorem of J. Eells and J.H. Sampson (see [68, p. 465]) on the nonexistence of harmonic mappings of compact Riemannian manifolds. The Bochner technique is used in hundreds of articles, monographs and analytical reviews (see, for example, [6, 32, 52, 60, 68–70, 74] and [44, pp. 333–363]), in which one will find a number of applications of the Bochner technique to Riemannian and Lorentzian geometries and even to General Relativity (e.g., [52, 62, 63]).

Distributions, as subbundles of the tangent bundle on a manifold, arise in such topics of mathematics and physics as fiber bundles, Lie groups actions, almost contact and sub-Riemannian manifolds, e.g., [57]. Singular distributions are important because many manifolds admit no smooth (for example, codimension-one) distributions, while they admit such distributions outside some set. In [50], a manifold endowed with a singular or regular distribution, determined as the image of the tangent bundle under a smooth endomorphism is studied and the Bochner's technique is extended to the case of a singular distribution.

1.2 The Generalized Bochner Technique

In this section, we consider the generalized Bochner technique that is a natural development of the classical Bochner technique. Since the 1970s, complete (noncompact) Riemannian manifolds have been included in the range of research carried out using the Bochner technique. For this, the methods of geometric analysis were developed and, in particular, partial differential equations techniques which provide approaches to many open problems in global differential geometry (for example, [68, pp. 361–394]; [25, 45, 46, 59]). As a result, vanishing theorems of the classical Bochner technique took the form of *Liouville type theorems*: vanishing theorems, which are proved in the case of complete noncompact (boundaryless) Riemannian manifolds and without any assumption on their curvature. The first outstanding result in this direction was theorem of S.-T. Yau and R. Schoen (see [59]), generalizing theorem of J. Eells and J. H. Sampson to the case of complete Riemannian manifolds. The new research method developed by S.-T. Yau, R. Shoen, R. Hamilton, H. Wu, P. Li and others, later was called the generalized Bochner technique (see, for example, [45]). This technique studies the relationships between the geometry of a complete Riemannian manifold and the global behavior of its subharmonic, superharmonic and convex functions under assumptions either on the curvature or on the volume growth of geodesic balls. Today, this research method is not as popular in geometry as the classical Bochner technique: there are monographs on it (e.g., [25, 45, 46, 59, 68]), but there are no so many geometric applications of his modern method. This shortcoming is corrected by numerous works of the authors of this paper (see, for example, [37–41, 53, 54, 63–65]).

Let's dwell on this method in more detail. The generalized Bochner technique includes the following directions. The first one is the application of generalized maximum principles. Namely, there are various forms of the maximum principle, starting from its classical formulation in the Hopf form up to generalizations of the Omori-Yau maximum principle at infinity in [46], where applications are given to a number of geometrical problems in the setting of complete Riemannian manifolds, under assumptions either on the curvature or on the volume growth of geodesic balls (for example, [61]). An easy application of the Hopf maximum principle shows that "every harmonic (subharmonic and superharmonic) function is constant on a compact Riemannian manifold", that coincides with the assertion of the Hopf's

Lemma from [70]. We recall that the condition $\Delta f \geq 0$ (respectively, $\Delta f \leq 0$) defines f as a subharmonic (respectively, superharmonic) function. Consequently, if (M, g) is complete and f is a nonnegative subharmonic function such that $f \in L^q(M)$ for any $1 < q < \infty$, then f is constant. Furthermore, in this case (M, g) is a manifold of finite volume, unless the constant is zero. On the other hand, a complete Riemannian manifold of finite volume does not carry nonconstant positive superharmonic functions (see also [30, 41]).

An application of the Hopf maximum principle to the theory of geodesic mappings of complete noncompact Riemannian manifolds can be found in [66]. Throughout the article, we consider these functions defined on complete noncompact manifolds. There are numerous generalizations of the Hopf maximum principle on complete Riemannian manifolds that are used in the generalized Bochner technique (see, for example, [46, 67]). We remark that there exist noncompact counterparts of the Green theorem. For example, such a result was obtained by generalized Bochner technique in [14], where it was proved that if the sign of div X remains unchanged for integrable vector fields X on a complete noncompact oriented Riemannian manifold (M, g), then div $X = 0$. Some applications of this theorem can be found in [2, 14, 61].

Also, there is the theory of parabolic manifolds, which is related to superharmonic functions and is a part of the generalized Bochner technique. The concept of a parabolic manifold is associated to a wide class of equivalent properties of Riemannian manifolds, including the Green kernel, linear capacity, Brownian motion, etc. Thus, there are several equivalent definitions of the parabolicity of a complete Riemannian manifold in various terms (see, for example, [24, pp. 164–165]). In particular, a complete Riemannian manifold of finite volume is an example of a parabolic manifold. Some applications of this theory can be found in articles [41, 61] on the generalized Bochner technique. Note that any parabolic manifold is stochastically complete. Recall that a diffusion process on a Riemannian manifold is conservative or stochastically complete if the associated stochastic process remains forever in the state space. Both parabolicity and stochastic completeness have been the subject of systematic study, see e.g., a survey by A. Grigor'yan [24]. In particular, any complete Riemannian manifold with the Ricci curvature bounded from below by a (possibly, negative) constant is stochastically complete. Moreover, if a manifold is stochastically complete, then any nonnegative integrable superharmonic function is constant on this manifold. Some applications of the theory of stochastically complete manifolds can be found in [1]. Third, there is the theory of subharmonic functions. A number of results on subharmonic functions on complete Riemannian manifolds have been obtained by R. Green and H. Wu, A. Huber, L. Karp, S.-T. Yau and others. In particular, it was proved that a complete manifold is parabolic if and only if any subharmonic function has finite Dirichlet integral. Some applications of the theory of subharmonic functions can be found in [54, 61], where we considered applications of the generalized Bochner technique to study of the geometry in the large of conformal diffeomorphisms and Laplacians. Note also that convex functions are an example of subharmonic functions because the Hessian of an arbitrary convex function is positive semi-definite. The famous Yau's theorem [71] states that a complete Riemannian manifold admitting a nonconstant convex function has infinite volume.

On the other hand, R.L. Bishop and B. O'Neill [8] proved that if a Riemannian manifolds is a connected and complete manifold of finite volume, then any convex function on this manifold is constant. An example of application of the theory of convex functions was published in [41], where we considered applications of the generalized Bochner technique to study of metrics projectively and holomorphically projectively equivalent to metrics of parabolic Riemannian and Kähler manifolds.

The Ricci flow, introduced by R. Hamilton, refers to the generalized Bochner technique (for detail, see [12]). Using this theory, R. Hamilton proved that a closed three-manifold (M, g) admitting a Riemannian metric of strictly positive Ricci curvature also admits a metric of constant positive sectional curvature. In dimension four, R. Hamilton showed (see review on this topic [12]) that a closed four-manifold (M, g) with positive curvature operator is diffeomorphic to the sphere $\mathbb{S}^4$ or the real projective space $\mathbb{R}\mathbb{P}^4$. We recall that the real projective space $\mathbb{R}\mathbb{P}^4$ admits a constant positive scalar curvature metric. The continuation of these studies can be found in monographs [16, 42] and in a survey on this topic [12]. This also includes G. Perelman's proof of the Poincaré conjecture (for details, see [42, Introduction]). Beside that a compact manifold of positive constant sectional curvature for even n is either a sphere $\mathbb{S}^n$ or a real projective space $\mathbb{R}\mathbb{P}^n$ with canonical metric; and for each odd $n \geq 3$, there are infinitely many spherical space forms that were completely classified by group-theoretical methods (see [12] for more details).

We conclude our overview with resent articles [21, 47, 48] on the geometry of noncompact Ricci solitons and almost Ricci solitons, where the results obtained earlier are generalized using the generalized Bochner technique.

2 The Lichnerowicz Type Laplacians

Here, we consider the Lichnerowicz type Laplacians using the Bochner technique. By now, the most general point of view on the Weitzenböck expansion of the Laplace operator has been presented in [27].

2.1 *The Lichnerowicz Type Laplacian and its Properties*

Let (M, g) be an n-dimensional connected complete Riemannian manifold with the Levi-Civita connection ∇. Let $E \to M$ be a smooth vector bundle of rank r over (M, g) with a scalar product in each fiber E_x, depending C^∞ on $x \in M$, and with a compatible connection. We recall that the rank r of a vector bundle E is the dimension of its fiber. The Riemannian metric g and the Levi-Civita connection ∇ induce on E a scalar product and a compatible connection, respectively. In addition, the scalar products and compatible connections of E, as well as of M, will be denoted by the same symbols g and ∇. Moreover, we can define the L^2-global scalar product on C^∞-sections of E by the formula $\langle \theta, \vartheta \rangle = \int_M g(\theta, \vartheta)\, \Omega_g$ for $\theta, \vartheta \in C^\infty(E)$ and introduce the associated Hilbert space $L^2(E)$. Using the L^2-structures on $C^\infty(E)$, we can define the rough Laplacian by the formula $\bar{\Delta} = \nabla^*\nabla$, where ∇^* is the formally adjoint operator w.r.t. the L^2-global scalar product of the compatible connection $\nabla : C^\infty(E) \to C^\infty(T^*M \otimes E)$ (see [7, p. 52]; [6, p. 378]). Denote by R^∇ the curvature tensor of the compatible connection ∇ as the 2-form on M with values in $E^* \otimes E$, see [7, p. 24]:

$$R^\nabla_{X,Y} = \nabla_X \nabla_Y - \nabla_Y \nabla_X - \nabla_{[X,Y]}.$$

By above definitions and notations, we can define the Lichnerowicz type Laplacian $\Delta_L : C^\infty(E) \to C^\infty(E)$ by the Weitzenböck decomposition formula (see [44, p. 344]; [27] and [6] for $t = 1$)

$$\Delta_L = \bar{\Delta} + t\,\mathfrak{R}(R^\nabla), \tag{1}$$

where $t \in \mathbb{R}$ and $\mathfrak{R}(R^\nabla)$ is the *Weitzenböck curvature operator* that depends linearly on curvature R^∇ and defines a symmetric endomorphism $\mathfrak{R}(R^\nabla) : E_x \to E_x$ for all $x \in M$. Various geometrical problems give rise to different values of $t \in \mathbb{R}$. Namely, the Laplacian Δ_L is a second-order elliptic linear differential operator on $C^\infty(E)$ which is symmetric with respect to the L^2-global scalar product. On a closed (i.e., compact without boundary) manifold (M, g), we have the orthogonal (with respect to the L^2-global scalar product) decomposition formula

$$C^\infty(E) = \ker \Delta_L \oplus \operatorname{Im} \Delta_L \tag{2}$$

(see [7, p. 464]), where the first component $\ker \Delta_L$ is the kernel of the Lichnerowicz type Laplacian Δ_L. Since Δ_L is elliptic, the dimension of the kernel of Δ_L is finite for the case of a closed (compact without boundary) Riemannian manifold (M, g). In accordance with [45, p. 104], its smooth sections will be called Δ_L*-harmonic*.

Define the vector space of Δ_L-harmonic C^∞-sections of $E \to M$ by

$$\ker \Delta_L = \left\{\varphi \in C^\infty(E) : \Delta_L \varphi = 0\right\},$$

and the vector space of $L^q(E)$-sections of $E \to M$ by the condition

$$L^q(\ker \Delta_L) = \left\{\varphi \in \ker \Delta_L : \|\varphi\| \in L^q(M)\right\},$$

where $\|\varphi\|^2 = g(\varphi, \varphi)$. Furthermore, Δ_L-harmonic C^∞-sections of $E \to M$ satisfy the unique continuation property. In coordinates, the condition $\Delta_L \varphi = 0$ becomes a system of elliptic equations satisfying the structural assumptions of Aronszajn-Cordes (see [45, Appendix A]). Consequently, the following theorem holds (see [45, p. 104]).

Theorem 1 *Let Δ_L be the Lichnerowicz type Laplacian on C^∞-sections of a smooth vector bundle $E \to M$ over (M, g) satisfying* (1). *Suppose that $\varphi \in \ker \Delta_L$ is a Δ_L-*

harmonic section of $E \to M$. If φ has a zero of infinite order at some point of M, then φ vanishes on M.

If (M, g) is a closed manifold, then the following integral formula holds:

$$\begin{aligned}(\Delta_L \varphi, \varphi)_{L^2} &= (\nabla^* \nabla \varphi, \varphi)_{L^2} + t\,(\mathfrak{R}(R^\nabla)\,\varphi, \varphi)_{L^2} \\ &= (\nabla \varphi, \nabla \varphi)_{L^2} + t\,(\mathfrak{R}(R^\nabla)\,\varphi, \varphi)_{L^2},\end{aligned} \tag{3}$$

where φ is a smooth section of the vector bundle $E \to M$. In this case, if $t > 0$ and $\mathfrak{R}(R^\nabla)$ is a nonnegative endomorphism at each point of M, then the right-hand-side is the sum of two nonnegative terms. Hence $\Delta_L \varphi = 0$ implies both $\nabla\varphi = 0$ and $\mathfrak{R}(R^\nabla)\,\varphi = 0$. Furthermore, if $\mathfrak{R}(R^\nabla)$ is strictly positive at some point of M, then clearly $\varphi = 0$. Therefore, we can formulate the following theorem (see also [6, p. 379]).

Theorem 2 *Let Δ_L be the Lichnerowicz type Laplacian on C^∞-sections of a smooth vector bundle $E \to M$ of rank r over a closed Riemannian manifold (M, g) satisfying* (1). *If $t > 0$ and the symmetric endomorphism $\mathfrak{R}(R^\nabla)$ satisfies $\mathfrak{R}(R^\nabla) \geq 0$,*
(i) then $\nabla\varphi = 0$ for an arbitrary $\varphi \in \ker \Delta_L$ and $\dim_{\mathbb{R}} \ker \Delta_L \leq r$;
(ii) and $\mathfrak{R}(R^\nabla) > 0$ at some point of M, then $\dim_{\mathbb{R}} \ker \Delta_L = 0$.

Remark 1 If $t < 0$ and $\mathfrak{R}(R^\nabla)$ is a nonpositive symmetric endomorphism at each point of M, then $\dim_{\mathbb{R}} \ker \Delta_L \leq r$ and $\nabla\varphi = 0$ for an arbitrary $\varphi \in \ker \Delta_L$. Furthermore, if $\mathfrak{R}(R^\nabla)$ is strictly negative at some point of M, then $\varphi = 0$.

Let (M, g) be an oriented manifold. As observed by Hitchin [21], the curvature operator of the first kind $\hat{R} : \Lambda^2 M \to \Lambda^2 M$ is positive (see [44, p. 82–83]) if and only if the symmetric endomorphism $\mathfrak{R}(R^\nabla) : E_x \to E_x$ is also positive for all nontrivial irreducible representations of the orthogonal group $SO(n) \to \mathrm{End}\, E_x$ for all $x \in M$. Thus, we can formulate the following corollary.

Corollary 1 *Let Δ_L be the Lichnerowicz-type Laplacian on C^∞-sections of a smooth vector bundle $E \to M$ over a closed and oriented Riemannian manifold (M, g), satisfying* (1). *If $t > 0$ and the curvature operator $\hat{R} : \Lambda^2 M \to \Lambda^2 M$ is positive, then* $\dim_{\mathbb{R}} \ker \Delta_L = 0$.

The above arguments refer to the classical Bochner technique. For the generalized Bochner technique these arguments have the following form. Namely, we get

$$\frac{1}{2}\Delta\,\|\varphi\|^2 = -g(\nabla^*\nabla\varphi, \varphi) + \|\nabla\varphi\|^2 = -g(\Delta_L\,\varphi, \varphi) + t\,g(\mathfrak{R}(R^\nabla)\,\varphi, \varphi) + \|\nabla\varphi\|^2. \tag{4}$$

In addition, by direct calculation we find the equality

$$\frac{1}{2}\Delta\,\|\varphi\|^2 = \|\varphi\| \cdot \Delta\,\|\varphi\| + \|d\,\|\varphi\|\|^2.$$

Then (4) can be rewritten in the form

$$\|\varphi\| \cdot \Delta \|\varphi\| = -g(\Delta_L \varphi, \varphi) + t\, g(\mathfrak{R}(R^\nabla)\varphi, \varphi) + \|\nabla\varphi\|^2 - \|d\,\|\varphi\|\|^2. \quad (5)$$

Using the first T. Kato's inequality $\|\nabla\varphi\|^2 \geq \|d\,\|\varphi\|\|^2$ (see [6, p. 380]), we obtain from (5) the following inequality:

$$\|\varphi\| \cdot \Delta \|\varphi\| \geq -g(\Delta_L \varphi, \varphi) + t\, g(\mathfrak{R}(R^\nabla)\varphi, \varphi). \quad (6)$$

If φ is a Δ_L-harmonic section, $t > 0$ and $\mathfrak{R}(R^\nabla)$ is a nonnegative symmetric endomorphism at each point of M, then the right-hand side of the above inequality is nonnegative. Applying the S.-T. Yau theorem to the above inequality, we conclude that for the any real $q > 1$, either $\int_M \|\varphi\|^q\, \Omega_g = \infty$ or $\|\varphi\| = C$ for some constant $C \geq 0$ (see [73]). In particular, if $\|\varphi\| \in L^q(M)$ for some real $q > 1$ and the volume of (M, g) is infinite, then the constant $C = 0$. On the other hand, for $\|\varphi\| = C$, $t > 0$ and $\mathfrak{R}(R^\nabla) \geq 0$ (nonnegative), from (5) we get $\nabla\varphi = 0$. Thus, the following statement holds.

Theorem 3 *Let (M, g) be a complete noncompact Riemannian manifold and Δ_L the Lichnerowicz type Laplacian on C^∞-sections of a smooth vector bundle $E \to M$, satisfying* (1). *If $t > 0$ and $\mathfrak{R}(R^\nabla)$ is a nonnegative symmetric endomorphism on M, then $\nabla\varphi = 0$ for any $\varphi \in L^q(\ker \Delta_L)$ and any $q > 1$. In particular, if (M, g) has infinite volume, then $L^q(\ker \Delta_L)$ is trivial for any $q > 1$.*

Remark 2 If $t < 0$ and $\mathfrak{R}(R^\nabla)$ is a nonpositive symmetric endomorphism on M, then $\nabla\varphi = 0$ for any $\varphi \in L^q(\ker \Delta_L)$ and for some real $q > 1$.

On an n-dimensional ($n \geq 2$) Riemannian manifold (M, g), one can introduce several Lichnerowicz-type Laplacians related to the metric tensor g that give rich information about the geometry of (M, g). Therefore, the Bochner technique has a wide range of applications.

2.2 Estimates of Eigenvalues of the Lichnerowicz Type Laplacian

Let (M, g) be an n-dimensional ($n \geq 2$) closed Riemannian manifold. An eigensection of the Lichnerowicz type Laplacian Δ_L is $\varphi \in C^\infty(E)$ satisfying the condition $\Delta_L \varphi_x = \lambda \varphi_x$ at each point $x \in M$, where λ is a constant called the eigenvalue of Δ_L corresponding to φ.

Since Δ_L is an elliptic operator, we conclude that on a closed manifold (M, g), the space $E(\lambda)$ of eigensection of Δ_L associated with the eigenvalue λ has a finite dimension, see [6, p. 389]. Moreover, the set of all eigenvalues $\{\lambda_a\}_{a\geq 1}$ of the Lichnerowicz Laplacian Δ_L is discrete and form a nondecreasing sequence of the eigenvalues counted with multiplicities (see [6]). We denote also by $\{\bar\lambda_a\}_{a\geq 1}$ a nondecreasing sequence of the eigenvalues of the Bochner Laplacian $\bar\Delta$ counted with multiplicities. Observe that $\bar\lambda_a \geq 0$ for all $a \geq 1$.

For a closed Riemannian manifold (M, g), we define the following numbers:

$$\mathfrak{R}_{\min} = \inf\{\mathfrak{R}_x(R^\nabla) : x \in M\}, \quad \mathfrak{R}_{\max} = \sup\{\mathfrak{R}_x(R^\nabla) : x \in M\}.$$

At the same time, suppose that $\{\lambda_a\}_{a\geq 1}$ is a nondecreasing sequence of the nonzero eigenvalues of Δ_L counted with multiplicities. Then $\mathfrak{R}_{\min} \leq \lambda_a - \bar{\lambda}_a$ for all $a \geq 1$ (see [6, p. 393]). In this case, if $t > 0$, then from (4) we obtain

$$0 = \int_M (-\lambda_a \|\varphi\|^2 + t\, g(\mathfrak{R}(R^\nabla)\varphi, \varphi) + \|\nabla\varphi\|^2)\, \Omega_g$$

$$\geq -\lambda_a \int_M \|\varphi\|^2\, \Omega_g + t \int_M g(\mathfrak{R}(R^\nabla)\varphi, \varphi)\, \Omega_g \geq (t\,\mathfrak{R}_{\min} - \lambda_a) \int_M \|\varphi\|^2\, \Omega_g.$$

On the other hand, if $t < 0$, then from (4) we deduce

$$(t\,\mathfrak{R}_{\max} - \lambda_a) \int_M \|\varphi\|^2\, \Omega_g \leq 0.$$

From the last integral inequality we conclude that $\lambda_a \geq t\,\mathfrak{R}_{\min}$ for $t > 0$ and $\lambda_a \geq t\,\mathfrak{R}_{\max}$ for $t < 0$. Thus, the following theorem holds.

Theorem 4 *Let $\Delta_L : C^\infty(E) \to C^\infty(E)$ be the Lichnerowicz type Laplacian on a closed Riemannian manifold (M, g) satisfying the Weitzenböck formula $\Delta_L = \bar{\Delta} + t\,\mathfrak{R}(R^\nabla)$, where $t \in \mathbb{R}$ and $\mathfrak{R}(R^\nabla)$ is the Weitzenböck curvature operator, which defines a symmetric endomorphism $\mathfrak{R}(R^\nabla) : E_x \to E_x$ for all $x \in M$. Suppose that $\{\lambda_a\}_{a\geq 1}$ is a nondecreasing sequence of nonzero eigenvalues of Δ_L counted with multiplicities. Then*

$$(i)\ \lambda_a \geq t\,\mathfrak{R}_{\min} \ \text{ for } \ t > 0, \quad (ii)\ \lambda_a \geq t\,\mathfrak{R}_{\max} \ \text{ for } \ t < 0.$$

2.3 Examples and Properties of the Lichnerowicz Type Laplacians

Here, we give an overview of various types of the Lichnerowicz type Laplacian, their elementary properties, and discuss the geometry of their kernels.

Example 1 (The ordinary Lichnerowicz Laplacian) Let $t > 0$ and $E = \bigotimes^p T^*M$, then the Lichnerowicz type Laplacian is the Lichnerowicz Laplacian defined in [7, p. 344]. In the special case when $t = 1$, the Lichnerowicz type Laplacian is the usual Lichnerowicz Laplacian defined in [33, p. 315] and [7, p. 54] by the Weitzenböck decomposition formula

$$\Delta_L = \bar{\Delta} + \mathfrak{R}(R^\nabla), \tag{7}$$

where the Weitzenböck curvature operator $\mathfrak{R}(R^\nabla)$ depends linearly on the Riemann curvature tensor and the Ricci tensor of (M, g) with the Levi-Civita connection ∇. In this case, Weitzenböck curvature operator $\mathfrak{R}(R^\nabla)$ is defined by equalities (1.143) from [68, p. 54],

$$\mathfrak{R}(R^\nabla)\,\varphi_{i_1\ldots i_q} = \sum\nolimits_{a=1}^{q} g^{jk} R_{i_a j}\,\varphi_{i_1\ldots i_{a-1}k i_{a+1}\ldots i_q}$$
$$-2\sum\nolimits_{a,b=1,\,a<b}^{q} g^{jk} g^{lm} R_{i_a j, i_b jl}\,\varphi_{i_1\ldots i_{a-1}k i_{a+1}\ldots i_{b-1} l i_{b-1}\ldots i_q}$$

for any $i, j, k, l, \ldots \in \{1, 2, \ldots, n\}$ and for local components g_{ij}, $\varphi_{i_1\ldots i_q}$, R_{ik} and R_{ikjl} of the metric tensor g, $\varphi \in C^\infty(\bigotimes^q T^*M)$, the Ricci tensor Ric and the curvature tensor R. In particular, $\mathfrak{R}(R^\nabla) = Ric$ for the case $p = 1$; thus, (7) has the form $\Delta_L = \bar{\Delta} + Ric$, and this Laplacian is called the Hodge Laplacian.

Example 2 **(The Hodge-de Rham Laplacian)** Let $t = 1$ and $E = C^\infty(\Lambda^p M)$ for an arbitrary $1 \le p \le n-1$, where $\Lambda^p M$ is the vector bundle of differential p-forms on M, then the Lichnerowicz type Laplacian Δ_L is the well-known Hodge-de Rham Laplacian Δ_H acting on C^∞-sections of $\Lambda^p M$ on (M, g). It admits the Weitzenböck decomposition (1) with $t = 1$, where $\mathfrak{R}(R^\nabla)$ is an algebraic operator, which is the restriction of the Weitzenböck curvature operator to differential forms. Moreover, Δ_H is a nonnegative and self-adjoint elliptic operator with respect to the L^2-global scalar product on C^∞-sections of $\Lambda^p M$ (see [44]).

Example 3 **(The Sampson Laplacian)** Let $t = -1$ and $E = C^\infty(S^p M)$ for any $1 \le p < \infty$, where the vector bundle $S^p M$ of covariant symmetric p-tensors on M (see details in [7]), then the Lichnerowicz type Laplacian Δ_L is the well-known Sampson Laplacian Δ_S acting on C^∞-sections of $S^p M$ over (M, g) (see [9, 35, 36, 54, 59]). In this case, the Weitzenböck decomposition formula has the form (1) with $t = -1$, where $\mathfrak{R}(R^\nabla)$ is an algebraic operator, which is the restriction of the Weitzenböck curvature operator to symmetric p-tensors (see also [19, 21]). In addition, the kernels of Δ_S for any $1 \le p < \infty$ are finite-dimensional vector spaces on a closed manifold (M, g). An arbitrary Δ_S*-harmonic section* of $S^p M$ is called a harmonic symmetric p-tensors (see also [35, 36]).

Example 4 **(The Lichnerowicz Laplacian acting on symmetric tensors)** Let $t > 0$ and $E = C^\infty(S^p M)$, then Δ_L is the Lichnerowicz Laplacian defined on symmetric covariant tensor fields. In the special case when $t = 1$ this differential operator, introduced by Lichnerowicz in [33], is self-adjoint, elliptic and respects the symmetries of tensor fields. Moreover, it is defined by Weitzenböck decomposition formulas

$$\Delta_L = \bar{\Delta} + \mathfrak{R}(R^\nabla) = \Delta_S + 2\,\mathfrak{R}(R^\nabla),$$

where the Weitzenböck curvature operator $\mathfrak{R}(R^\nabla)$ is an algebraic operator, which is the restriction of the Weitzenböck curvature operator to symmetric p-tensors (see also [10, 29]). The Lichnerowicz Laplacian acting on symmetric covariant tensor fields is of fundamental importance in mathematical physics (see, for instance, [22,

49]). Note also that the Lichnerowicz Laplacian acting on symmetric covariant 2-tensor fields appears in many problems in Riemannian geometry (for example, [7, 10, 29]).

Example 5 (The Bourguignon Laplacian) Let $t = 1$ and $E = C^{\infty}(S^2M)$. The Lichnerowicz type Laplacian Δ_L is the well-known Bourguignon Laplacian Δ_B acting on C^{∞}-sections of S^2M on (M, g) (see [11, p. 273]; [55]). The operator Δ_B admits the Weitzenböck decomposition with $t = 1$, where $\mathfrak{R}(R^{\nabla})$ is an algebraic operator, which is the restriction of the Weitzenböck curvature operator to symmetric 2-tensors. Also, Δ_B is a fundamental operator in the context of Ricci flow, it seems to be more natural than the connection or Bochner Laplacian $\bar{\Delta}$. An example of such naturalness is the appearance of Δ_B in the linearized Ricci flow equation (e.g., the evolution formula of the Ricci tensor under the Ricci flow in [16, p. 112]).

Example 6 (The spinor Laplacian) Let $t = \frac{1}{4}$ and $E \to M$ be the bundle of spinors over a closed Riemannian manifold (M, g). The Lichnerowicz type Laplacian Δ_L is the spinor Laplacian Δ_D and the Weitzenböck decomposition formula (3) has the form

$$\Delta_D = \bar{\Delta} + \frac{1}{4}\,\mathrm{Scal},$$

where $\mathfrak{R}(R^{\nabla}) = \mathrm{Scal}$ for the scalar curvature Scal of (M, g) (see [44, p. 352]; [34]; [7, p. 55]). In addition, Δ_D is the square D^2 of the Dirac operator D acting on C^{∞}-sections of $E = \Sigma\, M$ (see also [7, p. 55]; [34]). Elements of $\ker \Delta_D$ are called *harmonic spinors* (e.g., [7, p. 170]; [34]). Let us use formula (3), then, if $\mathrm{Scal} > 0$, the condition $\Delta_D\, \varphi = 0$ implies that the harmonic spinor φ is equal to zero, since

$$0 = \int_M g(\Delta_D\, \varphi,\ \varphi)\, \Omega_g = \int_M \|\nabla\varphi\|^2\, \Omega_g + \frac{1}{4}\int_M \mathrm{Scal}\cdot \|\varphi\|^2\, \Omega_g.$$

Therefore, on a closed manifold (M, g) with nonnegative scalar curvature, every harmonic spinor field is parallel. Furthermore, if the scalar curvature is positive at some point of (M, g), then every harmonic spinor field vanishes. In addition, Gromov-Lawson showed that any metric on a torus with $\mathrm{Scal} \geq 0$ is flat (see [44, p. 352]). Note that Dirac operators and their Weitzenböck decomposition formulas were also extremely important in physics and the theory of four-dimensional manifolds (see also [44, p. 352]).

Example 7 (The Einstein Laplacian) Let $t = 1$, $E = S^2M$ and $\mathfrak{R}(R^{\nabla}) = -2\,\overset{o}{R}$, where $\overset{o}{R} : S^2M \to S^2M$ is the symmetric *curvature operator of the second kind* (see [7, pp. 51–52]). The Einstein Laplacian is denoted as $\Delta_E = \bar{\Delta}\,\varphi - 2\,\overset{o}{R}$. This is a self-adjoint elliptic operator, and on a compact (M, g) it has a discrete spectrum. In particular, the Einstein Laplacian Δ_E maps the infinite-dimensional vector space of TT-tensors to itself (see also [28]). We recall that a TT-tensor (Transverse Traceless tensor) is a divergence free and traceless covariant symmetric 2-tensor field. If a *TT*-tensor φ belongs to ker Δ_E, then it can be seen as an *infinitesimal Einstein deformation* through g (see details in [30, pp. 54 and Chap. 12]; [28]; [7, pp. 346–348]). Many authors have studied infinitesimal Einstein deformations, that are symmetric 2-tensors φ which are *TT*-tensors and satisfy the *linearized Einstein equation* $\Delta_E\,\varphi = \bar{\Delta}\,\varphi - 2\overset{o}{R} = 0$. By ellipticity of the involved operator, the space of infinitesimal Einstein deformations is finite dimensional since (M, g) is closed, i.e., $\dim_{\mathbb{R}} \ker \Delta_E < \infty$. Moreover, if $\varphi \in \ker \Delta_E$, then from (6) we obtain the following inequality: $\|\varphi\| \cdot \Delta\, \|\varphi\| \geq g(\overset{o}{R}\varphi, \varphi)$. Remark that the sectional curvature *sec* of a Riemannian manifold (M, g) with the nonnegative curvature operator of the second kind $\overset{o}{R}$ is also nonnegative since

$$g(\overset{o}{R}(X \odot Y), X \odot Y) = 2\, g(R(X, Y)Y, X),$$

where $X \odot Y = \frac{1}{2}\,(X \otimes Y + Y \otimes X)$ for any $X, Y \in T_x M$ at any $x \in M$. Recall that a complete noncompact Riemannian manifold with nonnegative sectional curvature has infinite volume (see [23]).

Therefore, based on Theorem 3, we get the following.

Corollary 2 *Let Δ_E be the Einstein Laplacian defined on a complete noncompact Riemannian manifold (M, g) with nonnegative curvature operator of the second kind $\overset{o}{R}$. Then there is no nonzero $\varphi \in L^q\,(\ker \Delta_E)$ for any $q > 1$.*

3 The Lichnerowicz Laplacian

Here, we consider the classical Lichnerowicz Laplacian using the Bochner technique, using, in particular, Lichnerowitz's article [33] and some of our articles, for example, [64].

3.1 Vanishing Theorems for the Kernel of the Lichnerowicz Laplacian

An example of the Lichnerowicz type Laplacian on (M, g) is the ordinary Lichnerowicz Laplacian defined by the Weitzenböck decomposition formula (1). In this case, the Weitzenböck curvature operator $\mathfrak{R}(R^\nabla) : \bigotimes^p T^*M \to \bigotimes^p T^*M$ satisfies the following equalities:

$$g(\mathfrak{R}(R^\nabla)\,\varphi, \varphi') = g(\varphi, \mathfrak{R}(R^\nabla)\,\varphi'), \tag{8}$$

$$\operatorname{trace}_g\left(\mathfrak{R}(R^\nabla)\,\varphi\right) = \mathfrak{R}(R^\nabla)\left(\operatorname{trace}_g \varphi\right), \tag{9}$$

for any $\varphi, \varphi' \in C^\infty\left(\bigotimes^p T^*M\right)$, see [33, p. 315]. From (8)–(9) we get the equality

$$\mathrm{trace}_g\,(\Delta_L\,\varphi) = \Delta_L\left(\mathrm{trace}_g\,\varphi\right). \tag{10}$$

Important examples of the Lichnerowicz Laplacian are the Hodge-de Rham Laplacian and the Bourguignon Laplacian, which we will consider below. We recall that the Riemann curvature tensor Rm of (M, g) defines a self-adjoint operator $\widehat{R} : \Lambda^2(T_xM) \to \Lambda^2(T_xM)$, where $\Lambda^2(T_xM)$ is the second power of T_xM at any $x \in M$, by the explicit formula (see [44, p. 82–83])

$$g(\widehat{R}(X \wedge Y),\, Z \wedge W) = Rm(X, Y, W, Z), \quad X, Y, Z, W \in TM.$$

This operator is called the *curvature operator of the first kind*. It is said to be nonnegative (resp. nonpositive) if all its eigenvalues are nonnegative (resp. nonpositive) at each point $x \in M$. By the above,

$$g(\Re(R^\nabla)\,\varphi_x,\, \varphi_x) = \sum\nolimits_\alpha \Lambda_\alpha\,\|\Xi_\alpha\,\varphi_x\|^2, \tag{11}$$

see [44, p. 346]. Here, Ξ_α is an orthonormal basis of eigenvectors for $\widehat{R}$, i.e., $\widehat{R}\,(\Xi_\alpha) = \Lambda_\alpha \Xi_\alpha$. Thus, if $\widehat{R}$ is nonnegative on (M, g), then $g(\Re(R^\nabla)\,\varphi,\, \varphi)$ is also nonnegative. More generally, if $\widehat{R} \geq \gamma > 0$ for some real $\gamma > 0$ at some point $x \in M$, then $g(\Re(R^\nabla)\,\varphi_x,\, \varphi_x) \geq \gamma \sum_\alpha \|\Xi_\alpha\,\varphi_x\|^2 > 0$ for $\varphi_x \neq 0$. Thus, using (3) and Theorem 2, we can formulate the following.

Theorem 5 *Let Δ_L be the Lichnerowicz Laplacian on C^∞-sections of a smooth tensor bundle $\bigotimes^p T^*M \to M$ over a closed Riemannian manifold (M, g). If $\widehat{R} \geq 0$ holds,*

(i) then $\nabla\varphi = 0$ for any $\varphi \in \ker \Delta_L$ and $\dim_{\mathbb{R}} \ker \Delta_L \leq p\,n$;
(ii) and $\widehat{R} \geq \gamma > 0$ at some point of M for some constant $\gamma > 0$, then $\dim_{\mathbb{R}} \ker \Delta_L = 0$.

Remark 3 It is well known (see, for example, [7, p. 282]) that if there exists on (M, g) a tensor field φ which has zero covariant derivative, i.e., $\nabla\,\varphi = 0$, then φ is invariant of the holonomy representation $\mathrm{Hol}(g)$ at each point $x \in M$. The converse is also true. Therefore, finding on a Riemannian manifold (M, g) parallel forms is equivalent to the algebraic problem of finding invariant forms of the holonomy representation $\mathrm{Hol}(g)$.

Let, in particular, the Lichnerowicz Laplacian Δ_L be defined on C^∞-sections of a smooth tensor bundle $\bigotimes^2 T^*M \to M$ over (M, g). In this case, by (10), $\mathrm{trace}_g\,\varphi$ is a harmonic function for any $\varphi \in \ker \Delta_L$. Therefore, the following theorem holds.

Theorem 6 *Let Δ_L be the Lichnerowicz Laplacian on C^∞-sections of a smooth tensor bundle $\bigotimes^2 T^*M \to M$ over a closed Riemannian manifold (M, g). Then $\mathrm{trace}_g\,\varphi = const$ for any $\varphi \in \ker \Delta_L$.*

It is known that a Riemannian symmetric space of compact type has nonnegative sectional curvature. For such a symmetric space, the curvature operator is nonnegative definite (see [14, p. 113–114]). Hence, we obtain the following.

Corollary 3 *Let (M, g) be a Riemannian symmetric space of compact type and Δ_L the Lichnerowicz Laplacian on C^∞-sections of a smooth tensor bundle $\bigotimes^p T^*M \to M$. Then $\nabla\varphi = 0$ for any $\varphi \in \ker \Delta_L$.*

Let (M, g) be a complete Riemannian manifold with nonnegative curvature operator $\widehat{R}$. An elementary linear algebra argument shows that if the curvature operator is nonnegative, then all sectional curvatures of (M, g) are also nonnegative. At the same time, a connected complete noncompact Riemannian manifold with nonnegative sectional curvature has infinite volume (see [23]).

Thus, using Theorem 3, we can formulate the following.

Theorem 7 *Let (M, g) be a connected complete noncompact Riemannian manifold and Δ_L the Lichnerowicz Laplacian on C^∞-sections of a smooth tensor bundle $\bigotimes^p T^*M \to M$. If the curvature operator $\widehat{R}$ is nonnegative on M, then $L^q(\ker \Delta_L)$ is trivial for any $q > 1$.*

If $p = 1$, then $\mathfrak{R}(R^\nabla) = Ric$ for the Ricci tensor Ric of (M, g). In addition, every complete noncompact Riemannian manifold with nonnegative Ricci curvature has infinite volume (see [72]). Thus, from Theorem 3 obtain the following.

Corollary 4 *Let (M, g) be a connected complete noncompact Riemannian manifold and Δ_L be the Lichnerowicz Laplacian on C^∞-sections of T^*M. If the Ricci curvature is nonnegative on M, then $L^q(\ker \Delta_L)$ is trivial for any $q > 1$.*

3.2 Estimates of Eigenvalues of the Lichnerowicz Laplacian

An eigentensor of Δ_L is a tensor field $\varphi \in C^\infty(\bigotimes^p T^*M)$ satisfying the condition $\Delta_L \varphi_x = \lambda^{[p]} \varphi_x$ at each point $x \in M$, where $\lambda^{[p]}$ is a constant called the eigenvalue of Δ_L corresponding to φ. Note that the superscript $[p]$ indicates the degree of the eigentensor φ. Since Δ_L is an elliptic operator, we conclude that on a closed manifold (M, g), the subspace of p-eigentensors of Δ_L corresponding to an eigenvalue $\lambda^{[p]}$ has finite dimension; moreover, the set of all eigenvalues of the Lichnerowicz Laplacian is discrete.

Let $\Delta_L \varphi_x = \lambda^{[p]} \varphi_x$ for $\lambda^{[p]} \in \mathbb{R}$ at any point $x \in M$, then from (4) we get

$$\frac{1}{2}\,\Delta\,\|\varphi\|^2 = -\lambda^{[p]}\,\|\varphi\|^2 + t\,g(\mathfrak{R}(R^\nabla)\,\varphi, \varphi) + \|\nabla\varphi\|^2\,. \tag{12}$$

If the curvature operator $\widehat{R}$ of (M, g) satisfies the condition $\widehat{R} \geq \gamma$ for some $\gamma < 0$, then there is a constant $C(n, p) > 0$ depending only on the type of the eigentensor φ and $\dim M$ such that (see [44, Corollary 9.3.4])

$$g(\mathfrak{R}(R^{\nabla})\,\varphi_x, \varphi_x) \geq \gamma\, C(p, n)\, \|\varphi_x\|^2 .$$

In this case, from (12) we conclude that

$$\frac{1}{2}\Delta\ \|\varphi\|^2 \geq \left(t\,\gamma C(p, n) - \lambda^{[p]}\right) \|\varphi_x\|^2 .$$

Using this inequality, we can compose the following integral inequality:

$$\left(\lambda^{[p]} - t\,\gamma\, C(p, n)\right) \int_M \|\varphi\|^2\ \Omega_g \geq 0.$$

As a result we have the inequality $\lambda^{[p]} \geq t\,\gamma\, C(p, n)$ for any $t > 0$, $\gamma < 0$ and some $C(p, n) > 0$. Let us now turn to the first nonzero eigenvalue $\lambda_1^{[p]}$ of Δ_L.

Thus, we have proved the following theorem.

Theorem 8 *Let* $\Delta_L : C^\infty(\bigotimes^p T^*M) \to C^\infty(\bigotimes^p T^*M)$ *be the Lichnerowicz Laplacian on a closed Riemannian manifold* (M, g)*. Then there exists a constant* $C(n, p) > 0$ *such that* $\lambda_1^{[p]} \geq C(n, p)\,\gamma$ *for a lower bound* $\gamma < 0$ *of the eigenvalues of the curvature operator* $\widehat{R}$ *of* (M, g)*.*

Many other properties, in particular, the spectral properties of the Lichnerowicz Laplacian can be found in [44] and, for example, in [64].

4 The Hodge-de Rham Laplacian

The Laplace operator acting on smooth functions has a natural extension to differential forms. In the theory of this operator, also called the Hodge-de Rham Laplacian, one can see the interplay between analysis, topology and geometry. Here, we provide an overview of this theory.

4.1 Vanishing Theorems for the Kernel of Hodge-de Rham Laplacian

The well-known example of the Lichnerowicz type Laplacian Δ_L is the Hodge-de Rham Laplacian Δ_H acting on C^∞-sections of $\Lambda^p M$ on (M, g). It is defined by $\Delta_H = d\delta + \delta d$, where d is the exterior derivative and $\delta = d^*$ is its formal adjoint with respect to L^2-product on forms. The Hodge decomposition theorem states that we have the following orthogonal direct sum:

$$C^\infty(\Lambda^p M) = \ker \Delta_H \oplus d(\Lambda^{p-1} M) \oplus \delta(\Lambda^{p+1} M)$$

for the space of smooth p-forms with compact support. An arbitrary Δ_H-harmonic section of $\Lambda^p M$ is called a *harmonic p-form.* By Hodge theory, the p-th Betti number of a closed manifold M is precisely $\beta_p(M) = \dim_{\mathbb{R}} \ker \Delta_H$—the dimension of the space of harmonic p-forms on (M, g). It is well known that a closed Riemannian manifold with positive curvature operator $\widehat{R} : \Lambda^2 M \to \Lambda^2 M$ has vanishing p-th Betti number $\beta_p(M)$, and if the curvature operator is nonnegative, then the p-th Betti number is less than or equal to $\beta_p(M) \leq \binom{n}{p}$ for $p = 1, \dots, n-1$, see [44]. Thus, the following statement (see [44, p. 351]) is a corollary of Theorem 3.

Corollary 5 *Let Δ_H be the Hodge-de Rham Laplacian on C^∞-sections of the smooth bundle $\Lambda^p M \to M$ of differential p-forms over a closed Riemannian manifold (M, g). If $\widehat{R} \geq 0$ holds,*

(i) then $\nabla\varphi = 0$ for any $\varphi \in \ker \Delta_L$, and $\dim_{\mathbb{R}} \ker \Delta_L = \beta_p(M) \leq \binom{n}{p}$ for $p = 1, \dots, n-1$.
(ii) and $\widehat{R} \geq \gamma > 0$ at some point on M and for some real $\gamma > 0$, then $\dim_{\mathbb{R}} \ker \Delta_L = \beta_p(M) = 0$ for $p = 1, \dots, n-1$.

These facts refer to the classical Bochner technique and were established using formula (3) and Theorem 5. On the other hand, on a closed n-dimensional Riemannian manifold (M, g) the normalized Ricci flow evolves a Riemannian metric with positive curvature operator to a limit metric with constant positive sectional curvature (see [9]); in this case, $\beta_p(M) = 0$ for any $p = 1, \dots, n-1$. In particular, if $p = 1$, then $\mathfrak{R}(R^\nabla) = Ric$ for the Ricci tensor Ric of (M, g). Using (2) and Theorem 2 for $t = 1$, it was proved that if (M, g) is closed and satisfies $Ric \geq 0$, then $\beta_1(M) \leq n = \dim M$. In addition, the equality $\beta_1(M) = n$ holds if and only if (M, g) is a flat torus (see [44, p. 338]). On the other hand, if the Ricci curvature of (M, g) is nonnegative and positive definite at some point $x \in M$, then the first Betti number $\beta_1(M)$ is equal to zero (see [6, p. 375]).

Consider a complete Riemannian manifold (M, g) with nonnegative curvature operator $\widehat{R}$. In this case, (11) can be rewritten in the form

$$g(\mathfrak{R}(R^\nabla)\,\varphi_x,\ \varphi_x) = \sum\nolimits_\alpha \Lambda_\alpha \, \|\Xi_\alpha \, \varphi_x\|^2 \geq 0,$$

where Ξ_α is an orthonormal basis of eigenvectors for the curvature operator $\widehat{R}$, i.e., $\widehat{R}\,(\Xi_\alpha) = \Lambda_\alpha \Xi_\alpha$ at each point $x \in M$. In turn, (6) can be rewritten as

$$\|\varphi\| \cdot \Delta\, \|\varphi\| \geq 0$$

for harmonic p-forms $\varphi \in \ker \Delta_H$. Thus, the following theorem, which follows from Theorem 3 and Theorem 7, holds.

Theorem 9 *Let (M, g) be a connected complete noncompact Riemannian manifold and Δ_H be the Hodge-de Rham Laplacian on C^∞-sections of a smooth vector bundle*

$\Lambda^p M$ of differential p-forms over (M, g). If the curvature operator $\widehat{R}$ of (M, g) is nonnegative on M, then L^q (ker Δ_H) is trivial for any $q > 1$.

If $p = 1$, then $\mathfrak{R}(R^\nabla) = Ric$ for the Ricci tensor Ric of (M, g). Also, a complete noncompact Riemannian manifold with nonnegative Ricci curvature has infinite volume (see [72]). Thus, from Theorem 3 we get the following (see [23]).

Corollary 6 *Let (M, g) be a connected complete noncompact Riemannian manifold and Δ_H the Hodge-de Rham Laplacian on C^∞-sections of T^*M. If the Ricci curvature of (M, g) is nonnegative, then L^q (ker Δ_H) is trivial for any $q > 1$.*

We recall that the description of all complete locally irreducible Riemannian manifolds with a form whose covariant derivative vanishes is given in Corollary 10.110 from [7, pp. 306–307]. Namely, the following statement holds:

"Let (M, g) be a complete locally irreducible Riemannian manifold admitting a nonzero and not decomposable form φ whose covariant derivative is vanishing. Then (M, g) should fall in one of the following categories: locally symmetric manifolds, Kähler manifolds, quaternionic-Kähler manifolds and manifolds with holonomy group G_2 or Spin(7)".

Next, we consider a complete noncompact locally irreducible Riemannian manifold with nonnegative curvature operator, which admits a nonzero form from $L^2(\ker \Delta_H)$. Recall that a complete, simply-connected locally symmetric space is globally symmetric, and a Riemannian globally symmetric space with nonnegative sectional curvature is a symmetric space of compact type and at the same time it is a compact manifold. A Riemannian globally symmetric space of noncompact type is complete and has a nonpositive sectional curvature. Therefore, we exclude these two cases from consideration.

Note that manifolds with holonomy group G_2 or Spin(7) are Ricci-flat manifolds. On the other hand, every complete noncompact Riemannian manifold with nonnegative Ricci curvature has infinite volume (see [72, p. 667]). Therefore, a complete noncompact Riemannian manifold with holonomy group G_2 or Spin(7) has infinite volume and has no a nonzero parallel form φ (since a parallel form has a constant length $\|\varphi\|$, see the proof of Theorem 6). If φ is an L^2-form, i.e., $\int_M \|\varphi\|^2 \, \Omega_g < \infty$, then $\mathrm{Vol}(M, g) < \infty$. The last inequality contradicts the assumption about infinite volume of the manifold (M, g).

These remarks complete the proof of the following.

Corollary 7 *Let (M, g) be a complete noncompact locally irreducible Riemannian manifold with nonnegative curvature operator. If (M, g) admits a nonzero and not decomposable form from $L^2(\ker \Delta_H)$, then it is a Kähler or quaternionic Kähler manifold.*

4.2 Estimates of Eigenvalues of the Hodge-de Rham Laplacian

Let us now turn to the first positive eigenvalue of the Hodge Laplacian Δ_H, which we will denote by $\lambda_1^{[p]}$. The estimate by Gallot and Meyer in [20] shows that

$$g(\mathfrak{R}(R^\nabla)\,\varphi,\varphi) \geq p(n-p)\,\gamma\,\|\varphi\|^2,$$

where γ is a lower bound of the eigenvalues of the curvature operator $\widehat{R}$ of (M,g). In this case, from [15, Theorem C1] we conclude that $\lambda_1^{[p]} \geq p(n-p)\,\gamma$ on $(M,\bar{g})$. Moreover, the Gallot-Meyer estimate and (3) are the main tools in the proof of the following lower bound.

Theorem 10 (see *[20]*) *Let (M,g) be an n-dimensional closed Riemannian manifold and $\gamma>0$ be the lowest eigenvalue of its curvature operator. Then the inequality $\lambda_1^{[p]} \geq C(n,\ p)\,\gamma$ holds for all $p=1,\dots,n-1$, where $C(n,\ p)=\min\{p(n-p+1),\ (p+1)(n-p)\}$ for the first eigenvalue of the Hodge Laplacian Δ_H. The equality holds for the Euclidean sphere $\mathbb{S}^n$.*

Remark 4 A closed Riemannian manifold (M,g) with positive curvature operator is diffeomorphic to a spherical space form, see [9]. The proof, in all dimensions, uses the normalized Ricci flow to evolve an initial metric with positive curvature operator into a limit metric with constant positive sectional curvature. This result makes assertion (ii) of Corollary 5 obvious. Other facts of the spectral theory of the Hodge-de Rham Laplacian can be found in [18].

Recall that an n-dimensional ($n\geq 2$) connected manifold M with a Riemannian metric g is called an Einstein manifold with Einstein constant α, if its Ricci tensor satisfies $Ric=\alpha\,g$; moreover, we have $\alpha=s/n$ for its scalar curvature s. Therefore, any Einstein manifold of dimensions two and three is a space form (i.e., has constant sectional curvature). The study of Einstein manifolds is more complicated in dimension higher than three (for example, [7, p. 44]). In particular, we proved the following theorem, see [56].

Theorem 11 *Let (M,g) be a compact Einstein manifold with positive Einstein constant α and sectional curvature bounded from below by real $\delta>0$ such that $\delta>\alpha/(n+2)$. Then the first eigenvalue $\lambda_1^{[p]}$ of the Hodge Laplacian Δ_H satisfies the following inequality:*

$$\lambda_1^{[p]} \geq (1/3)((n+2)\delta-\alpha)(n-p).$$

5 The Sampson Laplacian

Here, we consider the little-known Sampson operator that is strongly elliptic and self-adjoint second order differential operator acting on covariant symmetric tensors on Riemannian manifolds. We study this operator using the analytical method, due to Bochner, of proving vanishing theorems for the null space of a Laplace operator admitting a Weitzenböck decomposition. Furthermore, we estimate operator's lowest eigenvalue.

5.1 Vanishing Theorems for the Kernel of the Sampson Laplacian

The Sampson Laplacian $\Delta_S : C^\infty(S^pM) \to C^\infty(S^pM)$ is the second order differential operator defined by the formula $\Delta_S := \delta\delta^* - \delta^*\delta$ (for example, [35]), where $\delta^* = (p+1)\mathrm{Sym}\,\nabla$ for the linear symmetrization operator $Sym : T^*M \bigotimes S^pM \to S^{p+1}M$ and $\delta : S^{p+1}M \to S^pM$ its formal adjoint operator with respect to the L^2-global product, which is called the *divergence operator* (see [7, p. 356]). Notice that δ is the $\bigotimes^{p+1} T^*M$ restriction of ∇^* to $S^{p+1}M$. The Weitzenböck decomposition formula of Δ_S has the form (1) with $t = -1$, where $\mathfrak{R}(R^\nabla)$ is an algebraic operator, which is the restriction of the Weitzenböck curvature operator to symmetric p-tensors S^pM (see [58]), i.e.,

$$\Delta_S = \bar{\Delta} - \mathfrak{R}(R^\nabla). \tag{13}$$

For particular case of 1-forms, we have $\Delta_S = \bar{\Delta} - Ric$. On a closed Riemannian manifold (M, g) the following L^2-orthogonal decomposition holds:

$$C^\infty(S^pM) = \mathrm{Im}\,\Delta_S \oplus \ker \Delta_S,$$

where the space $\ker \Delta_S$ consists of Δ_S-harmonic symmetric p-tensors on (M, g).

Consider an example with Δ_S-harmonic symmetric p-tensor fields. Namely, a tensor field $\varphi \in C^\infty(S^pM)$, which satisfies the equation $\delta^*\varphi = 0$, is well known in the theory of General Relativity as a *symmetric Killing tensor* (see, for example, [26]). This is a natural generalization of the *Killing vector field* equation for $p = 1$. A divergence-free symmetric Killing p-tensor is determined by two conditions $\delta^*\varphi = 0$ and $\delta\,\varphi = 0$. It is obvious that a divergence-free symmetric Killing p-tensor φ belongs to $\ker \Delta_S$; thus, this tensor is an example of Δ_S-harmonic symmetric tensors. If a Riemannian manifold (M, g) is closed, then the converse is also true (see [35]). A traceless symmetric Killing p-tensor ($p \geq 2$) is a smooth section of a subbundle $S_0^pM \subset S^pM$ of a covariant symmetric p-tensor fields, which are totally traceless, that is, $\mathrm{trace}_g\,\varphi = 0$ on any pair of indices at any point $x \in M$. Obviously, every traceless Killing tensor is a divergence-free tensor. Thus, any traceless Killing p-tensor ($p \geq 2$) is an example of Δ_S-harmonic symmetric tensors. From (3) we obtain the following integral equation:

$$\int_M g(\Delta_L\,\varphi,\ \varphi)\,\Omega_g = \int_M \|\nabla\varphi\|^2\ \Omega_g - \int_M g(\mathfrak{R}(R^\nabla)\,\varphi, \varphi)\,\Omega_g.$$

Furthermore, if $\Delta_L \varphi = 0$, then we refer to the Berger-Ebin Theorem (see also [7, p. 436]) and formulate the following theorem based on it.

Theorem 12 *Let Δ_S be the Sampson Laplacian defined on C^∞-sections of a smooth tensor bundle $S^2M \to S^2M$ over a closed Riemannian manifold (M, g). If the sectional curvature of (M, g) is nonpositive, then $\nabla\varphi = 0$ for any $\varphi \in \ker \Delta_S$ and $\dim_{\mathbb{R}} \ker \Delta_S \le \binom{n+1}{2}$. Moreover, if (M, g) has negative sectional curvature at some point, then $\varphi = \lambda\, g$ for some $\lambda \in \mathbb{R}$.*

The Sampson Laplacian Δ_S maps $S_0^p M$ to itself (see [35]). Moreover, the inequality $g(\mathfrak{R}(R^\nabla)\,\varphi, \varphi) \le 0$ holds at every point of (M, g) for every tensor field $\varphi \in C^\infty(S_0^p M)$ on a manifold (M, g) with nonpositive sectional curvature (see [26]). Moreover, if $\Delta_S \varphi = 0$ for any $\varphi \in C^\infty(S_0^p M)$, then from (13) we obtain the following integral inequality:

$$0= \int_M \|\nabla\varphi\|^2\, \Omega_g - \int_M g(\mathfrak{R}(R^\nabla)\,\varphi, \varphi)\, \Omega_g \ge 0.$$

Thus, using Theorem 2 we can formulate the following theorem.

Theorem 13 *Let Δ_S be the Sampson Laplacian on C^∞-sections of a smooth tensor bundle $S_0^p M \to S_0^p M$ over a closed Riemannian manifold (M, g). If the sectional curvature of (M, g) is nonpositive, then $\nabla\varphi = 0$ for any $\varphi \in \ker \Delta_S$ and*

$$\dim_{\mathbb{R}} \ker \Delta_S \le \binom{n+p-1}{p} - \binom{n+p-3}{n-1}.$$

For $p = 1$, the kernel of Δ_S consists of infinitesimal harmonic transformations of (M, g) (see [35, 36]), holomorphic vector fields on nearly Kählerian manifolds and the vector fields that transform Riemannian metrics into a Ricci soliton metrics (see [33, 72]). In addition, if $p = 1$, then $\mathfrak{R}(R^\nabla) = Ric$ for the Ricci tensor Ric of (M, g). In this case, using (1) and Theorem 2 for $t = -1$, the Sampson Laplacian Δ_S has a trivial kernel on a closed Riemannian manifold (M, g) with quasi-negative Ricci curvature.

Let us consider a Hadamard manifold, which, by definition, is a complete simply connected nonpositive curved manifold (M, g) (see [32, p. 381]). A Hadamard manifold has infinite volume. In addition, for every symmetric two-tensor field φ, the inequality $g(\mathfrak{R}(R^\nabla)\,\varphi, \varphi) \le 0$ holds at every point of a Riemannian manifold (M, g) with nonpositive sectional curvature for (see [7, p. 436]). Using the above arguments, we proved the following theorem in [35].

Theorem 14 *Let (M, g) be an Hadamard manifold, then there are no nonzero Δ_S-harmonic symmetric tensors $\varphi \in C^\infty\left(S^2M\right)$ such that $\varphi \in L^q$ (ker Δ_S) for at least one $q \in (0, \infty)$.*

Let $\varphi \in C^{\infty}(S_0^p M)$, on a complete manifold (M, g) with nonpositive sectional curvature, then the inequality $g(\mathfrak{R}(R^{\nabla})\,\varphi, \varphi) \leq 0$ holds on (M, g) (see [26]), and from (4) we get

$$\frac{1}{2}\,\Delta\,\|\varphi\|^2 = -g(\mathfrak{R}(R^{\nabla})\,\varphi, \varphi) + \|\nabla\varphi\|^2 \geq 0.$$

Recall that a nonnegative subharmonic L^q-function for $q \in (0, \infty)$ on a complete simply connected manifold (M, g) of nonpositive sectional curvature is constant (see [31]). In particular, if the traceless symmetric p-tensor $\varphi \in L^q\,(\ker \Delta_S)$ for $q \in (0, \infty)$ is defined on such a manifold, then $\|\varphi\| = const$. Thus, the following our theorem for Hadamard manifolds holds.

Theorem 15 *Let (M, g) be an Hadamard manifold, then there is no nonzero Δ_S-harmonic tensor $\varphi \in C^{\infty}(S_0^p M)$ such that $\varphi \in L^q\,(\ker \Delta_S)$ for at least one $q \in (0, \infty)$.*

5.2 Spectral Properties of the Sampson Laplacian

Here, we consider the Sampson Laplacian $\Delta_S : C^{\infty}(S^2M) \to C^{\infty}(S^2M)$ on a closed manifold (M, g). Denote by $\lambda^{[2]}$ an arbitrary eigenvalue of Δ_S acting on symmetric 2-tensor fields, i.e., $\Delta_S\,\varphi = \lambda^{[2]}\,\varphi$ for some $\varphi \in C^{\infty}(S^2M)$. We have the following theorem (see [35]).

Theorem 16 *Let (M, g) be an n-dimensional $(n \geq 2)$ closed connected Riemannian manifold with negative sectional curvature and $\Delta_S : C^{\infty}(S^2M) \to C^{\infty}(S^2M)$ the Sampson Laplacian. Then all eigenvalues of Δ_S are positive and Spec $\Delta_S = \{0 < \lambda_1^{[2]} \leq \lambda_2^{[2]} \leq \ldots \to \infty\}$.*

We will estimate the first eigenvalue $\lambda_1^{[2]}$ of the Sampson Laplacian when its eigentensor φ has $\operatorname{trace}_g\varphi \neq 0$. Thus, the following theorem is true (see [35]).

Theorem 17 *Let (M, g) be an n-dimensional $(n \geq 2)$ closed connected Riemannian manifold with diameter $d(M)$ and the Ricci curvature $Ric \geq (n-1)K$ for some $K \in \mathbb{R}$. If the eigentensor $\varphi \in C^{\infty}(S^2M)$ corresponding to the first nonzero eigenvalue $\lambda_1^{[2]}$ of Sampson Laplacian has a nonconstant* $\operatorname{trace}_g \varphi$*, then*

1. $\lambda_1^{[2]} \geq nK$ *when* $K > 0$;
2. $\frac{\pi^2}{4d(M)^2} \leq \lambda_1^{[2]} \leq \frac{n\pi^2}{d(M)^2}$ *when* $K = 0$;
3. $\lambda_1^{[2]} \geq \frac{\exp -1-\sqrt{1-4(n-1)^2 d(M)^2 K}}{2(n-1)^2 d(M)^2}$ *when* $K < 0$.

In particular, if (M, g) is a compact hyperbolic manifold $(\mathbb{H}^n, g_0)$ with $n \geq 3$ and standard metric g_0 having constant sectional curvature equal to -1, then there exists a constant $C = C(n) > 0$ such that $\lambda_1 \geq C \cdot \mathrm{Vol}(M, g)^{-2}$.

6 The Bourguignon Laplacian

Here, we show that the kernel of the little-known Bourguignon Laplacian is a finite-dimensional vector space of harmonic symmetric bilinear forms on a closed Riemannian manifold (see [61, p. 350]). We also prove that every harmonic symmetric bilinear form on a compact Riemannian manifold with nonnegative sectional curvature is invariant under parallel translations. In addition, we investigate the spectral properties of the little studied Bourguignon Laplacian.

6.1 Vanishing Theorems for the Kernel of the Bourguignon Laplacian

We will consider a symmetric bilinear form $\varphi \in C^\infty(S^2M)$ as a one-form with values in the cotangent bundle T^*M over M. This bundle comes equipped with the Levi-Civita covariant derivative ∇; thus, there is an induced exterior differential $d^\nabla : C^\infty(S^2M) \to C^\infty(\Lambda^2 M \otimes T^*M)$ on T^*M-valued differential one-forms such that

$$d^\nabla \varphi(X, Y, Z) := (\nabla_X \varphi)(Y, Z) - (\nabla_Y \varphi)(X, Z)$$

for any vector fields $X,\ Y,\ Z$ on M and any $\varphi \in C^\infty(S^2M)$.

Remark 5 The theory on T^*M-valued differential one-forms can be found in the following monographs: [7, p. 133–134; 355], [44, p. 349–350].

J.-P. Bourguignon defined in [46, p. 273] the Laplacian $\Delta_B : C^\infty(S^2M) \to C^\infty(S^2M)$ by the formula $\Delta_B := \delta^\nabla d^\nabla + d^\nabla \delta^\nabla$, where $\delta^\nabla : C^\infty(\Lambda^2 M \otimes T^*M) \to C^\infty(S^2M)$ is the formal adjoint operator of the exterior differential d^∇. If (M, g) is a compact Riemannian manifold, then by direct computations we obtain the following integral formula:

$$\langle \Delta_B \varphi,\ \varphi \rangle = \langle d^\nabla \varphi,\ d^\nabla \varphi \rangle + \langle \delta^\nabla \varphi,\ \delta^\nabla \varphi \rangle.$$

Based on this formula, we conclude that the *Bourguignon Laplacian* Δ_B is a nonnegative operator. Recall that a symmetric bilinear form $\varphi \in C^\infty(S^2M)$ on a Riemannian manifold (M, g) is Δ_B-harmonic if and only if $\varphi \in \ker d^\nabla \cap \ker \delta^\nabla$. Therefore, a symmetric bilinear form $\varphi \in C^\infty(S^2M)$ on (M, g) is harmonic if and only if it is a *Codazzi tensor* with constant trace (see [44, p. 350]). An easy computation yields the Weitzenböck decomposition formula, $\Delta_B \varphi = \bar{\Delta} \varphi + \mathfrak{R}(R^\nabla)\varphi$, see also [7, p. 355]. Moreover, it has the form $\mathfrak{R}(R^\nabla)\varphi := \varphi \circ Ric - \overset{o}{R}\varphi$, where $\circ$ is a composition of endomorphisms and the curvature operator of the second kind $\overset{o}{R}$. Thus, the classical Berger-Ebin Theorem holds (see [7, p. 436]).

Theorem 18 *Every* Δ_B*-harmonic form* $\varphi \in C^\infty(S^2M)$ *on a closed Riemannian manifold* (M, g) *with nonnegative sectional curvature is parallel. Moreover, if the sectional curvature of* (M, g) *is positive at some its point, then* $\varphi = \lambda g$ *for some nonzero* $\lambda \in \mathbb{R}$.

The following theorem supplements the Berger-Ebin Theorem for the case of a complete noncompact Riemannian manifold.

Theorem 19 *Let* (M, g) *be a complete simply connected Riemannian manifold with nonnegative sectional curvature. Then* L^q (ker Δ_B) *is trivial for any* $q > 1$.

Namely, if a complete simply connected Riemannian manifold (M, g) has nonnegative sectional curvature, then (M, g) has infinite volume and $\|\varphi\| \, \Delta \, \|\varphi\| \geq 0$ holds for any $\varphi \in \ker \Delta_B$. To show this, we should turn to Theorem 3.

6.2 Spectral Properties of the Bourguignon Laplacian

Let $\Delta_B \, \varphi = \lambda \, \varphi$ for some nonzero $\lambda \in \mathbb{R}$ and $\varphi \in C^\infty(S^2M)$. Then we can rewrite $\Delta_B \, \varphi = \bar{\Delta} \, \varphi + \mathfrak{R}(R^\nabla) \, \varphi$ in the form

$$\lambda \, \varphi = \bar{\Delta} \, \varphi + \mathfrak{R}(R^\nabla) \, \varphi.$$

In this case, direct verification yields the equality $\bar{\Delta} \left(\text{trace}_g \, \varphi\right) = \lambda \left(\text{trace}_g \, \varphi\right)$. Using the above, we get the following theorem (see [55]).

Theorem 20 *Let* (M, g) *be an n-dimensional* $(n \geq 2)$ *compact Riemannian manifold and* $\Delta_B \, \varphi = \lambda \, \varphi$ *holds for* $\Delta_B : C^\infty(S^2M) \to C^\infty(S^2M)$ *and for its nonzero eigenvalue* λ. *If* $\text{trace}_g \, \varphi$ *is not constant, then* $\text{trace}_g \, \varphi$ *is an eigenfunction of the Bochner Laplacian* $\bar{\Delta} : C^\infty(M) \to C^\infty(M)$ *such that* $\bar{\Delta} \left(\text{trace}_g \, \varphi\right) = \lambda \, \text{trace}_g \, \varphi$. *Furthermore, if the Ricci curvature* (M, g) *satisfies the inequality* $Ric \geq (n-1)k > 0$ *for some real* $k > 0$, *then* λ *has the lower bound* $\lambda \geq nk$. *Moreover, if* $Ric \geq (n-1)k \geq 0$ *for some real* $k \geq 0$, *then* λ *satisfies the following lower bound depending on the diameter* $d(M)$ *of* (M, g):

$$\lambda \geq \frac{1}{4}(n-1)k + \frac{\pi^2}{d(M)^2}.$$

Other interesting spectral properties of the Bourguignon Laplacian can be found in [55].

7 The Hodge Type Laplacian for Singular Distributions

Singularities play a crucial role in mathematics (e.g., in real and complex analysis, algebraic geometry and differential topology, theory of dynamical systems), and its applications (e.g., in theories of nonlinear control systems and relativity). Here, we consider the Hodge type Laplacian and Weitzenböck type curvature operator in the Bochner technique for singular distributions on Riemannian manifolds with the Levi-Civita and statistical connectrions.

7.1 Singular Distributions

A *singular distribution* $\mathcal{D}$ on a manifold M assigns to each point $x \in M$ a linear subspace $\mathcal{D}_x$ of the tangent space T_xM in such a way that, for any $v \in \mathcal{D}_x$, there exists a smooth vector field V defined in a neighborhood U of x and such that $V(x) = v$ and $V(y) \in \mathcal{D}_y$ for all $y \in U$. A priori, the dimension $\dim \mathcal{D}_x$ depends on $x \in M$. If $\dim \mathcal{D}_x = \text{const}$, then $\mathcal{D}$ is a regular distribution. We consider singular distributions as images of TM under smooth endomorphisms.

Definition 1 An image $\mathcal{D} = P(TM)$ of a smooth endomorphism $P \in \text{End}(TM)$ is called here a *singular distribution* or a *generalized vector subbundle* of TM.

The distribution $\mathcal{D} = P(TM)$ in above definition is singular in the sense of P. Stefan and H.J. Sussmann. Indeed, for any vector $v \in P(T_xM)$ there exists a vector $\tilde{v} \in T_xM$ such that $P(\tilde{v}) = v$. Take a smooth vector field $\tilde{V}$ in a neighborhood U of x such that $\tilde{V}(x) = \tilde{v}$. Then $V = P(\tilde{V})$ is a smooth vector field in U and such that $V(x) = v$ and $V(y) \in P(T_yM)$ for all $y \in U$.

Example 8 Let $P \in \text{End}(TM)$ on (M, g) be of constant rank satisfying conditions $P^2 = P$ and $P^* = P$, where P^* is adjoint endomorphism to P, i.e., $g(P^*X, Y) = g(X, PY)$, then we have an almost-product structure on (M, g). In this case, P and $P^\perp := \text{id} - P$ are orthoprojectors onto regular distributions $P(TM)$ and $P^\perp(TM)$, which are complementary orthogonal, but none of which is in general integrable. If $f \in C^\infty(M)$ has isolated zeros in M, then $P = f\, \text{id}_{TM}$ defines a singular distribution, but $P^\perp(TM)$ is not a singular distribution.

7.2 The Covariant P-Derivative and its L^2-adjoint Operator

Given $P \in \text{End}(TM)$, define a *P-connection* $\nabla^P : \mathcal{X}_M \times \mathcal{X}_M \to \mathcal{X}_M$ (which generally is not a linear connection on TM) by

$$\nabla^P_X Y := \nabla_{PX} Y \quad (X, Y \in \mathcal{X}_M), \tag{14}$$

and set $\nabla^P_X f = (PX)f$ for $f \in C^\infty(M)$. Such ∇^P satisfies Koszul conditions

$$\nabla^P_X(fY + Z) = (PX)(f)Y + f\nabla^P_X Y + \nabla^P_X Z, \quad \nabla^P_{fX+Z} Y = f\nabla^P_X Y + \nabla^P_Z Y. \tag{15}$$

We will use ∇^P to define the covariant P-derivative and its L^2-adjoint operator on tensors on (M, g) endowed with a singular distribution.

Definition 2 The *covariant P-derivative on tensors* is defined by the following:
—if $f \in C^\infty(M)$, then $(\nabla^P f)(X) = \nabla^P_X f$, and $\nabla^P f$ is called the *P-gradient* of f.
—if $Y \in \mathcal{X}_M$, then $(\nabla^P Y)(X) = \nabla^P_X Y$, see (14).
—if φ is a (s, k)-tensor, where $s = 0, 1$, then

$$(\nabla^P\varphi)(Y, X_1, \ldots, X_k) = \nabla^P_Y(\varphi(X_1, \ldots, X_k)) - \sum\nolimits_{i=1}^k \varphi(X_1, \ldots, \nabla^P_Y X_i, \ldots, X_k). \tag{16}$$

This makes sense for all $s, k \geq 0$, using the product rule with respect to tensors:

$$\nabla^P_X(\varphi_1 \otimes \varphi_2) = \nabla^P_X \varphi_1 \otimes \varphi_2 + \varphi_1 \otimes \nabla^P_X \varphi_2.$$

The metric and volume form are P-parallel, since ∇^P has the metric property:

$$\nabla^P_X g(Y, Z) = g(\nabla_{PX}Y, Z) + g(Y, \nabla_{PX}Z) = g(\nabla^P_X Y, Z) + g(Y, \nabla^P_X Z)$$

for all $X, Y, Z \in \mathfrak{X}_M$, and commutes with contractions for tensors. Define the P-divergence of a vector field X using a local orthonormal frame $\{e_i\}$ by

$$\operatorname{div}_P X = \operatorname{trace}(Y \to \nabla^P_Y X) = \sum\nolimits_i g(\nabla^P_{e_i} X, e_i), \tag{17}$$

and note that $(\operatorname{div} P)X = \sum_i g((\nabla_{e_i} P)X, e_i)$.

Lemma 1 *Given $P \in \operatorname{End}(TM)$, the condition*

$$div\, P = 0 \tag{18}$$

is valid if and only if the equality $div_P\, X = div(PX)$ holds for all $X \in \mathcal{X}_M$.

Proof Note that

$$\sum\nolimits_i g(\nabla_{Pe_i} X, e_i) = \sum\nolimits_{i,j} g(Pe_i, e_j)g(\nabla_{e_j} X, e_i) = \sum\nolimits_{i,j} g(e_i, P^* e_j)g(\nabla_{e_j} X, e_i)$$
$$= \sum\nolimits_j g(\nabla_{e_j} X, P^* e_j) = \sum\nolimits_j g(P\nabla_{e_j} X, e_j) = \operatorname{div}(PX) - (\operatorname{div} P)(X).$$

Using this, we get $\operatorname{div}_P X = \sum_i g(\nabla_{Pe_i} X, e_i) = \operatorname{div}(PX) - (\operatorname{div} P)(X)$. □

Lemma 1 allows us to extend the Divergence Theorem as follows:
if (18) *holds and a vector field X has compact support on M, then*

$$\int_M (\mathrm{div}_P X)\,\Omega_g = \int_{\partial M} g(X, P(\nu))\,\omega_{\hat g}, \tag{19}$$

where ν is the unit inner normal to the boundary ∂M and $\omega_{\hat g}$ is the volume form on ∂M with induced metric $\hat g$. We will use the operator of *contraction* ι_Y: if φ is an $(s, k+1)$-tensor and Y is a vector field, then $\iota_Y\,\varphi$ is an (s, k)-tensor given by $\iota_Y\,\varphi(X_2,\dots,X_k) = \varphi(Y, X_2,\dots,X_k)$, where $X_i \in \mathfrak{X}_M$.

The next proposition shows that, under certain conditions, the *L^2-adjoint to the P-covariant derivative* on tensors, denoted here by ∇^{*P}, is given by

$$(\nabla^{*P}\varphi)(X_2,\dots,X_k) = -\sum\nolimits_i (\nabla^P_{e_i}\varphi)(e_i, X_2,\dots,X_k). \tag{20}$$

Proposition 1 *If* (18) *holds, then ∇^{*P} is L^2-adjoint to ∇^P on tensors; namely, for any compactly supported (s,t)-tensor φ_1 and an $(s,t+1)$-tensor φ_2 we have*

$$(\nabla^{*P}\varphi_2,\ \varphi_1)_{L^2} = (\varphi_2,\ \nabla^P\varphi_1)_{L^2}. \tag{21}$$

Proof Define a compactly supported 1-form ω by $\omega(Y) = g(\iota_Y\,\varphi_2,\ \varphi_1)$ for $Y \in \mathcal{X}_M$. Take an orthonormal frame (e_i) such that $\nabla_Y e_i = 0$ for all $Y \in T_x M$. To simplify calculations, assume that $s = t = 1$, then at $x \in M$,

$$\begin{aligned}
-\nabla^{*P}\omega &= \sum\nolimits_j (\nabla^{*P}_{e_j}\,\omega)(e_j) = \sum\nolimits_{i,j} g(Pe_j, e_i)\,(\nabla_{e_i}\omega)(e_j)\\
&= \sum\nolimits_{i,j,c} g(Pe_j, e_i)\,\nabla_{e_i} g(\varphi_2(e_j, e_c), \varphi_1(e_c))\\
&= \sum\nolimits_{i,j,c} g(Pe_j, e_i)\big(g(\nabla_{e_i}\varphi_2(e_j, e_c), \varphi_1(e_c)) + g(\varphi_2(e_j, e_c), \nabla_{e_i}\varphi_1(e_c))\big)\\
&= \sum\nolimits_{i,j,c} (g(\,g(Pe_j, e_i)\nabla_{e_i}\varphi_2(e_j, e_c), \varphi_1(e_c)) + g(\varphi_2(e_j, e_c), g(Pe_j, e_i)\nabla_{e_i}\varphi_1(e_c)))\\
&= \sum\nolimits_{j,c} \big(g(\nabla^P_{e_j}\varphi_2(e_j, e_c), \varphi_1(e_c)) + g(\varphi_2(e_j, e_c), \nabla^P_{e_j}\varphi_1(e_c))\big)\\
&= -g(\nabla^{*P}\varphi_2, \varphi_1) + g(\varphi_2, \nabla^P\varphi_1).
\end{aligned}$$

The ∇^{*P} is related to the *P-divergence* of a vector field $X \in \mathcal{X}_M$ by $\mathrm{div}_P X = -\nabla^{*P} X^\flat$. Indeed, at $x \in M$, we calculate

$$\begin{aligned}
\mathrm{div}_P X &= \sum\nolimits_j g(\nabla^P_{e_j} X,\ e_j) = \sum\nolimits_{i,j} g(Pe_j, e_i)\,g(\nabla_{e_i} X,\ e_j)\\
&= \sum\nolimits_{i,j} g(Pe_j, e_i)\,e_i(g(X,\ e_j)) = \sum\nolimits_{i,j} g(Pe_j, e_i)\,e_i(X^\flat(e_j))\\
&= \sum\nolimits_{i,j} g(Pe_j, e_i)(\nabla_{e_i} X^\flat)(e_j) = \sum\nolimits_j (\nabla_{Pe_j} X^\flat)(e_j) = -\nabla^{*P} X^\flat.
\end{aligned}$$

By the above with $X^\flat = \omega$ and (19), we obtain (21). □

We apply (21): $0 \ge (\nabla^{*P}\nabla^P\varphi,\ \varphi)_{L^2} = (\nabla^P\varphi,\ \nabla^P\varphi)_{L^2} \ge 0$; and get the following maximum principle for the Bochner P-Laplacian $\nabla^{*P}\nabla^P$.

Corollary 8 *Let* (18) *hold for $P \in \mathrm{End}(TM)$ on a closed Riemannian manifold (M, g) and φ a smooth tensor field such that $g(\nabla^{*P}\nabla^P\varphi,\ \varphi) \le 0$. Then, φ is P-parallel: $\nabla^P\varphi = 0$.*

Define a skew-symmetric *P-bracket* $[\,\cdot\,,\cdot\,]_P : \mathcal{X}_M \times \mathcal{X}_M \to \mathcal{X}_M$ by

$$[X, Y]_P = \nabla^P_X Y - \nabla^P_Y X. \tag{22}$$

Define the following operator:

$$\mathfrak{D}^P(X, Y) = [PX, PY] - P[X, Y]_P. \tag{23}$$

Proposition 2 *Condition* $\mathfrak{D}^P = 0$ *is equivalent to the following symmetry on covariant components of the* $(1, 2)$*-tensor* $\nabla^P P$*:*

$$(\nabla^P P)(X, Y) = (\nabla^P P)(Y, X), \quad X, Y \in \mathcal{X}_M. \tag{24}$$

Proof Generally, we have

$$\begin{aligned}
&(\nabla^P_X P)(Y) = (\nabla_{PX} P)(Y) = \nabla_{PX} PY - P\nabla_{PX} Y,\\
&[PX, PY] - P[X, Y]_P = (\nabla_{PX} P)(Y) - (\nabla_{PY} P)(X).
\end{aligned}$$

Thus, $\mathfrak{D}^P(X, Y) = (\nabla^P_X P)(Y) - (\nabla^P_Y P)(X)$, and the claim follows. □

If (24) holds, then, by (22), the covariant P-derivative is torsion free.

Example 9 For an almost complex structure $P = J$ on TM, the property (24) describes a class of *almost Hermitian manifolds* which includes Kählerian manifolds, i.e., $\nabla J = 0$. Differentiating $J^2 = -\,\mathrm{id}_{TM}$, we obtain $(\nabla_X J)J = -J(\nabla_X J)$ for $X \in \mathfrak{X}_M$. Thus, our class contains a wider class of *nearly Kählerian manifolds*, which are defined by $(\nabla_X J)X = 0$. Indeed, if (M, J, g) is nearly Kählerian, then (24) follows from

$$\begin{aligned}
0 = &\ (\nabla_{J(X+Y)} J)J(X + Y)\\
= &\ (\nabla_{JX} J)JX + (\nabla_{JX} J)JY + (\nabla_{JY} J)JX + (\nabla_{JY} J)JY\\
= &\ (\nabla_{JX} J)JY + (\nabla_{JY} J)JX = -J\big((\nabla_{JX} J)Y + (\nabla_{JY} J)X\big).
\end{aligned}$$

There are many nearly Kählerian manifolds that are not Kählerian.

One can show that if $P \in \mathrm{End}(TM)$ defines an almost-product structure on (M, g), see Example 8, then (18) holds when both distributions are harmonic.

7.3 The Hodge P-Laplacian Δ^P_H

The *exterior P-derivative* of a differential form $\omega \in \Lambda^k M$ as a $(k + 1)$-form

$$d^P\omega(X_0, \dots, X_k) = \sum\nolimits_i (-1)^i (\nabla^P_{X_i} \omega)(X_0, \dots, \widehat{X_i}, \dots X_k).$$

For a differential form ω, the form $\nabla^P\omega$ given in (16) is not skew-symmetric, but $d^P\omega$ is skew-symmetric. For a function f on M, we have $d^P f = \nabla^P f$, see Definition 2. Put $\delta^P = \nabla^{*P} : \Lambda^{k+1}M \to \Lambda^k M$ for forms. By (20), we have

$$(\delta^P\omega)(X_2,\dots,X_k) = -\sum\nolimits_i (\nabla^P_{e_i}\omega)(e_i, X_2,\dots,X_k).$$

Similarly to Proposition 1, we have the following.

Proposition 3 *If* (18) *holds on a closed* (M,g) *then* δ^P *is* L^2*-adjoint to* d^P,

$$(\delta^P\omega_2,\,\omega_1)_{L^2} = (\omega_2,\,d^P\omega_1)_{L^2},\quad \omega_1\in\Lambda^k M,\ \omega_2\in\Lambda^{k+1}M.$$

We have

$$\begin{aligned}
d^P\omega(X_0,\dots,X_k) &= \sum\nolimits_{i=0}^k(-1)^i(\nabla_{PX_i}\omega)(X_0,\dots,\widehat{X_i},\dots,X_k)\\
&= \sum\nolimits_{i=0}^k(-1)^i PX_i(\omega(X_0,\dots,\widehat{X_i},\dots,X_k))\\
&+ \sum\nolimits_{i=0}^k(-1)^i\Big(\sum\nolimits_{j=0}^{i-1}\omega(X_0,\dots,\nabla_{PX_i}X_j,\dots,\widehat{X_i},\dots,X_k)\\
&+ \sum\nolimits_{j=i+1}^k\omega(X_0,\dots,\widehat{X_i},\dots,\nabla_{PX_i}X_j,\dots,X_k)\Big)\\
&= \sum\nolimits_{i=0}^k(-1)^i PX_i(\omega(X_0,\dots,\widehat{X_i},\dots,X_k))\\
&+ \sum\nolimits_{0\le i<j\le k}(-1)^{i+j}\omega\big(\nabla_{PX_i}X_j-\nabla_{PX_j}X_i, X_0,\dots,\widehat{X_i},\dots,\widehat{X_j},\dots,X_k\big).
\end{aligned}$$

The use of (14) and (22) shows that $d^P : \Lambda^k M \to \Lambda^{k+1}M$ $(1\le k\le n)$ is an exterior derivation, that is,

$$\begin{aligned}
d^P\omega(X_0,\dots,X_k) = &\sum\nolimits_{i=0}^k(-1)^i PX_i(\omega(X_0,\dots,\widehat{X_i},\dots,X_k))\\
&+\sum\nolimits_{0\le i<j\le k}(-1)^{i+j}\omega\big([X,Y]_P, X_0,\dots,\widehat{X_i},\dots,\widehat{X_j},\dots,X_k\big).
\end{aligned}$$

Define the *Hodge type Laplacian* Δ^P_H for differential forms by

$$\Delta^P_H = d^P\,\delta^P + \delta^P d^P. \tag{25}$$

We say that a differential form ω is *P-harmonic* if $\Delta^P_H\omega = 0$. The P-harmonic forms have similar properties as harmonic forms, see, e.g., [44, Lemma 9.1.1]. Let (18) hold on a closed (M,g). For $\omega\in\Lambda^k M$, by Proposition 3 and (25), we get

$$(\Delta^P_H\omega,\,\omega)_{L^2} = (d^P\omega,\,d^P\omega)_{L^2} + (\delta^P\omega,\,\delta^P\omega)_{L^2};$$

thus, a differential form ω is P-harmonic if and only if $d^P\omega = 0$ and $\delta^P\omega = 0$. Moreover, if ω is P-harmonic and $\omega = d^P\theta$, then $\omega = 0$.

We call $\Delta^P_B f = \operatorname{div}^P(\nabla^P f)$ the *P-Laplacian for functions*. A function $f \in C^\infty(M)$ is *P-subharmonic* if $\Delta^P_B f \le 0$. If $\Delta^P_B f \ge 0$, then f is *P-superharmonic*. Modifying Stokes theorem on a complete open (M, g), yields the following generalization of [14, Proposition 1].

Lemma 2 *Let (M, g) be a complete open Riemannian manifold endowed with a vector field X such that $div_P\, X \ge 0$ (or $div_P\, X \le 0$), where $P \in \operatorname{End}(TM)$ such that* (18) *and $\|PX\|_g \in \mathrm{L}^1(M, g)$ hold. Then, $div_P\, X \equiv 0$.*

Proof Let ω be the $(n-1)$-form in M given by $\omega = \iota_{PX}\,\Omega_g$, i.e., the contraction of the volume form Ω_g in the direction of PX. If $\{e_1, \dots, e_n\}$ is an orthonormal frame on an open set $U \subset M$ with coframe $\omega_1, \dots, \omega_n$, then

$$\iota_{PX}\,\Omega_g = \sum\nolimits_{i=1}^{n} (-1)^{i-1} g(PX, e_i)\,\omega_1 \wedge \ldots \wedge \widehat{\omega}_i \wedge \ldots \wedge \omega_n.$$

Since the $(n-1)$-forms $\omega_1 \wedge \ldots \wedge \widehat{\omega}_i \wedge \ldots \wedge \omega_n$ are orthonormal in $\Lambda^{n-1}M$, we get $\|\omega\|^2_g = \sum_{i=1}^n g(PX, e_i)^2 = \|PX\|^2_g$. Thus, $\|\omega\|_g \in \mathrm{L}^1(M, g)$ and $d\omega = d(\iota_{PX}\,\Omega_g)$ $= (\operatorname{div}_P X)\,\Omega_g$. There exists a sequence of domains B_i on M such that $M = \bigcup_{i\ge1} B_i$, $B_i \subset B_{i+1}$ and $\lim_{i\to\infty} \int_{B_i} d\omega = 0$, see [72]. Then

$$\int_{B_i} (\operatorname{div}_P X)\,\Omega_g = \int_{B_i} \operatorname{div}(PX)\,\Omega_g = \int_{B_i} d\omega \to 0.$$

But since $\operatorname{div}_P X \ge 0$ on M, it follows that $\operatorname{div}_P X = 0$ on M. □

Consider the system of singular distributions $\mathcal{D}_1 = \mathcal{D}$, $\mathcal{D}_2 = \mathcal{D}_1 + [\mathcal{D}, \mathcal{D}_1]$, etc. The distribution $\mathcal{D}$ is said to be *bracket-generating* of the step $r \in \mathbb{N}$ if $\mathcal{D}_r = TM$, see, e.g., [13].

The condition $\nabla^P f = 0$ means that f is constant along the integral curves of $\mathcal{D} = P(TM)$; moreover, $f = \text{const}$ on M when $\mathcal{D}$ is bracket-generating. The next theorem for $P = \mathrm{id}_{TM}$ reduces to the classical result on subharmonic functions.

Theorem 21 *Let* (18) *hold and $f \in C^2(M)$ satisfy $\Delta^P_B f \ge 0$ or $\Delta^P_B f \le 0$. Suppose that (M, g) is open complete, $\|P\,\nabla^P f\| \in L^1(M, g)$ and $\|f\,P\,\nabla^P f\| \in L^1(M, g)$. Then $\nabla^P f = 0$; and if $P(TM)$ is bracket-generating, then $f =$* const.

Proof Set $X = \nabla^P f$, then $\Delta^P_B f = \operatorname{div}_P X$. By Lemma 2 with $X = \nabla^P f$ and $\|P\,\nabla^P f\| \in L^1(M, g)$, we get $\Delta^P_B f \equiv 0$. Using the equality with $Y = \nabla^P f$:

$$\operatorname{div}_P(f \cdot Y) = f \cdot \operatorname{div}_P Y + g(\nabla^P f,\ Y).$$

By Lemma 2 with $X = f\nabla^P f$ and $\|f\,P\nabla^P f\| \in L^1(M, g)$, we get $(\nabla^P f, \nabla^P f) =$ 0; thus, $\nabla^P f = 0$. If $P(TM)$ is bracket-generating, then we use Chow's theorem [17]. □

7.4 The P-curvature Tensor and the Weitzenböck Type Curvature Operator

Define the *P-curvature* of P-connection by

$$R^P_{X,Y} Z = \nabla^P_X \nabla^\rho_Y Z - \nabla^P_Y \nabla^P_X Z - \nabla^P_{[X,Y]_P} Z.$$

The *Ricci type tensor* Ric^P is defined by $\mathrm{Ric}^P(X,Y) = \sum_i R^P(X, e_i, e_i, Y)$. The *Jacobiator* of the bracket is defined by

$$\mathcal{J}_P(X,Y,Z) = \sum\nolimits_{\rm cycl.} [X, [Y,Z]_P\,]_P.$$

Since ∇^P is torsionless, we get $\sum_{\rm cycl.} R^P_{X,Y} Z = \mathcal{J}_P(X,Y,Z)$. Generally, R^P acts on $(0,k)$-tensors by

$$\begin{aligned}(R^P_{X,Y}\,\varphi)(X_1,\ldots,X_k) &= \mathfrak{D}^P(X,Y)(\varphi(X_1,\ldots,X_k)) \\ &\quad -\sum\nolimits_i \varphi(X_1,\ldots R^P_{X,Y}X_i,\ldots,X_k).\end{aligned} \tag{26}$$

Observe that the action of $R^P_{X,Y}$ on tensors is $C^\infty(M)$-linear if $\mathfrak{D}^P = 0$, i.e., (24) holds; we conclude the same about the $C^\infty(M)$-homogeneity of $R^P_{X,Y}\,\omega$ in ω.

Based on the above arguments, we assume (18) and (24) in the rest of this article. The P-curvature has some symmetry properties, similar to the classical case $P = \mathrm{id}_{TM}$.

Proposition 4 *For $X, Y, Z, W \in \mathcal{X}_M$, $\omega \in \Lambda^1 M$ and $f \in C^\infty(M)$, we have*

1. $R^P_{X,Y} Z = R_{PX,PY} Z$; $(R^P_{X,Y}\,\omega)(Z) = -\omega(R^P_{X,Y} Z)$;
2. $R^P_{X,Y}\, f = 0$; $R^P_{X,Y}\, g = 0$;
3. *for every $(1,k)$-tensor T we have*

$$(R^P_{X,Y}\,T)(Z_1,\ldots,Z_k) = (R_{PX,PY}\,T)(Z_1,\ldots,Z_k).$$

4. $g(R^P_{X,Y} Z, W) = g(R_{PX,PY} Z, W)$;
5. $g(R^P_{X,Y} Z, W) = -g(R^P_{Y,X} Z, W) = -g(R^P_{X,Y} W, Z)$, *where* $X, Y, Z, W \in \mathcal{X}_M$, $\omega \in \Lambda^1 M$ *and* $f \in C^2(M)$.

Proof 1. Since $P[X,Y]_P = [PX, PY]$, see definition of $\mathfrak{D}^P$, we have

$$R^P_{X,Y} Z = R_{PX,PY} Z, \quad (R^P_{X,Y}\,\omega)(Z) = -\omega(R_{PX,PY} Z).$$

2. We calculate

$$R^P_{X,Y}\, f = PX(PY(f)) - PY(PX(f)) - (P[X,Y]_P) f = \mathfrak{D}^P(X,Y) f = 0.$$

Using 1. we get $g(R^P_{X,Y}Z, W) = g({}_{PX,PY}Z, W)$. Similarly, $g(R^P_{X,Y}W, Z) = g(R_{PX,PY}W, Z)$. By this and (26), we get

$$\begin{aligned}(R^P_{X,Y}\, g)(Z, W) &= -g(R^P_{X,Y}Z, W) - g(Z, R^P_{X,Y}W)\\ &= (R_{PX,PY}\, g)(Z, W) - (R_{PX,PY}\, g)(Z, W).\end{aligned}$$

Using $R_{PX,PY}\, g = 0$ (for the curvature tensor), we obtain $R^P_{X,Y}\, g = 0$.
3. From the above and (26) the claim follows.
4. The equality follows from 1.
5. Since $R^P_{X,Y}Z = -R^P_{Y,X}Z$, see 1., the first equality follows. For the second equality we use 2:

$$0 = (R^P_{X,Y}\, g)(Z, Z) = -2\, g(R^P_{X,Y}Z,\ Z);$$

thus, the claim follows from the equality $g(R^P_{X,Y}(Z+W),\ Z+W) = 0$. □

One can show that $P \in \mathrm{End}(TM)$ induces endomorphisms of $\Lambda^2 M$:

$$\mathcal{P}(X \wedge Y) = PX \wedge PY, \quad \mathcal{P}^*(X \wedge Y) = P^*X \wedge P^*Y.$$

The curvature tensor $R_{X,Y}$ can be seen as a self-adjoint linear operator $\mathcal{R}$ on bivectors $\Lambda^2 M$, called the *curvature operator* (note the reversal of Z and W):

$$g(\mathcal{R}(X \wedge Y), Z \wedge W) = g({}_{X,Y}W, Z), \quad \mathcal{R}^* = \mathcal{R}.$$

Similarly, we can consider $R^P_{X,Y} = R_{PX,PY}$ as a linear operator $\mathcal{R}^P = \mathcal{R} \circ \mathcal{P}$ or as a corresponding bilinear form on $\Lambda^2 M$, that is,

$$\begin{aligned}\mathcal{R}^P(X \wedge Y) &= \mathcal{R} \circ \mathcal{P}(X \wedge Y) = \mathcal{R}(PX \wedge PY),\\ \mathcal{R}^P(X \wedge Y, Z \wedge W) &= g(\mathcal{R}^P(X \wedge Y), Z \wedge W).\end{aligned}$$

Using known properties of $\mathcal{R}$ and R^P, we have

$$g(\mathcal{R}^P(X \wedge Y), Z \wedge W) = g(\mathcal{R}(PX \wedge PY), Z \wedge W) = g(R^P_{X,Y}W, Z).$$

Note that $\mathcal{R}^P$ on $\Lambda^2 M$ generally is not self-adjoint:

$$(\mathcal{R}^P)^* = (\mathcal{R} \circ \mathcal{P})^* = \mathcal{P}^* \circ \mathcal{R} \neq \mathcal{R} \circ \mathcal{P} = \mathcal{R}^P.$$

The next definition allows us to continue the series of examples in Sect. 2.3.

Definition 3 The *Weitzenböck type curvature operator* $\mathfrak{R}^P$ on $(0, k)$-tensors is defined by

$$\mathfrak{R}^P(\varphi)(X_1,\ldots,X_k)=\sum\nolimits_{\alpha=1}^{k}\sum\nolimits_{i=1}^{n}(R^P_{e_i,X_\alpha}\varphi)(\underbrace{X_1,\ldots,e_i}_{\alpha},\ldots,X_k)$$

$$=-2\sum\nolimits_{i,m,a;b<a}g(R^P_{e_i,X_a}e_m,X_b)\cdot\varphi(\underbrace{X_1,\ldots,e_m}_{b},\underbrace{\ldots,e_i}_{a-b},\ldots,X_k)$$

$$+\sum\nolimits_{i,a}\mathrm{Ric}^P(e_i,X_a)\cdot\varphi(\underbrace{X_1,\ldots,e_i}_{a},\ldots,X_k),$$

or, in coordinates (similarly to the Weitzenböck curvature operator $\mathfrak{R}(R^\nabla)$, see Sect. 2.3),

$$\mathfrak{R}^P(\varphi)_{i_1,\ldots,i_k}=\sum\nolimits_a\mathrm{Ric}^P_{i_a j}\,\varphi_{i_1\ldots\ \ldots i_k}^{\ \ \ \ j}-2\sum\nolimits_{a<b}R^P_{j\,i_a s\,i_b}\varphi_{i_1\ldots\ \ldots\ \ldots i_k}^{\ \ \ \ j\ \ \ s}.$$

For a differential form ω, the $\mathfrak{R}^P(\omega)$ is skew-symmetric. For vector fields and 1-forms, $\mathfrak{R}^P$ reduces to the kind of Ricci curvature, $\mathrm{Ric}^P=\mathrm{trace}_{1,3}R^P$.

Proposition 5 *The following Weitzenböck type decomposition formula is true:*

$$\Delta^P_H=\nabla^{*P}\,\nabla^P+\mathfrak{R}^P. \tag{27}$$

Proof For $\omega\in\Lambda^k M$, we obtain

$$d^P\delta^P\omega(X_1,\ldots,X_k)=-\sum\nolimits_{i,\alpha}(-1)^{\alpha+1}(\nabla^{P,2}_{X_\alpha,e_i}\omega)(e_i,\underbrace{X_1,\ldots,\widehat{X_\alpha}}_{i},\ldots,X_k)$$

$$=-\sum\nolimits_{i,\alpha}(\nabla^{P,2}_{X_\alpha,e_i}\omega)(\underbrace{X_1,\ldots,e_i}_{\alpha},\ldots,X_k),$$

$$\delta^P d^P\omega(X_1,\ldots,X_k)=-\sum\nolimits_i(\nabla^{P,2}_{e_i,e_i}\omega)(X_1,\ldots,X_k)$$

$$-\sum\nolimits_{i,\alpha}(-1)^\alpha(\nabla^{P,2}_{e_i,X_\alpha}\omega)(e_i,\underbrace{X_1\ldots,\widehat{X_\alpha}}_{i},\ldots,X_k)$$

$$=(\nabla^{*P}\nabla^P\omega)(X_1,\ldots,X_k)+\sum\nolimits_{i,\alpha}(\nabla^{P,2}_{e_i,X_\alpha}\omega)(\underbrace{X_1\ldots,e_i}_{\alpha},\ldots,X_k),$$

where $1\le i\le n$ and $1\le\alpha\le k$. Substituting the above into (25) yields (27). □

Remark 6 The Weitzenböck type decomposition formula (27) allows to extend the Hodge type Laplacian to arbitrary tensors and is important in the study of interactions between the geometry and topology of manifolds equipped with distributions. The Bochner technique works for tensors lying in the kernel of Δ^P_H on a closed manifold: using maximum principles, we can prove that such tensors are P-parallel. For this, we need $g(\mathfrak{R}^P(\varphi),\varphi)\ge 0$, which by (27) is equivalent to $g(\nabla^{P*}\nabla^P\varphi,\varphi)\le 0$. There is a simple formula for $\mathfrak{R}^P(\varphi)$, which shows that $g(\mathfrak{R}^P(\varphi),\varphi)\ge 0$ when the P-curvature operator on $\Lambda^2 M$ is nonnegative.

Proposition 6 *The Bochner–Weitzenböck type formula for k-forms is true:*

$$\frac{1}{2}\,\Delta^P_B(\|\omega\|^2) = -g(\Delta^P_H\,\omega,\omega) + \|\nabla^P\omega\|^2 + g(e^P(\omega),\omega). \tag{28}$$

Proof Take an orthonormal frame (e_i) such that $\nabla_Y\,e_i = 0\,(Y \in T_xM)$ and, in order to simplify calculations, assume that $s = t = 1$. By $\Delta^P_B\,f(x) = \sum_i (Pe_i)^2(f)(x)$ we get at $x \in M$,

$$\begin{aligned}\frac{1}{2}\,\Delta^P_B\,(\,\|\omega\|^2\,) &= \frac{1}{2}\sum\nolimits_i (Pe_i)^2 g(\omega,\,\omega) = \sum\nolimits_i (Pe_i)g(\nabla_{Pe_i}\omega,\,\omega)\\ &= \sum\nolimits_i \big[g(\nabla_{Pe_i}\nabla_{Pe_i}\omega,\,\omega) + g(\nabla_{Pe_i}\omega,\,\nabla_{Pe_i}\omega)\big]\\ &= -g(\nabla^{*P}\,\nabla^P\,\omega,\omega) + \|\nabla^P\,\omega\|^2.\end{aligned}$$

Then, using Proposition 5, we obtain (28). □

Next, we rewrite $\mathfrak{R}^P$ using an orthonormal basis (ξ_a) of skew-symmetric transformations $\mathfrak{so}(TM)$ and give some applications. For every bivector $X\wedge Y \in \Lambda^2M$, we build a map $\mathcal{R}^P(X\wedge Y) : \mathcal{X}_M \to \mathcal{X}_M$, given by

$$g(\mathcal{R}^P(X\wedge Y)Z,W) = g(\mathcal{R}^P(X\wedge Y),W\wedge Z) = g(R_{PX,PY}Z,W).$$

Since bivectors generate Λ^2M, we obtain in this way a map $\mathcal{R}^P(\xi) : \mathcal{X}_M \to \mathcal{X}_M$ similarly to algebraic curvature operator $\mathcal{R}(\xi)$. Using Proposition 4, we get

$$\begin{aligned}&g(\mathcal{R}^P(X\wedge Y)Z,W) = g(R_{PX,PY}Z,W)\\ =\ &-g(R_{PX,PY}W,Z) = -g(\mathcal{R}^P(X\wedge Y)W,Z).\end{aligned}$$

Thus, the map $\mathcal{R}^P(\xi)$, where $\xi \in \Lambda^2M$, is skew-symmetric: $g(\mathcal{R}^P(\xi)W,Z) = -g(\mathcal{R}^P(\xi)Z,W)$. The associated *P-curvature operator* is given by $g(\mathcal{R}^P(X\wedge Y),Z\wedge W) = g(R_{PX,PY}W,Z)$. We are based on the fact that, if X and Y are orthonormal, then $X\wedge Y$ is a unit bivector, while the corresponding skew-symmetric operator (a counterclockwise rotation of $\pi/2$ in the plane span(X,Y)) has Euclidean norm $\sqrt{2}$. To simplify calculations, assume that metric on $\mathfrak{so}(TM)$ is induced from Λ^2M, e.g., [50]. If $L \in \mathfrak{so}(TM)$, then

$$(L\,\varphi)(X_1,\ldots,X_k) = -\sum\nolimits_i \varphi(X_1,\ldots,L(X_i),\ldots,X_k). \tag{29}$$

Let $\{\xi_a\}$ be an orthonormal base of skew-symmetric transformations such that $(\xi_u)_x \in \mathfrak{so}(T_xM)$ for x in an open set $U \subset M$. By (29), for any $(0,k)$-tensor φ,

$$(\xi_\alpha\varphi)(X_1,\ldots,X_k) = -\sum\nolimits_i \varphi(X_1,\ldots,\xi_\alpha(X_i),\ldots,X_k);$$

Proposition 7 *If φ is a $(0,k)$-tensor on (M,g), then*

$$\Re^P(\varphi) = -\sum\nolimits_{\alpha} \mathcal{R}^P(\xi_a)(\xi_a\,\varphi), \qquad (\Re^P(\varphi))^* = \Re^{P^*}(\varphi).$$

In particular, if P is self-adjoint, then $\Re^P$ is self-adjoint too.

For a $(0,k)$-tensor φ on M, define a $(0,k)$-tensor $\widetilde{\varphi}$ with values in $\Lambda^2 M$ by

$$g(\,L,\ \widetilde{\varphi}(X_1,\dots,X_k)\,) = (L\,\varphi)(X_1,\dots,X_k), \quad L \in \mathfrak{so}(TM).$$

Using a local orthonormal base $\{\xi_\alpha\}$ of $\Lambda^2 M$, we get

$$\begin{aligned} g(\,\Re^P(\varphi_1),\varphi_2) &= \sum\nolimits_{\alpha} g(\,\mathcal{R}^P(\xi_\alpha)(\xi_\alpha\varphi_1),\ \varphi_2) = \sum\nolimits_{\alpha} g(\,\xi_\alpha\varphi_1,\ \mathcal{R}^{P*}(\xi_\alpha)\varphi_2) \\ &= \sum\nolimits_{\alpha,i} (\xi_\alpha\varphi_1)(e_{i_1},\dots,e_{i_k})(\mathcal{R}^{P*}(\xi_\alpha)\varphi_2)(e_{i_1},\dots,e_{i_k}) \\ &= \sum\nolimits_{\alpha,i} g(\,\xi_\alpha,\widetilde{\varphi}_1(e_{i_1},\dots,e_{i_k})\,)g(\,\mathcal{R}^{P*}(\xi_\alpha),\widetilde{\varphi}_2(e_{i_1},\dots,e_{i_k})\,) \\ &= \sum\nolimits_{\alpha,i} g(\,\mathcal{R}^{P*}((\,\xi_\alpha,\widetilde{\varphi}_1(e_{i_1},\dots,e_{i_k})\,)\,\xi_\alpha),\widetilde{\varphi}_2(e_{i_1},\dots,e_{i_k})\,) \\ &= \sum\nolimits_{i} g(\,\mathcal{R}^{P*}(\widetilde{\varphi}_1(e_{i_1},\dots,e_{i_k})),\widetilde{\varphi}_2(e_{i_1},\dots,e_{i_k})\,) \\ &= \sum\nolimits_{i} g(\,\widetilde{\varphi}_1(e_{i_1},\dots,e_{i_k}),\mathcal{R}^{P}(\widetilde{\varphi}_2(e_{i_1},\dots,e_{i_k}))\,). \end{aligned}$$

Thus, the following is true.

Proposition 8 *If φ_1,φ_2 are $(0,k)$-tensors, then*

$$g(\,\Re^P(\varphi_1),\varphi_2\,) = g(\,\widetilde{\varphi}_1,\ \mathcal{R}^P(\widetilde{\varphi}_2)\,).$$

Proposition 9 *If $g(\mathcal{R}^P(\varphi),\varphi) \ge 0$ for any $(0,k)$-tensor φ, then $g(\Re^P(\varphi),\varphi) \ge 0$. Moreover, if $g(\mathcal{R}^P(\varphi),\varphi) \ge k\,\|\varphi\|^2$ for any $(0,k)$-tensor φ, where $k<0$, then $g(\Re^P(\varphi),\varphi) \ge k\,C\,\|\varphi\|^2$, where a constant C depends only on the type of φ.*

Theorem 22 *Let $g(\mathcal{R}^P(\omega),\omega) \ge 0$ for any k-form ω on an open complete Riemannian manifold (M,g). If $\|\,P\,\nabla^P(\|\omega\|^2)\| \in L^1(M,g)$ for a P-harmonic form ω, then ω is P-parallel.*

Proof By conditions and Proposition 9(i), we get $g(\Re^P(\omega),\omega) \ge 0$. By the Bochner–Weitzenböck type formula (28) with $\varphi=\omega$, since $\Delta^P_H\,\omega = 0$, we obtain $\Delta^P_B(\|\omega\|^2) \ge 0$. By Lemma 2 with $X = \nabla^P(\|\omega\|^2)$, we get $\Delta^P_B(\|\omega\|^2) = 0$. Hence, by (28) again, $\nabla^P\omega = 0$. □

Corollary 9 *If $g(\mathcal{R}^P(\omega),\omega) \ge 0$ for any k-form ω on a closed Riemannian manifold (M,g), then any P-harmonic differential form on M is P-parallel.*

Proof By conditions and Proposition 9(i), we get $g(\Re^P(\omega),\omega) \ge 0$. By Proposition 5, since $\Delta^P_H\,\omega = 0$, we obtain $g(\nabla^{*P}\,\nabla^P\omega,\ \omega) \le 0$. By Proposition 8 with $\varphi=\omega$, we get $\nabla^P\omega = 0$. □

If $P(TM)$ in Theorem 22 and Corollary 9 is bracket-generating, then $\|\omega\| =$ const.

7.5 Statistical P-structure for Singular Distributions

In [51], we extended the Bochner's technique discussed in Sects. 7.1–7.4 to a Riemannian manifold endowed with a singular (or regular) distribution and a statistical type connection.

Recall, e.g., [3], that a linear connection $\widetilde{\nabla}$ on a Riemannian manifold (M, g) is called *statistical* if it is torsion-free and $\widetilde{\nabla} g$ is symmetric in all its entries:

$$(\widetilde{\nabla}_X\, g)(Y, Z) = (\widetilde{\nabla}_Y\, g)(X, Z) = (\widetilde{\nabla}_Z\, g)(X, Y).$$

Such a pair $(g, \widetilde{\nabla})$ is called a *statistical structure* on M. We generalize this concept for singular distributions.

The covariant P-derivative and the P-divergence. Given $P \in \mathrm{End}(TM)$, define the map $\bar{\nabla}^P : \mathfrak{X}_M \times \mathfrak{X}_M \to \mathfrak{X}_M$ satisfying Koszul conditions (15),

$$\bar{\nabla}^P_X\, Y = \nabla_{PX}\, Y + K_X Y, \tag{30}$$

called *affine P-connection*. The $\bar{\nabla}^P$ depends on the P-connection ∇^P, see (14), and the $(1, 2)$-tensor K (contorsion tensor), but generally is not a linear connection on M. Set $\bar{\nabla}^P_X f = (PX)f$ for $f \in C^1(M)$ (P-*gradient* of f). The P-derivative $\bar{\nabla}^P \varphi$ of an (s, k)-tensor φ, where $s = 0, 1$, is defined similarly to (16). A tensor φ is called P-*parallel* if $\bar{\nabla}^P \varphi = 0$. An affine P-connection $\bar{\nabla}^P$ on (M, g) will be called *statistical P-connection* if the cubic form $g(K_X Y, Z)$ is symmetric, or, equivalently,

$$K^*_X = K_X, \quad K_X Y = K_Y X \quad (X, Y \in TM). \tag{31}$$

In this case, the pair $(g, \bar{\nabla}^P)$ is called a *statistical P-structure* on M. We calculate

$$(\bar{\nabla}^P_X\, g)(Y, Z) = (\nabla_{PX}\, g)(Y, Z) - g(K_X Y, Z) - g(K_X Z, Y). \tag{32}$$

By (32), (31) and the property $\nabla g = 0$, we have $(\bar{\nabla}^P_X\, g)(Y, Z) = -2\, g(K_X Y, Z)$; thus, if $\bar{\nabla}^P$ is a statistical P-connection for g, then the (3,0)-tensor $\bar{\nabla}^P g$ is symmetric in all its entries.

The P-*bracket* $\overline{[\cdot, \cdot]}_P : \mathfrak{X}_M \times \mathfrak{X}_M \to \mathfrak{X}_M$ is defined similarly to (22) by

$$\overline{[X, Y]}_P = \bar{\nabla}^P_X\, Y - \bar{\nabla}^P_Y\, X.$$

By this, $\bar{\nabla}^P$ is torsion free. Similarly to (23), we define the following operator:

$$\mathfrak{D}^P(X, Y) = [PX, PY] - P\overline{[X, Y]}_P.$$

By (31), the P-bracket of a statistical P-structure does not depend on K:

$$\overline{[X, Y]}_P = \nabla_{PX} Y - \nabla_{PY} X.$$

If $\bar\nabla^P$ is statistical, then ∇^P_X has the same P-bracket and $\mathfrak{D}^P = \bar{\mathfrak{D}}^P$.

Proposition 10 *Condition $\bar{\mathfrak{D}}^P = 0$ is equivalent to the symmetry on covariant components of the $(1,2)$-tensor $(\nabla_{PX}P)(Y) - P(K_X Y)$. For a statistical P-structure, condition $\bar{\mathfrak{D}}^P = 0$ is equivalent to*

$$(\nabla_{PX}P)(Y) = (\nabla_{PY}P)(X), \quad X, Y \in \mathcal{X}_M.$$

The notion of dual connection is important for statistical manifolds, see [43]. For a P-connection $\bar\nabla^P$ on (M, g), its *dual P-connection* $\breve\nabla^P$ is defined by the following equality:

$$PX(g(Y,Z)) = g(\bar\nabla^P_X Y, Z) + g(Y, \breve\nabla^P_X Z).$$

For a statistical P-connection $\bar\nabla^P$, its dual connection $\breve\nabla^P$ is given by $\breve\nabla^P_X = \nabla^P_X - K_X$. In turn, the statistical P-connection $\bar\nabla^P$ is dual to $\breve\nabla^P$. Note that $\bar\nabla^P + \breve\nabla^P = 2\,\nabla^P$. For a dual statistical P-connection $\breve\nabla^P$, we define the P-bracket and the tensor $\breve{\mathfrak{D}}^P$ by

$$\breve{[X, Y]}_P = \breve\nabla^P_X Y - \breve\nabla^P_Y X, \quad \breve{\mathfrak{D}}^P(X,Y) = [PX, PY] - P\breve{[X, Y]}_P.$$

By (31), we have $\breve{[\cdot\,,\cdot\,]}_P = \overline{[\cdot\,,\cdot\,]}_P$ and $\breve{\mathfrak{D}}^P = \bar{\mathfrak{D}}^P$. Thus, we obtain the following.

Corollary 10 *$(g, \bar\nabla^P)$ and $(g, \breve\nabla^P)$ are simultaneously statistical P-structures.*

For the rest of this article we restrict ourselves to statistical P-structures, see (31), and we assume (24).

For the vector field $E := \sum_i K_{e_i} e_i$ on M, using (31), we get

$$g(E, X) = \mathrm{trace}_g K_X \quad (X \in \mathfrak{X}_M).$$

Lemma 3 (see Lemma 6.3 in *[43]*) *For any local orthonormal frame $\{e_i\}$, for any $(k+1)$-form, $k \ge 1$, and an index $a \in \{1, \ldots, k\}$ be fixed, we have*

$$\sum\nolimits_i \omega(e_i, X_1, \ldots, K_{e_i} X_a, \ldots, X_k) = 0. \tag{33}$$

The P-divergence for $\bar\nabla^P$ is defined similarly to (17): $\overline{\mathrm{div}}_P\, X = \sum_i g(\bar\nabla^P_{e_i} X,\, e_i)$.

Lemma 4 *On a Riemannian manifold (M, g) with a statistical P-structure, the condition*

$$(div\, P)(X) = trace_g K_X, \quad X \in \mathcal{X}_M \tag{34}$$

is equivalent to the following equality: $\overline{div}_P\, X = div(PX)$ for $X \in \mathcal{X}_M$.

Lemma 4 allows us to extend the Divergence Theorem and (19) as follows:

Theorem 23 *Let a statistical P-structure on a compact Riemannian manifold (M, g) with boundary satisfy* (34). *Then for any $X \in \mathfrak{X}_M$ we have*

$$\int_M (\overline{div}_P\, X)\, \Omega_g = \int_{\partial M} g(X, P(\nu))\, \omega_{\hat{g}},$$

where ν is the unit inner normal to ∂M. In particular, on (M, g) without boundary, for any $X \in \mathfrak{X}_M$ with compact support, we get $\int_M (\overline{div}_P X)\, \Omega_g = 0$.

For example, if $\operatorname{div} P = 0$, then the tensor $K = c\, \nabla P$ with any $c \in \mathbb{R}$ satisfies (34). The L^2-adjoint operator $\bar{\nabla}^{*P}$ for $\bar{\nabla}^P$ is defined similarly to (20), and similarly for $\breve{\nabla}^{*P}$. Using (33), we relate $\bar{\nabla}^{*P}$ and ∇^{*P} for any k-form ω:

$$\bar{\nabla}^{*P}\omega = \nabla^{*P}\omega + \iota_E\, \omega, \quad \breve{\nabla}^{*P}\omega = \nabla^{*P}\omega - \iota_E\, \omega.$$

Thus, $\breve{\nabla}^{*P}\omega = \bar{\nabla}^{*P}\omega - 2\, \iota_E\, \omega$. To simplify the calculations, we will consider statistical P-structures with stronger conditions than (34),

$$a)\ \operatorname{div} P = 0, \qquad b)\ E = 0. \tag{35}$$

The next result shows that $\breve{\nabla}^{*P}$ is L^2-adjoint to $\bar{\nabla}^P$ on k-forms.

Proposition 11 *If a statistical P-connection $\bar{\nabla}^P$ satisfy* (35), *then for any compactly supported k-form ω_1 and $(k+1)$-form ω_2, we have*

$$(\breve{\nabla}^{*P}\omega_2,\ \omega_1)_{L^2} = (\omega_2,\ \bar{\nabla}^P\omega_1)_{L^2}.$$

The operator $\breve{\nabla}^{*P}\, \bar{\nabla}^P$ will be called the *Bochner P-Laplacian* for a statistical P-structure.

Proposition 12 *Let a statistical P-connection $\bar{\nabla}^P$ on a closed Riemannian manifold (M, g) satisfy* (34). *If $g(\breve{\nabla}^{*P}\, \bar{\nabla}^P\omega,\ \omega) \le 0$ holds for a k-form ω, then $\bar{\nabla}^P\omega = 0$, i.e., ω is P-parallel.*

The Hodge P-Laplacians $\bar{\Delta}^P_H$ and $\breve{\Delta}^P_H$. Using a statistical P-connection $\bar{\nabla}^P$, we define the *exterior P-derivative* of $\omega \in \Lambda^k(M)$ similarly as for ∇^P in Sect. 7.2. For a function f on M, we have $\bar{d}^P f = \bar{\nabla}^P f$ and $\breve{d}^P f = \breve{\nabla}^P f$. The next proposition shows that $\breve{d}^P = \bar{d}^P$.

Proposition 13 *The P-derivative $\bar{d}^P : \Omega^k(M) \to \Omega^{k+1}(M)$ is a 1-degree derivation, i.e.,*

$$\bar{d}^P\omega(X_0, \dots, X_k) = \sum\nolimits_{i=0}^k (-1)^i P X_i(\omega(X_0, \dots, \widehat{X}_i, \dots, X_k))$$
$$+ \sum\nolimits_{0 \le i < j \le k} (-1)^{i+j} \omega\big([X, Y]_P, X_0, \dots, \widehat{X}_i, \dots, \widehat{X}_j, \dots, X_k\big).$$

Put $\bar\delta^P = \bar\nabla^{*P}$ for the P-codifferential $\bar\delta^P : \Lambda^k(TM) \to \Lambda^{k-1}(TM)$. Similarly, we define

$$\breve\delta^P\omega(X_2,\dots,X_k) = -\sum\nolimits_i (\breve\nabla^P_{e_i}\omega)(e_i, X_2,\dots,X_k).$$

Proposition 14 *On a closed manifold (M,g) with a statistical P-structure, $\breve\delta^P$ is L^2-adjoint to $\bar d^P$, i.e., for any differential forms $\omega_1 \in \Lambda^k(TM)$, $\omega_2 \in \Lambda^{k+1}(TM)$, we have*

$$(\breve\delta^P\omega_2,\,\omega_1)_{L^2} = (\omega_2,\,\bar d^P\omega_1)_{L^2}.$$

Definition 4 Define the *Hodge P-Laplacians* $\bar\Delta^P_H$ and $\breve\Delta^P_H$ for a differential form ω by

$$\bar\Delta^P_H\omega = \bar d^P\breve\delta^P\omega + \breve\delta^P\bar d^P\omega, \quad \breve\Delta^P_H\omega = \bar d^P\bar\delta^P\omega + \bar\delta^P\bar d^P\omega. \tag{36}$$

A form ω is called *P-harmonic* if $\bar\Delta^P_H\omega = 0$ and $\|\omega\|_{L^2} < \infty$ (similarly for $\breve P$).

P-harmonic forms have similar properties as in the classical case, e.g., [44]. Let (34) hold on a closed (M,g). For $\omega \in \Lambda^k(TM)$, using Proposition 14 and (36), we have

$$(\bar\Delta^P_H\omega,\,\omega)_{L^2} = (\bar d^P\omega,\,\bar d^P\omega)_{L^2} + (\breve\delta^P\omega,\,\breve\delta^P\omega)_{L^2},$$

thus, ω is P-harmonic (and similarly for $\breve P$-harmonic) if and only if $\bar d^P\omega = 0$ and $\breve\delta^P\omega = 0$. If $\bar\Delta^P_H\omega = 0$ and $\omega = \bar d^P\theta$, then $\breve\delta^P\bar d^P\theta = \breve\delta^P\omega = 0$. Thus,

$$(\omega,\,\omega)_{L^2} = (\bar d^P\theta,\,\bar d^P\theta)_{L^2} = (\theta,\,\breve\delta^P\bar d^P\theta)_{L^2} = (\theta,\,\breve\delta^P\omega)_{L^2} = 0.$$

Moreover, if $\omega \in \Lambda^k(TM)$ is P-harmonic and $\omega = \bar d^P\theta$ for some $\theta \in \Lambda^{k-1}(TM)$, then $\omega = 0$.

Lemma 5 *For a statistical P-structure the following equalities are satisfied:*

$$a)\ d^P = \bar d^P = \breve d^P,\quad b)\ \delta^P = \bar\delta^P - \iota_E = \breve\delta^P + \iota_E,\quad c)\ \Delta^P_H = \bar\Delta^P_H + \bar{\mathcal L}^P_E = \breve\Delta^P_H - \bar{\mathcal L}^P_E,$$

where $\bar{\mathcal L}^P := \bar d^P\iota - \iota\,\bar d^P$ is the Lie P-derivative.

The following proposition extends Lemma 2.

Proposition 15 *Let (M,g) be a complete noncompact Riemannian manifold endowed with a vector field X such that $\overline{\operatorname{div}}_P X \ge 0$ (or $\overline{\operatorname{div}}_P X \le 0$), where $P \in$* End(TM) *such that conditions* (34) *and $\|PX\|_g \in \mathrm{L}^1(M,g)$ hold. Then, $\overline{\operatorname{div}}_P X \equiv 0$.*

We call $\bar\Delta^P f = \overline{\operatorname{div}}_P(\bar\nabla^P f)$ the *P-Laplacian for functions*. By (30), we obtain

$$\bar\Delta^P f = \Delta^P f + (PE)(f).$$

The condition $\bar\nabla^P f = 0$ means that $f \in C^2(M)$ is constant along the integral curves of $\mathcal D$; moreover, if $\mathcal D$ is bracket-generating then $f = \mathrm{const}$ on M. The next theorem extends the well-known classical result on subharmonic functions.

Theorem 24 *Let* (34) *hold for a statistical P-connection $\bar\nabla^P$, and let $f \in C^2(M)$ satisfy either $\bar\Delta^P f \ge 0$ or $\bar\Delta^P f \le 0$. If (M, g) is complete noncompact, $\|P\bar\nabla^P f\|$ and $\|f P\bar\nabla^P f\|$ belong to $L^1(M, g)$. Then, $\bar\nabla^P f = 0$; moreover, if $P(TM)$ is bracket-generating, then $f =$ const.*

We omit results on the statistical P-curvature tensor and its Weitzenböck type curvature operator (see [51]), which extend the results in Sect. 7.4.

References

1. Alexandrova I.A., Mikeš J., Stepanov, S.E., Tsyganok I.I.: Liouville-type theorems in the theory of mappings of complete Riemannian manifolds, J. of Mathematical Sciences (N.Y.), 221:6 (2017), 737–744
2. Alexandrova I.A., Stepanov S.E., Tsyganok I.I.: Liouville-type theorems for theories of Riemannian almost product structures and submersions, J. of Mathematical Sciences (N.Y.), 230:1 (2018), 1–9
3. Amari S.-I.: *Information geometry and its applications*. Applied Math. Sciences, 194. Springer, 2016
4. Barros A., Gomes J.N., Rebeiro E.: A note on rigidity of almost Ricci soliton, Archiv der Mathematik, 100, 481–490 (2013)
5. Barros A., Batista R., Ribeiro jr. E.: Compact almost Ricci solitons with constant scalar curvature are gradient, Monatshefte für Mathematik, 174, 29–39 (2014)
6. Berard P.H.: From vanishing theorems to estimating theorems: the Bochner technique revisited, Bull. of the AMS, 1988, 19:2, 371–406
7. Besse A.L.: Einstein Manifolds, Springer-Verlag, Berlin, 1987
8. Bishop R.L., O'Neill B.: Manifolds of negative curvature. TAMS, 145 (1969), 1–9
9. Böhm C., Wilking B.: Manifolds with positive curvature operators are space forms, Annals of Mathematics, 2008, 167, 1079–1097
10. Boucetta M.: Spectra and symmetric eigentensors of the Lichnerowicz Laplacian on S^n, Osaka J. Math., 2009, 46, 235–254
11. Bourguignon J.-P.: Les variétés de dimension 4 á signature non nulle dont la courbure est harmonique sont d'Einstein, Invent. Math., 1981, 63, 263–286
12. Brendle S., Schoen S.: Curvature, sphere theorem, and the Ricci flow, Bull. AMS, 2011, 48, 1–32
13. Calin O. and Chang D.-C.: Sub-Riemannian geometry. General theory and examples. Encyclopedia of Math. and its Applications, 126. Cambridge Univ. Press, 2009
14. Caminha A., Souza P., Camargo F.: Complete foliations of space forms by hypersurfaces, Bull. Braz. Math. Soc., 2001, 41:3, 339–353
15. Chavel I.: Eigenvalues in Riemannian Geometry, New York, Academic Press, 1984
16. Chow B., Lu P., Ni L.: Hamilton's Ricci flow, AMS, Providence, RI (2006)
17. Chow W.-L.: Über Systeme von linearen partiellen Differentialgleichungen erster Ordnung. Math. Ann. 1939, 117, 98–105
18. Craioveanu M., Puta M., Rassias T.M.: Old and New Aspects in Spectral Geometry, Kluwer Acad Publ, Cambridge, 2001
19. Deshmukh S.: Almost Ricci solitons isometric to spheres. Int. J. of Geom. Methods in Modern Physics, 16 (5), 1950073 (2019)
20. Gallot S., Meyer D.: Opérateur de courbure et laplacien des formes différentielles d'une variété riemannienne, J. Math. Pures Appl., 1975, 54:3, 259–284
21. Ge H., Zhang S.: Liouville-type theorems on the complete gradient shrinking Ricci solitons, Differential Geometry and its Applications, 2018, 56, 42–53

22. Gibbons G.W., Perry M.J.: Quantizing gravitational instantons, Nuclear Phys., 1978, B 146, 90–108
23. Greene R.E., Wu H.: Harmonic forms on noncompact Riemannian and Kähler manifolds, Michigan Math. J., 1981, 28:1, 63–81
24. Grigor'yan A.: Analytic and geometric background of recurrence and non-explosion of the Brownian motion on Riemannian manifolds, Bull. Amer. Math. Soc., 36:2 (1999), 135–249
25. Grigor'yan A.: Heat kernel and analysis on manifolds, AMS/IP, Boston, 2009
26. Heil K., Moroianu A., Semmelmann U.: Killing and conformal Killing tensor, J. of Geometry and Physics, 2016, 106, 383–400
27. Hitchin, N.: A note on vanishing theorems, In: Geometry and Analysis on Manifolds, Progr. Math., 2015, 308, 373–382
28. Kröncke K.: On the stability of Einstein manifolds, Annals of Global Analysis and Geometry, 2015, 47:1, 81–98
29. Kröncke K.: Stability of Einstein metrics under Ricci flow, Communications in Analysis and Geometry, 2020, 28:2, 351–394
30. Kröncke K.: On infinitesimal Einstein deformations, Differential Geometry and its Applications, 2015, 38, 41–57
31. Li P., Schoen R.: L^q and mean value properties of subharmonic functions on Riemannian manifolds, Acta Mathematica, 1984, 153, 279–301
32. Li P.: Geometric Analysis, Cambridge University Press, Cambridge, 2012
33. Lichnerowicz A.: Propagateurs et commutateurs en relativité générale. Publ. Math., Inst. Hautes Étud. Sci. 10 (1961), 293–344
34. Lichnerowicz A.: Spineurs harmoniques, C.R. Acad. Sci. Paris, Ser. A-B, 1963, 257, 7-9
35. Mikeš J., Stepanov S.E., Tsyganok I.I.: On the Sampson Laplacian, Filomat, 2019, 33:4, 1059–1070
36. Mikeš J., Stepanov S.E.: The spectral theory of the Yano rough Laplacian with some of its applications, Ann. Glob. Anal. Geom., 2015, 48, 37–4
37. Mikeš J., Rovenski V., Stepanov S.E.: A Contribution of Liouville-Type Theorems to the Geometry in the Large of Hadamard Manifolds, Mathematics, 2022, 10(16), article number 2880
38. Mikeš J., Stepanov S.E.: What is the Bochner technique and where is it applied, Lobachevskii J. of Mathematics, 2022, 43(3), 709–719
39. Mikeš J., Rovenski V., Stepanov S.E., Tsyganok I.I.: Application of the generalized Bochner technique to the study of conformally flat Riemannian manifolds, Mathematics, 2021, 9(9), article number 927
40. Mikeš J., Rovenski V., Stepanov S.E.: An example of Lichnerowicz-type Laplacian, Ann. Global Anal. Geom. 58:1 (2020), 19–34
41. Mikeš J., Stepanov S.E., Tsyganok I.I.: On metrics projectively and holomorphically projectively equivalent to metrics of parabolic Riemannian and Kähler manifolds, Filomat, 2023, 37:3, 949–956
42. Morgan J., Tian G.: Ricci flow and Poincare conjecture, Clay Math. Monographs, 3. AMS, Providence, RI; Clay Mathematics Institute, Cambridge, MA (2007)
43. Opozda B.: Bochner's technique for statistical structures. Ann. Global Anal. Geom. 48, 357–395 (2015)
44. Petersen P.: Riemannian geometry, 3d ed. New York, Springer AG, 2016
45. Pigola S., Rigoli M., Setti A.G.: Vanishing and finiteness results in geometric analysis: A generalization of the Bochner technique, Birkhäuser, 2008
46. Pigola S., Rigoli M., Setti A.G.: Maximum principles on Riemannian manifolds and applications, AMS, Providence, 2005
47. Pigola S., Rigoli M., Rimoldi M., Setti A.G.: Ricci almost solitons, Ann. Scuola Norm. Sup. Pisa Cl. Sci. 10 (4), 757–799 (2011)
48. Pigola S., Rimoldi M., Setti A.G.: Remarks on non-compact gradient Ricci solitons, Mathematische Zeitschrift, 2011, 268, 777–790

49. Pilch K., Schellekens A.N.: Formulas for the eigenvalues of the Laplacian on tensor harmonics on symmetric coset spaces, J. Math. Phys., 1984, 25, 34
50. Popescu P. Rovenski V. and Stepanov S.E.: The Weitzenböck type curvature operator for singular distributions, Mathematics, MDPI, 2020, 18 pp.
51. Popescu P. Rovenski V. and Stepanov S.E.: On singular distributions with statistical structure, Mathematics, MDPI, 2020, 8, 1825
52. Romero A.: The introduction of Bochner's technique on Lorentzian manifolds, Nonlinear Analysis, 47, 2001, 3047–3059
53. Rovenski V., Stepanov S.E., Tsyganok I.I.: On evolution equations under the Hamilton's Ricci flow, Results in Mathematics, 2020, 75(4), article number 166
54. Rovenski V., Stepanov S.E., Tsyganok I.I.: On the geometry in the large of Lichnerowicz type Laplacians and its applications, Balkan J. Geom. Appl. 25:2 (2020), 76–93
55. Rovenski V., Stepanov S.E., Tsyganok I.I.: The Bourguignon Laplacian and harmonic symmetric bilinear forms, Mathematics, 2020, 8, 83
56. Rovenski V., Stepanov S.E., Tsyganok I.I.: On the Betti and Tachibana numbers of compact Einstein manifolds, Mathematics, 2019, 7, 1210
57. Rovenski V. and Walczak P.: Extrinsic geometry of foliations, Birkhäuser, Progress in Mathematics, Vol. 339, 2021
58. Sampson J.H.: On a theorem of Chern, Transactions of the American Mathematical Society, 1973, 177, 141–153
59. Schoen R., Yau S.T.: Lectures on harmonic maps, Boston, Internat. Press, 1994
60. Stepanov S.E.: Vanishing theorems in affine, Riemannian, and Lorentz geometries, J. of Mathematical Sciences, 2007, 141: 1, 929–964
61. Stepanov S.E.: A contribution to the geometry in the large of conformal diffeomorphisms, J. of Geometry and Physics, 2019, 143, 1–10
62. Stepanov S.E., Denezhkina I.E., Ovchinnikov A.V.: On geometric analysis of the dynamics of volumetric expansion and its applications to general relativity, J. Math. Sci. (New York), 2020, 245:5, 659–668
63. Stepanov S.E., Mikeš J.: The generalized Landau–Raychaudhuri equation and its applications, International J. of Geometric Methods in Modern Physics, 2015, 12:08, article number 1560026
64. Stepanov S.E., Tsyganok I.I.: On the Lichnerowicz Laplacian, J. of Mathematical Sciences, Vol. 263, No. 3, 2022, 416–422
65. Stepanov S.E., Aleksandrova I.A., Tsyganok I.I.: From harmonic mappings to Ricci flows due to the Bochner technique, J. of Mathematical Sciences (New York), 2022, 263: 3, 423–435
66. Stepanov S.E., Mikeš J.: Application of the Hopf maximum principle to the theory of geodesic mappings, Kragujevac J. of Mathematics, 2021, 45:5, 781–786
67. Suh Y.J.: Generalized maximum principles and their applications to submanifolds and S. S. Chern's conjectures, Proc. of the 11-th Int. Workshop on Diff. Geom. 2007, 11, 135–152
68. Wu H.-H.: The Bochner technique in differential geometry, Harwood Academic Publ., 1988
69. Yano K.: Integral formulas in Riemannian geometry, N.Y., Marcel Dekker, 1970
70. Yano K., Bochner S.: Curvature and Betti numbers, Princeton University Press, 1953
71. Yau S.T.: Non-existence of continuous convex functions on certain Riemannian manifolds, Math. Ann., 207 (1974), 269–270
72. Yau S.-T.: Some function-theoretic properties of complete Riemannian manifold and their applications to geometry, Indiana Univ. Math. J., 1976, 25:1, 659–670
73. Yau S.-T.: Erratum: Some function-theoretic properties of complete Riemannian manifold and their applications to geometry, Vol. 25 (1976), 659–670, Indiana Univ. Math. J., 1982, 31:4, 607
74. Yau S.T.: Review of geometry and analysis, Asian J. Math., 2000, 4:1, 235–278

On General Solutions of Sinyukov Equations on Two-Dimensional Equidistant (pseudo-)Riemannian Spaces

Patrik Peška, Lenka Vítková, Josef Mikeš, and Irina Kuzmina

Abstract The paper is devoted to study of Sinyukov equations on two-dimensional equidistant (pseudo-) Riemannian spaces. The general solution of Sinyukov equations is found beyond these spaces under minimal requirements for the differentiability of the studied objects.

Keywords Equidistant space · (pseudo-) Riemannian space · Sinyukov equations · Geodesic mapping

1 Introduction

The geometric properties of equidistant (pseudo-) Riemannian manifolds are frequently studied in connection with the existence of special equations. For completeness, we start here with a fairly general class of Sinyukov equations of geodesic mappings.

It is well-known that equidistant spaces introduced Sinyukov [35, 36], see [28, 37]. On these spaces, there exists a concircular vector field that Yano studied [50]. These vector fields were known earlier, for example, in papers by Brinkmann [2], Kagan [19], Shirokov [43], or Fialkov [6]. These vector fields were intensively studied under different names even later, e.g., Vries [48], Shapiro [42].

P. Peška · L. Vítková · J. Mikeš (✉)
Palacky University, 77146 Olomouc, Czech Republic
e-mail: Josef.Mikes@upol.cz
URL: https://kag.upol.cz

P. Peška
e-mail: Patrik.Peska@upol.cz

L. Vítková
e-mail: Lenka.Vitkova@upol.cz

I. Kuzmina
Tupolev Kazan National Research Technical University, Kazan 420111, Russia
e-mail: iranina@mail.ru

© The Author(s), under exclusive license to Springer Nature Switzerland AG 2024
V. Rovenski et al. (eds.), *Differential Geometric Structures and Applications*,
Springer Proceedings in Mathematics & Statistics 440,
https://doi.org/10.1007/978-3-031-50586-7_9

In our paper [32], we obtained general solutions of concircular vector fields on two-dimensional (pseudo-) Riemannian spaces with minimal requirements on the differentiability of the metric tensor in canonical coordinates. In [32], examples of isometric embeddings of these spaces in three-dimensional Euclidean or Minkowski spaces were given.

Ultimately, we notice that any equidistant spaces admit geodesic mapping (Sinyukov [36]). In his other monograph [37, pp. 97–98], there is a mathematical mistake in the formulas, which have been corrected, for example, in [21] or [28, pp. 333–334]. Considering the above, it is very interesting to find a general solution of Sinyukov equations of geodesic mappings on equidistant spaces.

2 Equidistant Spaces

An n-dimensional (pseudo-) Riemannian space V_n with a metric tensor g is called *equidistant* (N.S. Sinyukov [35, 36], see [37, p. 92], [28, p. 168]) if there exists a linear form $\varphi \neq 0$ satisfying the equations

$$\nabla\varphi = \rho \cdot g, \tag{1}$$

where ρ is a function and ∇ is a connection on V_n. Since the form φ is necessarily a locally gradient (locally, there exists a function Φ for which $\varphi(X) = \nabla_X \Phi$), it defines a normal congruence in V_n, which we will call *equidistant*. The origin of this term will be explained in what follows.

Spaces satisfying conditions (1) were introduced by Brinkmann [2] and later also, e.g. Kagan [19], Shirokov [43], Fialkov [6], Vries [48], Yano [50] etc., see [3, 18, 20]. The special significance of equidistant spaces in the theory of geodesic and holomorphically projective mappings was discovered by Sinyukov [35–37], Mikeš [21–28], Shandra [38–41], and others [4, 7–14, 16, 30, 31, 52].

Some equidistant Riemannian spaces can be used as cosmological models. Modern cosmological models are based on the so-called Copernican principle (Bondi [1], 1960), stating that no position in the universe is distinguished in any way. A suitable concrete realization of this principle is the assumption of spatial homogeneity. At a sufficiently large scale, the universe's structure is essentially the same everywhere. Such a space is invariant under translations. The latter forms an isometry group, i.e., a group of transformations that leaves the metric invariant. A further interpretation of the Copernican principle is isotropy: the universe appears in every direction approximately the same. The associated isometry group is the group of rotations. Isotropy at every point ascertains homogeneity. The reverse is not true. Homogenous and isotropic models are the simplest cosmological models characterized by constant spatial curvature. They were introduced and studied by Friedmann, Lemaître, Robertson and Walker.

The vector field ξ, which is associated with the equidistant field φ: $g(\xi, X) = \varphi(X)$ for all tangent vector X, will also be called *equidistant*. This vector field is a special case of torse-forming and concircular vector fields introduced by Yano [49, 50]. *Torse-forming* vector fields ξ are characterized by the following equation $\nabla\xi = a\cdot\xi + \rho\cdot Id$, where a is a linear form, and ρ is a function on V_n. If a is a gradient, the vector field ξ is called *concircular*. We note that for any concircular vector field ξ, there exists a function f on V_n such that $f\cdot\xi$ is equidistant. Therefore, equidistant and concircular vector fields are often equated. In detail are the torse-forming, concircular and equidistant vector fields described in monographs [28, p. 168]. A lot of works are devoted to the above vector fields in various directions, for example, [4, 10, 31, 38, 41, 44]. Most of these works concern spaces of dimension $n > 2$. We devote our work to the study of equidistant 2-dimensional spaces V_2.

Next, we will suppose that equidistant vector field ξ is not isotropic, i.e. the length $\|\xi\| \neq 0$. For these equidistant spaces there exists a local coordinate system $x = (x^1, x^2, \dots, x^n)$ for which the metric form of V_n has the following form (see [37, p. 96], [28, p. 179]): $ds^2 = {dx^1}^2 + f(x^1)\cdot d\tilde{s}^2$, where f is a differentiable function of x^1 and $d\tilde{s}^2$ is a metric form $d\tilde{s}^2 = \tilde{g}_{\alpha\beta}(x^2,\dots,x^n)\,dx^\alpha dx^\beta$ of an $(n-1)$-dimensional (pseudo-) Riemannian space $\bar{V}_{n-1}$. We can consider the metrics ds^2 and $-ds^2$ to be equivalent.

The metric of 2-dimensional equidistant space V_2 has the form

$$ds^2 = du^2 + f(u)\cdot dv^2. \tag{2}$$

For clarity, here and hereafter, $u \equiv x^1$ and $v \equiv x^2$. In local coordinate system Eq. (1) of equidistant field φ has the following shape:

$$\nabla_j\varphi_i = \rho\cdot g_{ij}, \tag{3}$$

where φ_i and g_{ij} are components of φ and g. In detail, the left-hand side of Eq. (3) is $\partial_j\varphi_i - \varphi_\alpha\Gamma^\alpha_{ij} = \rho\, g_{ij}$, where $\Gamma^h_{ij} = \Gamma_{ij\alpha}g^{\alpha h}$, $\|g^{ij}\| = \|g_{ij}\|^{-1}$ and $\Gamma_{ijk} = \frac{1}{2}\,(\partial_i g_{jk} + \partial_j g_{ik} - \partial_k g_{ik})$ are Christoffel symbols and $\partial_i \equiv \partial/\partial x^i$.

By direct substitution, we make sure that $\varphi = c\,(\sqrt{|f(u)|}, 0, 0, \ \dots\ , 0)$, $c =$ const, is a particular solution of the Eqs. (1), therefore φ is an equidistant vector field.

In [32] we proved the following theorem:

Theorem 1 *In $V_2 \in C^1$ of non-constant curvature the general solution of the equation* (1) *has at most one linearly independent solution φ. If $V_2 \in C^k$, $k \geq 1$, then $\varphi_i(x) \in C^k$ holds and the general solution has the form* const $\cdot\,\varphi$.

This theorem holds globally even if V_2 has local regions with constant curvature. Furthermore, we can assume a priori that $\varphi_i(x)$ are differentiable functions, and $\varrho(x)$ is continuous. Higher differentiability of $\varphi_i(x)$ and $\varrho(x)$ follows from the results of I. Hinterleitner and J. Mikeš, see [15, 17] and [28, p. 171].

3 Geodesic Mappings and Sinyukov Equations

Diffeomorphism V_n onto $\bar{V}_n$ is called *geodesic mappings* if any geodesic in V_n is mapped onto geodesic in $\bar{V}_n$. Levi-Civita obtained necessary and sufficient condition of geodesic mapping between Riemannian spaces in common coordinates x:

$$\bar{\Gamma}_{ij}^h(x) = \Gamma_{ij}^h(x) + \delta_i^h \psi_j(x) + \delta_j^h \psi_i(x), \tag{4}$$

$$\nabla_k \bar{g}_{ij} = 2\psi_k \bar{g}_{ij} + \psi_i \bar{g}_{jk} + \psi_j \bar{g}_{ik}, \tag{5}$$

where g_{ij} and $\bar{g}_{ij}$ are components of metric tensor, Γ_{ij}^h and $\bar{\Gamma}_{ij}^h$ are Christoffel symbols of V_n and $\bar{V}_n$, ψ_i are components of a gradient vector

$$\psi_i = \partial_i \Psi = \frac{1}{n+1}\, \partial_i \left(\frac{\det |\bar{g}_{ij}|}{\det |g_{ij}|} \right).$$

Here, $\partial_i = \partial/\partial x^i$, δ_i^h is the Kronecker delta, ∇ is the covariant derivative in V_n, see [5], [37, pp. 72–75] and [28, pp. 291, 307, 329].

Equations (4) and (5) are called the *Levi-Civita equations*. Sinyukov, see [46], [37, p. 121] and [28, p. 329], found that (4) and (5) are equivalent to the following equations:

$$\nabla_k a_{ij} = \lambda_i g_{jk} + \lambda_j g_{ik}, \tag{6}$$

where a_{ij} are components of a symmetric and regular tensor, and λ_i are components of the gradient vector, for which $\lambda_i = \frac{1}{2}\, \partial_i(a_{ij} g^{ij})$.

There is a relationship between g_{ij}, $\bar{g}_{ij}$ and a_{ij}:

$$a_{ij} = e^{-2\Psi} g_{i\alpha} g_{j\beta} \cdot \bar{g}^{\alpha\beta}, \quad \lambda_i = e^{-2\Psi} g_{i\alpha} \psi_\beta \cdot \bar{g}^{\alpha\beta},$$

where $\|\bar{g}^{ij}\| = \|\bar{g}_{ij}\|^{-1}$.

Equations (6) have been completed by Sinyukov [37, pp. 132–134] to a system of differential equations of Cauchy type in covariant derivatives. In principle, this system for the unknown tensors a_{ij}, λ_i and μ has the form (6) with

$$\begin{aligned} n \nabla_j \lambda_i &= \mu\, g_{ij} - a_{i\alpha} R_j^\alpha - a_{\alpha\beta} R^\alpha{}_{ij}{}^\beta, \\ (n-1)\ \ \nabla_i \mu &= 2(n+1) \lambda_\alpha R_i^\alpha + a_{\alpha\beta}(2R^\alpha{}_{i,}{}^\beta - R^{\alpha\beta}{}_{,i}). \end{aligned}$$

For $n = 2$, this system has the following form:

$$\begin{aligned} \nabla_k a_{ij} &= \lambda_i g_{jk} + \lambda_j g_{ik}, \\ \nabla_j \lambda_i &= \mu\, g_{ij} - K\, a_{ij}; \\ \nabla_i \mu &= -\ \ 2K\lambda_k + (a_{\alpha\beta} K_{,k} - K_\alpha a_{\beta k}) g^{\alpha\beta}, \end{aligned}$$

where K is the gaussian curvature on V_2.

The above mentioned systems were obtained at higher differentiability requirements, and exist for $g_{ij}(x) \in C^3$. A detailed analysis of the differentiability of these equations is carried out in the papers [15, 17].

These systems have no more than one solution for the initial conditions $a_{ij}(x_0) = a^{\circ}_{ij}$, $\lambda_i(x_0) = \lambda^{\circ}_i$ and $\mu(x_0) = \mu^{\circ}$. This means that their general solution depends on $r_{geo} \leq \frac{1}{2}\, n(n+1)$ essential parameters, when $n = 2$ is $r_{geo} \leq 6$. The number r_{geo} was called a *degree of mobility with respect to geodesic mappings*. The maximum r_{geo} is achieved in spaces of constant curvature and only in them, see [37, p. 137] and [28, p. 309].

4 Integration of Sinyukov Equation on Two-Dimensional Equidistant Space

4.1 Lemma of Independently Special Set Functions

First, we introduce for an arbitrary continuous function $f(u)$ ($\neq 0$) of variable $u \in I \subset \mathbb{R}$, the following integrals:

$$I(u) = \int_{u_0}^{u} \frac{1}{f(u)}\, du \quad \text{and} \quad J(u) = \int_{u_0}^{u} \frac{1}{f(u)}\, I(u)\, du, \qquad u_0 \in I.$$

Now, we will prove the following lemma.

Lemma 1 *The functions* 1, $I(u)$ *and* $J(u)$ *are linearly independent set on an interval I, respective real coefficients.*

Remark 1 The above means that for real constants α, β, γ and $u \in I \subset \mathbb{R}$ the following must be true:

$$\alpha \cdot 1 + \beta \cdot I(u) + \gamma \cdot J(u) = 0 \tag{7}$$

if and only if $\alpha = \beta = \gamma = 0$.

Proof Let (7) hold. After derivation and multiplication by function f, we get

$$\beta + \gamma \cdot J(u) = 0.$$

Derivation of this equation, we get $\gamma / f(u) = 0$. Now, it is obvious that $\gamma = 0$. It follows that $\beta = 0$, and finally from (7), we get $\alpha = 0$. □

4.2 Sinyukov Equations of Equidistant Spaces in Special Coordinates

In equidistant space V_2 we choose coordinate system x in which metric form has the form (2): $ds^2 = du^2 + f(u)\, dv^2$, where $f(u)$ is a non-zero differentiable function of variable u, coordinates $u \equiv x^1$ and $v \equiv x^2$.

In this case, the Christoffel symbols have the following form:

$$\Gamma^1_{22} = -1/2\ f' \ \text{ and } \ \Gamma^2_{12} = \Gamma^2_{21} = 1/2\ f'/f,$$

other components are vanishing. After substituting the Christoffel symbols in Sinyukov equation (6) we get

$$\partial_k a_{ij} - a_{\alpha j}\Gamma^\alpha_{ik} - a_{i\alpha}\Gamma^\alpha_{jk} = \lambda_i g_{jk} + \lambda_j g_{ik},$$

and then obtain the following system:

$$\begin{array}{ll}(a) & \partial_1 a_{11} = 2\,\lambda_1,\\ (b) & \partial_2 a_{11} = f'/f \cdot a_{12},\\ (c) & \partial_1 a_{12} = 1/2 \cdot f'/f \cdot a_{12} + \lambda_2,\\ (d) & \partial_2 a_{12} = -1/2 \cdot f' \cdot a_{11} + 1/2 \cdot f'/f \cdot a_{22} + f \cdot \lambda_1,\\ (e) & \partial_1 a_{22} = f'/f \cdot a_{22},\\ (f) & \partial_2 a_{22} = -f' \cdot a_{12} + 2f\,\lambda_2.\end{array} \tag{8}$$

4.3 Integration of Sinyukov Equations (8) on Equidistant Spaces

Integrating equation (8e) with respect to unknown function $a_{22}(u, v)$ and variable u, we obtain

$$a_{22} = f \cdot A(v), \tag{9}$$

where function $A(v)$ is integrating "constant" and function of variable v. Since $\partial_2 a_{22}$ exists, the derivative of function $A(v)$ also exists.

From (8f) it follows that

$$\lambda_2 = 1/2f \cdot (f \cdot A' + f' \cdot a_{12}) = 1/2 \cdot A' + 1/2 \cdot f'/f \cdot a_{12}.$$

Substituting λ_2 to (8c) we find

$$\partial_1 a_{12} = f'/f \cdot a_{12} + 1/2 \cdot A'.$$

Integrating above equation with respect to unknown function $a_{12}(u, v)$ and the corresponding variable u, we get

$$a_{12} = f \cdot B(v) + 1/2 \cdot A' \cdot I(u), \tag{10}$$

where function $B(v)$ is integrating "constant" which is a function of variable v. Using Lemma 1, we can prove that B and A' are differentiable, i.e. there exist B' and A''.

Namely, based on Lemma 1, it follows that 1 and $I(u)$ are independent. Thus, there exist two points $u_1, u_2 \in I$, for which we have $\begin{vmatrix} 1 & I(u_1) \\ 1 & I(u_2) \end{vmatrix} \neq 0$.

From (10), we obtain two equations for $u = u_1$ and $u = u_2$:

$$\frac{a_{12}(u_1, v)}{f(u_1)} = B(v) \cdot 1 + 1/2\, A'(v) \cdot I(u_1)\,,$$

$$\frac{a_{12}(u_2, v)}{f(u_2)} = B(v) \cdot 1 + 1/2\, A'(v) \cdot I(u_2)\,.$$

These equations are nonhomogeneous linear equations respective unknowns $B(v)$ and $A'(v)$. Since the above determinant is nozero, these unknowns are expressed as a linear combination of $a_{12}(u_1, v)$ and $a_{12}(u_2, v)$ with real parameters. Since these functions are differentiable, the functions $A'(v)$ and $B(v)$ are also differentiable. This implies the existence of A'' and B'.

Equation (8f) implies

$$\lambda_1 = 1/f \cdot \partial_2 a_{12} + 1/2 \cdot f'/f \cdot a_{11} - 1/2 \cdot f'/f^2 \cdot a_{22}.$$

Then from (8a), (9) and (10) it follows the equation

$$\partial_1 a_{11} = f'/f \cdot a_{11} + A''\, I(u) - A \cdot f'/f + 2B'.$$

We integrate this equation as an inhomogeneous linear equation with respect to a_{11} and find a solution in the form

$$a_{11} = A'' \cdot f \cdot J(u) + A + 2B' \cdot f \cdot J(u) + C \cdot f,$$

where function $C(v)$ is an integrating "constant" which is a function of variable v. By a similar analysis, using Lemma 1, one can verify that the existence of $\partial_2 a_{11}$ implies the existence of functions A''', B'' and C'.

As a result, we obtain the following.

Theorem 2 *On equidistant space V_2 in coordinates (2) the general solution of Eqs. (8) has the following form:*

$$
\begin{aligned}
a_{11} &= f \cdot C(v) + 2B' \cdot f \cdot I(u) + A'' \cdot f \cdot J(u) + A, \\
a_{12} &= f \cdot B(v) + 1/2 \cdot A' \cdot f \cdot I(u), \\
a_{22} &= f \cdot A(v),
\end{aligned}
\tag{11}
$$

where $A''(v)$, $B'(v)$, $C(v)$ are differentiable functions.

4.4 Trivial Solution

For equidistant spaces, Sinyukov [37, p. 123] found the construction of the solution of the Eq. (6) in the following form: $a_{ij} = \alpha \cdot a_{ij} + \beta \cdot \varphi_i \varphi_j$, where φ_i is concircular vector field (1) and α, β are constants. It follows that $r_{geo} \geq 2$.

The above corresponds to the trivial solution of the Eq. (6) for the values $A(v) = A_0$, $B(v) = 0$ and $C(v) = C_0$, where A_0, C_0 are constants. These solutions take the following form:

$$
a_{11} = A_0 + C_0 \cdot f(u), \qquad a_{12} = 0, \qquad a_{22} = A_0 \cdot f(u). \tag{12}
$$

The solutions (12) correspond to the solution of the Levi-Civita equations (5) (see [21], [28, pp. 333–334]):

$$
\bar{g}_{11} = \frac{\alpha}{(1 + \beta f(u))^2}, \qquad \bar{g}_{12} = 0, \qquad \bar{g}_{22} = \frac{\alpha f(u)}{1 + \beta f(u)},
$$

where α, β are constants, for which $\alpha \neq 0$ and $1 + \beta f \neq 0$. At the same time,

$$
\Psi = -\frac{1}{2} \ln |1 + \beta f(u)| \quad \text{and} \quad \psi_i = \Big(-\frac{\beta f'}{1 + \beta f(u)}, \; 0 \Big).
$$

These solutions exist under minimal requirements for the differentiability of the function $f \in C^1$. Thus, equidistant space admits geodesic mappings, even if they have bifurcations of geodesics, see [33, 34].

4.5 General Solution

On the other hand, in general the general solution (11) satisfies all Eq. (8) except (8b). Therefore, we substitute (11) in the Eq. (8b) and get

$$
A''' \cdot J(u) + 2B'' \cdot I(u) + C' \cdot 1 + A' \Big(\frac{1}{f} - \frac{f'}{2f} I(u) \Big) - B \, \frac{f'}{f} = 0. \tag{13}
$$

By Lemma 1, there exist three points $u_1, u_2, u_3 \in I$, whose determinant is

$$\begin{vmatrix} 1 & I(u_1) & J(u_1) \\ 1 & I(u_2) & J(u_2) \\ 1 & I(u_3) & J(u_3) \end{vmatrix} \neq 0.$$

Then, the three equations we obtained from (13) for $u = u_1, u_2, u_3 \in I$ imply the following solution:

$$A''' = \alpha_1 \cdot A' + \beta_1 \cdot B, \quad B'' = \alpha_2 \cdot A' + \beta_2 \cdot B, \quad C' = \alpha_3 \cdot A' + \beta_3 \cdot B, \qquad (14)$$

where α_i and β_i are constants, which are uniquely determined by the function f.

Substituting (14) to (13), we have:

$$G \cdot A' + H \cdot B = 0, \qquad (15)$$

where

$$G = \alpha_1 \cdot J(u) + \alpha_2 \cdot 2I(u) + \alpha_3 + \frac{1}{f} - \frac{f'}{2f} I(u),$$

$$H = \beta_1 \cdot J(u) + \beta_2 \cdot 2I(u) + \beta_3 - \frac{f'}{f}. \qquad (16)$$

(1) Case $H = 0$. Then from (16), we have $f'/f = \beta_1 J(u) + \beta_2 \cdot 2I(u) + \beta_3$. Since the right-hand-side is differentiable, there exists f'', and, evidently it holds

$$f'' = f'^2/f + \beta_1 I(u) + 2\beta_2. \qquad (17)$$

The existence of f''' follows from differentiability of the right-hand-side (17), and similarly, after differentiation we obtain that the function $f \in C^\infty$.

(a) Subcase $H = 0$ and $G = 0$. In this case, if $G = 0$, formulas (11) are all fulfilled, whereas for functions A, B, C system (14) hold. This system has a solution for six independent variables (they are, for example, Cauchy initial conditions $A_0, A_0', A_0'', B_0, B_0', C_0$). This solution corresponds to the maximal solution of system (6), which is possible only in the space of constant curvature K.

For equidistant metrics (2), the Gaussian curvature is $K = -\dfrac{f''}{2f} + \left(\dfrac{f'}{2f}\right)^2$. Therefore, the general form of the function $f(v)$ for spaces of constant curvature K form a solution of the differential equation $f'' = \dfrac{f'^2}{2f} - 2fK$. Using methods of solving ordinary differential equations (see [51, p. 374, 2.9.3, No. 1]), we can check that this equation is equivalent to its first integral

$$f' = \pm\sqrt{-Kf^2 + cf}, \qquad c = \text{const}. \qquad (18)$$

As it is well known, spaces with the same constant curvature K are locally isometric. Therefore, we can limit ourselves to the three cases: $K = 0$, $K > 0$, and $K < 0$, respectively. From $H = 0$, $C = 0$ and (14) it is easy to prove that the functions A, B and C have the form

(i) for $K = 0$, $f(u) = 1$:
$A = c_1 + c_2 v + c_3 v^2$, $B = c_4 + c_5 v$, $C = c_6 - A$,

(ii) for $K > 0$, $f(u) = \cos^2(\omega u)$, $\omega = \sqrt{K}$:
$A = c_1 + c_2 \cdot \cos(2\omega v) + c_3 \cdot \sin(2\omega v)$, $B = c_4 \cdot \cos(\omega v) + c_5 \cdot \sin(\omega v)$,
$C = c_6 + A$,

(iii) for $K < 0$, $f(u) = \cosh^2(\omega u)$, $\omega = \sqrt{-K}$:
$A = c_1 + c_2 \cdot \cos(2\omega v) + c_3 \sin(2\omega v)$, $B = c_4 \cdot \cosh(\omega v) + c_5 \cdot \sinh(\omega v)$,
$C = c_6 + A$,

for any constants $c_1, \ldots, c_6$, for which $\det \|a_{ij}\| \neq 0$.

(b) Subcase $H = 0$ and $G \neq 0$. It follows from (15) that $A' = 0$. Moreover, $B = 0$ leads to a particular solution, see Sect. 3.4, therefore we will assume $B \neq 0$. In this case, the formula (14) implies $\beta_1 = 0$. Then Eq. (17) simplifies to $f'' = \frac{f'^2}{f} + 2\beta_2$. Analogically using the methods of solving ODE (see [51, p. 374, 2.9.3, No. 1]), we can check that this equation is equivalent to its first integral

$$f' = \pm\sqrt{-4\beta f + C_0 f^2}, \qquad C_0 - \text{const.}$$

Comparing with (18), we find that V_2 is a space with constant curvature $K = -C_0$, which leads to the previous point (a).

(2) Case $H \neq 0$. It follows from (15) that $B = \beta \cdot A'$, β is a constant. Condition $A' = 0$, and also $B = 0$, leads to a particular solution, see point (1), therefore we will assume $A' \not\equiv 0$.

In this case, the formula (14) simplified

$$A''' = \alpha \cdot A', \quad B = \beta \cdot A', \quad C' = \gamma \cdot A', \qquad \alpha, \beta, \gamma - \text{const.} \tag{19}$$

The general solution of system (19) has the following form:
$C = c_0 + \gamma \cdot A$, and for

$\alpha = 0$: $A = a_0 + a_1 v + a_2 v^2$; $\quad B = a_1\beta + 2a_2 v$,
$\alpha < 0$: $A = a_0 + a_1 \sin \omega v + a_2 \cos \omega v \quad B = a_1\beta\omega \cos \omega v - a_2\beta\omega \sin \omega v$
$\alpha > 0$: $A = a_0 + a_1 \sinh \omega v + a_2 \cosh \omega v \quad B = a_1\beta\omega \cosh \omega v + a_2\beta\omega \sinh \omega v$,

where a_0, a_1, a_2, c_0 are constants, and $\omega = \sqrt{|\alpha|}$.

After substitution (19) into the formula (13), we get

$$\alpha\, J(u) + 2\alpha\beta\, I(u) + \gamma + \left(\frac{1}{f} - \frac{f'}{2f}\, I(u)\right) - \beta\,\frac{f'}{f} = 0. \tag{20}$$

In this case, when the conditions (20) are met for the function f, the solution of the Sinyukov equations has the form (11). Obviously, the number of the parameter $r_{geo} = 4$, in the case that the spaces do not have a constant curvature.

Formula (20) is an integro-differential equation with respect to the unknown function f. If we express the derivative of f' and then integrate, then we obtain recursive formulas for the function $f(u)$ which is equivalent to (20):

$$f(u) = f(u_*) + 2\int_{u_*}^{u} \frac{1 + f \cdot (\alpha\, J(u) + 2\alpha\beta\, I(u) + \gamma)}{I(u) + 2\beta}\, du. \qquad (21)$$

Obviously, (20) (and, equivalently, also (21)) has an unique solution for the given constants α, β, γ and $f(u_*)$, evidently, $f(u_*) \neq 0$, $I(u_*) \neq -2\beta$, $u_* \in I$. In case $\beta \neq 0$, we can choose $u_* \equiv u_0$. Among other things, the functional sequence constructed from the iteration of the function $f(u)$ with the help of the recursive formula (21) converges in a sufficiently small neighborhood of the point u_* to the solution $f(u)$, which is theoretically a real analytical function.

Differentiating the formula (20), after modification, we find

$$(2\alpha f - f''f + f'^2)\, I(u) + 4\alpha\beta f - 3f' - 2\beta f''f + 2\beta f'^2 = 0. \qquad (22)$$

Differentiating this condition, after eliminating the integral $I(u)$ with the help of (22), we get the 3rd order differential equation respective unknown function $f(u)$.

5 Main Theorem

Finally, we can formulate the following.

Theorem 3 *The equidistant space V_2 with metric (2) admits the following geodesic mappings of degree r_{geo}:*
$r_{geo} = 6$ if and only if V_2 has a constant curvature,
$r_{geo} = 4$ if and only if V_2 has f that satisfies Eq. (21),
$r_{geo} = 2$ in other cases.

Remark 2 In special cases, Eq. (21) are satisfied in spaces of constant curvature, for them, $r_{geo} = 6$ applies. The result (of Theorem 3) is interesting from the point of view that V_3 has a maximum $r_{geo} = 9$ for spaces of constant curvature, and for other spaces the maximum is only $r_{geo} = 2$. This fact was actually proven in Moldobayev's dissertation [29]. It is also cited in works [21].

The article is motivated by Vagner and Shulikovskij works [45, 47].

References

1. Bondi, H. et al.: Rival theories of cosmology. A symposium and discussion of modern theories of the structures of the universe. Oxford University Press (1960)
2. Brinkmann, H.W.: Einstein spaces which mapped conformally on each other. Math. Ann. 94 (1925)
3. Chepurna, O., Hinterleitner, I.: On the mobility degree of (pseudo-) Riemannian spaces with respect to concircular mappings. Miskolc Math. Notes 14:2, 561–568 (2013)
4. Ćirić, M.S., Zlatanović, M.Lj.; Stanković, M.S., Velimirović, L.S.: On geodesic mappings of equidistant generalized Riemannian spaces. Appl. Math. Comput. 218:12, 6648–6655 (2012)
5. Eisenhart, L.P.: Non-Riemannian geometry. Reprint of the 1927 original. Mineola, NY: Dover Publ. (2005)
6. Fialkow, A.: Conformals geodesics. Trans. AMS 45, 443–473 (1939)
7. Gorbatyi, E.Z.: On geodesic mapping of equidistant Riemannian spaces and first class spaces. Ukr. Geom. Sb. 12, 45–53 (1972)
8. Hall, G.: Four-Dimensional Manifolds and Projective Structure. New York etc.: CRC Press (2023)
9. Hinterleitner, I.: Conformally-projective harmonic mappings of equidistant manifolds with the Friedmann models as example. Proc. 6th Int. Conf. Aplimat, 97–102 (2007)
10. Hinterleitner, I.: Conformally-projective harmonic diffeomorphisms of equidistant manifolds. R. Soc. Mat. Esp. 11, 298–303 (2007)
11. Hinterleitner, I.: Selected special vector fields and mappings in Riemannian geometry. J. Appl. Math. 1:2, 30–37 (2008)
12. Hinterleitner, I.: Special mappings of equidistant spaces. J. Appl. Math. 2, 31–36 (2008)
13. Hinterleitner, I.: Special mappings of equidistant spaces. Vědecké spisy VUT Brno. Edice Ph.D. Thesis, 525, 20p. (2009)
14. Hinterleitner, I.: Geodesic mappings on compact Riemannian manifolds with conditions on sectional curvature. Publ. Inst. Math. 94 (108), 125–130 (2013)
15. Hinterleitner, I., Mikeš, J.: Geodesic mappings of (pseudo-)Riemannian manifolds preserve class of differentiability. Miskolc Math. Notes 14:2, 575–582 (2013)
16. Hinterleitner, I., Berezovski, V., Chepurna, E., Mikeš, J.: On the concircular vector fields of spaces with affine connection. Acta Math. Acad. Paedagog. Nyházi. (N.S.) 33:1, 53–60 (2017)
17. Hinterleitner, I., Mikeš, J.: Geodesic mappings onto Riemannian manifolds and differentiability. Geometry, Integrability and Quantization 18, 183–190 (2017)
18. Hinterleitner, I., Berezovski, V., Chepurna, E., Mikeš, J.: On the concircular vector fields of spaces with affine connection. Acta Math. Acad. Paedagog. Nyházi. (N.S.) 33:1, 53–60 (2017)
19. Kagan, V.F.: Subprojective spaces. Bibl. Russkoi Nauki, Moscow (1961)
20. Kühnel, W., Rademacher, H.-B.: Conformal diffeomorphisms preserving the Ricci tensor. Proc. AMS 123:9, 2841–2848 (1995)
21. Mikeš, J.: Geodesic mappings of affine-connected and Riemannian spaces. J. Math. Sci. (New York) 78:3, 311–333 (1996)
22. Mikeš, J.: Geodesic mappings of semisymmetric Riemannian spaces. Odessk. Univ. Moscow: Archives at VINITI, No. 3924-76, 19 p. (1976)
23. Mikeš, J.; Equidistant Kähler spaces. Math. Notes 38:4, 855–858 (1985)
24. Mikeš, J.: On Sasaki spaces and equidistant Kähler spaces. Sov. Math. Dokl. 34, 428–431 (1987)
25. Mikeš, J., Starko, G.A.: Hyperbolically Sasakian and equidistant hyperbolically Kählerian spaces. J. Sov. Math. 59:2, 756–760 (1992)
26. Mikeš, J.: Holomorphically projective mappings and their generalizations. J. Math. Sci. (New York) 89:3, 1334–1353 (1998)
27. Mikeš, J., Rachůnek, L.: T-semisymmetric spaces and concircular vector fields. Suppl. Rend. Circ. Mat. Palermo, II. Ser. 69, 187–193 (2002)
28. Mikeš, J. et al.: Differential geometry of special mappings. 2th ed. Palacky Univ. Press, Olomouc (2019)

29. Moldobayev, Dz.: Special transformations of Riemannian spaces, which preserved Einstein tensor. Ph.D. Thesis, Bishkek Univ. 1986
30. Najdanović, M.S., Zlatanović, M.Lj., Hinterleitner, I.: Conformal and geodesic mappings of generalized equidistant spaces. Publ. Inst. Math., Nouv. Sér. 98 (112), 71–84 (2015)
31. Najdanović M.S.: On the problem of geodesic mappings and deformations of generalized Riemannian spaces. J. Math. Anal. Appl. 452:1, 634–645 (2017)
32. Peška, P., Mikeš, J., Rýparová, L., Chepurna, O.: On general solutions of equidistant vector fields on two-dimensional (pseudo-) Riemannian spaces. Filomat 37:25, 8569–8574 (2023)
33. Rýparová, L., Mikeš, J.: On geodesic bifurcations. Geom., Integrability, Quantization 18, 217–224 (2017)
34. Rýparová, L., Mikeš, J., Sabykanov, A.: On geodesic bifurcations of product spaces. J. Math. Sci., New York 239, No. 1, 86–91 (2019)
35. Sinyukov, N.S.: On geodesic mappings of Riemannian manifolds onto symmetric spaces. Dokl. Akad. Nauk SSSR 98, 21–23 (1954)
36. Sinyukov, N.S.: On equidistant spaces. Vestn. Odessk. Univ., Odessa, 133–135 (1957)
37. Sinyukov, N.S.: Geodesic mappings of Riemannian spaces. Nauka, Moscow (1979)
38. Shandra, I.G.: Geodesic mappings of equidistant spaces and Jordan algebras of spaces $V_n(K)$. Diff. geom. mnogoobr. figur, Kaliningrad, 104–111 (1993)
39. Shandra, I.G.: On concircular tensor fields and geodesic mappings of pseudo-Riemannian spaces. Russ. Math. 45:1, 52–62 (2001)
40. Shandra, I.G.: Concircular vector fields on semi-Riemannian spaces. J. Math. Sci., New York 142:5, 2419–2435 (2007)
41. Shandra I. G., Mikeš, J.: Geodesic mappings of $V_n(K)$-spaces and concircular vector fields. Mathematics 7(8), 692 (2019)
42. Shapiro, Ya.L.: Geodesic fields of directions and projective systems of paths. Mat. Sb. (N.S.) 36(78), 125–148 (1955)
43. Shirokov, P.A.: Selected investigations on geometry. Kazan Univ. Press (1966)
44. Shiha, M., Mikeš, J.: On equidistant, parabolically Kählerian spaces. Tr. Geom. Semin. 22, 97–107 (1994)
45. Shulikovskij, V.I.: Invariant criterium of Liouvill surfaces. Dokl. AN SSSR 94:1, 29–32, 1954
46. Sinyukov, N.S.: A contribution to the theory fo geodesic mapping of Riemannian spaces. Sov. Math., Dokl. 7, 1004–1006 (1966)
47. Vagner, V.: On the problem of determining the invariant characteristics of Liouville surfaces. Tr. Semin. Vektorn. Tenzorn. Anal. 5, 246–249 (1941)
48. Vries, H.L:. Über Riemannsche Räume, die infinitesimal konforme Transformationen gestatten. Math. Z. 60:3, 328–347 (1954)
49. Yano, K.: On the torse-forming directions in Riemannian spaces. Proc. Imp. Acad. Tokyo 20, 340–345 (1944)
50. Yano, K.: Concircular geometry. Proc. Imp. Acad. Tokyo, 16, 195–200, 354–360, 442–448, 505–511 (1940)
51. Zajtsev, V.F., Polyanin, A.D.: Handbook of ordinary differential equations. (Spravochnik po obyknovennym differentsial'nym uravneniyam.) Moscow: Fizmatlit. (2001)
52. Zlatanović, M., Hinterleitner, I., Najdanović, M.: On equitorsion concircular tensors of generalized Riemannian spaces. Filomat 28:3, 463–471 (2014)

Fundamental Equations on Conformal Fedosov Spaces

Chakibek Almazbekov, Nadezda Guseva, and Josef Mikeš

Abstract The present work is devoted to the study of the basic equations of conformally Fedosov structures. These equations are obtained in the form of closed linear equations in covariant derivatives of the Cauchy type. It is established that the general solution depends on no more than $1/2n(n+1)$ numerical parameters. The maximum is achieved in projectively Euclidean spaces. Estimates are found for these equations' dependence on spaces that are not projectively Euclidean.

Keywords Conformally Fedosov structure · Conformally Fedosov space · Fundamental equations · Lacuna of degree

1 Introduction to the Theory of Conformally Fedosov Structures and Spaces

The present work is devoted to the study of the basic equations of conformally Fedosov structures. These equations are obtained in the form of closed linear equations in covariant derivatives of the Cauchy type. It is established that the general solution depends on no more than $\frac{1}{2}n(n+1)$ numerical parameters. The maximum is achieved in projectively Euclidean spaces. At the beginning of the paper, the notion of a conformally Fedosov structure is introduced.

On a Riemannian manifold, there is a unique torsion-free connection ∇ with respect to which the metric g is covariantly constant, i.e., $\nabla g = 0$. This is the

WWW home page: https://kag.upol.cz

C. Almazbekov
Kyrgyz National University, Bishkek, Kyrgyzstan

N. Guseva
Moscow Pedagogical State University, Moscow, Russia

J. Mikeš (✉)
Palacky University, Olomouc 77146, Czech Republic
e-mail: Josef.Mikes@upol.cz

© The Author(s), under exclusive license to Springer Nature Switzerland AG 2024

V. Rovenski et al. (eds.), *Differential Geometric Structures and Applications*,
Springer Proceedings in Mathematics & Statistics 440,
https://doi.org/10.1007/978-3-031-50586-7_10

Levi-Civita connection, it is the basis for computing and understanding geometry on Riemannian and pseudo-Riemannian manifolds. There is no preferred connection on a symplectic manifold. Instead, there are many symplectic connections, torsion-free connections that preserve the symplectic form. The choice of one of them determines what is called the Fedosov manifold [9].

On a conformally symplectic manifold [24], there exists a local equivalence class, the class of symplectic forms defined only up to scale. In the introduction, we show that it is possible to combine projective differential geometry with the concept of a Fedosov manifold and to obtain the so-called conformally Fedosov manifold.

Conformally symplectic manifold [3, 24] is an even-dimensional manifold M of at least four dimensions n endowed with a non-degenerate 2-form J such that

$$dJ = 2a \wedge J \tag{1}$$

for some closed 1-form a.

The non-degeneracy of J in (1) guarantees that $J : \Lambda^1 \to \Lambda^3$ is injective, hence the 1-form a is uniquely determined by J if such a 1-form exists. It is called the *form of Lee* [15] and in the case $\dim M \geq 6$ we see that $0 = d^2 J = 2da \wedge J$. In coordinates, the Eq. (1) have the form $\nabla_{[i} J_{jk]} = a_{[i} J_{jk]}$, where J_{ij} and a_i are the components of J and a, square brackets denote alternation by the indicated indices without division. The notion of conformality of manifolds is related, for example, to the papers [11–14, 23, 25].

Projective structure [25] on a manifold M is an equivalence class of torsion-free affine connections on M, where two connections ∇ and $\hat{\nabla}$ are *projectively equivalent* if and only if $\hat{\nabla}_a \varphi_b = \nabla_a \varphi_b - \nu_{(a} \varphi_{b)}$ for some 1-form ν_a. Parentheses denote symmetrization without dividing the corresponding indices. For the antisymmetric tensor J we have $\hat{\nabla}_{(a} J_{b)c} = \nabla_{(a} J_{b)c} - 3\nu_{(a} J_{b)c}$ and for projectively invariant J the condition [5] holds: $\nabla_{(a} J_{b)c} = \beta_{(a} J_{b)c}$.

The last condition, together with (1), is a formal definition of conformally Fedosov manifolds, see [5]. Moreover, based on these results, this definition can be formalized as follows.

Definition 1 *Conformally Fedosov* manifold $A_n = (M, [J], [\nabla])$ is a conformally symplectic manifold $(M, [J])$ in which the skew-symmetric form J satisfies the equations $\nabla_k J_{ij} = J_{i[j} a_{k]}$, where a_i are components of the linear form.

In [5], it was shown that these formulas are equivalent to the equations

$$\nabla_k J^{ij} = a^{[i} \delta_k^{j]}, \tag{2}$$

where J^{ij} are the components of the matrix inverse to $\|J_{ij}\|$, i.e. $J_{ac} J^{bc} = \delta_a^b$, δ_a^b is the Kronecker delta and $\varphi^a \equiv J^{ab} \varphi_b$, $\psi_b \equiv \psi^a J_{ab}$. The tensors J_{ij} and J^{ij} define Fedosov manifolds, we will call them the *Fedosov structure*.

Note that the basic equations in the Definition 1 are a system of nonlinear differential equations in covariant derivatives with respect to the unknown tensors J_{ij} and

a_i, while the Eq. (2) are linear with respect to J^{ij} and a^i. For this reason, they are preferable.

Moreover, by folding (2) we get $a^i = \frac{1}{n-1} \nabla_a J^{ia}$, and after excluding a^i from (2) we obtain a system of linear equations in covariant derivatives with respect to the unknown tensor J^{ij}. In this system, the components of the J^{ij} tensor and their first-order partial derivatives are presented in a linear fashion.

2 Basic Equations of Conformally Fedosov Spaces as a System of Differential Equations of Cauchy Type

As we said in the previous section, conformally Fedosov spaces A_n are characterized by the presence of a regular skew-symmetric tensor J^{ij} that satisfies the Eq. (2).

In our work [1], the following task was posed: Find a conformally Fedosov structure J in the space A_n with affine connection, i.e. make sure that the given space A_n is conformally Fedosov manifold.

The following theorem has been proved in [1].

Theorem 1 *The basic equations of the conformally Fedosov structure J can be written in the affine connection space A_n as a closed linear system of differential equations in covariant derivatives with respect to the unknown regular skew-symmetric tensor J^{ij} and the vector a^i in the form*

$$\begin{aligned} &(a) \qquad \nabla_k J^{ij} = a^i \delta^j_k - a^j \delta^i_k, \\ &(b)\ (n-2)\nabla_k a^j = -\frac{1}{n-1} J^{\alpha\beta} R_{\alpha\beta} \cdot \delta^j_k + J^{j\alpha} R_{[\alpha k]} + J^{\alpha\beta} R^j_{\alpha\beta k}, \end{aligned} \tag{3}$$

where R^h_{ijk} and R_{ij} $(\equiv R^\alpha_{i\alpha j})$ are the components of the Riemannian and Ricci curvature tensors, respectively.

Note that in equiaffine spaces, which are characterized by the symmetry of the Ricci tensor (in particular, Riemannian and pseudo-Riemannian spaces are equiaffine), the basic Eq. (3) are simplified as follows:

$$(a)\ \nabla_k J^{ij} = a^i \delta^j_k - a^j \delta^i_k, \quad (b)\ (n-2)\nabla_l a^i = -J^{\alpha p} R^i_{\alpha p l} - J^{i\alpha} R_{\alpha l}.$$

Based on the general theory of closed systems of linear equations in covariant derivatives, which is detailed in [17, p. 130–135], it follows that the system (3) and (4) has at most one solution with respect to the initial values of the unknown functions $J^{ij}(x_0)$ and $a^i(x_0)$ at the given point x_0.

Therefore, the general solution (3), resp. (4), depends on a finite number of numerical parameters, which could be, for example, these initial values $J^{ij}(x_0)$ and $a^i(x_0)$. Since $J^{ij}(x_0)$ is skew-symmetric, these initial values $\frac{1}{2}n(n-1) + n = \frac{1}{2}n(n+1)$. Thus, we obtain the following, see [1].

Theorem 2 *The general solution of the basic equations of the conformally Fedosov equations depends on $r \leq \frac{1}{2}n(n+1)$ essential numerical parameters.*

Similar results have been obtained for motions, conformal and projective motions, see [6, 22]. Basic equations in the form of systems of Cauchy type were found, for example, in the theory of geodesic, holomorphically projective, and rotary mappings [4, 7, 8, 18, 19, 22].

3 Lacuna in the Distribution of the Degree of Solutions

As was shown, the general solution of the fundamental Eq. (3) depends on a finite number of numerical parameters: $r \leq \frac{1}{2}n(n+1)$.

We note that in the affine coordinate system x in the flat space A_n the general solution of the Eq. (2) has the form $J^{ij} = x^i\alpha^j - x^j\alpha^i + \beta^{ij}$, where α^i, β^{ij} are constants such that $\alpha^i \not\equiv 0, \beta^{ij} + \beta^{ji} = 0$, in the domain $\Omega \subset \mathbb{R}^n$ where $\det \|x^i\alpha^j - x^j\alpha^i + \beta^{ij}\| \neq 0$, see [2]. Note that $0 \in \Omega$ in the case when $\det \|\beta^{ij}\| \neq 0$. This solution is obtained by integrating the Eq. (2) with minimal requirements on the differentiability of the unknown functions $J^{ij} \in C^1$ and $a^i \in C^0$.

Based on the results of [1, 10], the above solution generates a general solution (2) in equiaffine projectively Euclidean spaces. In these spaces, the maximum number of parameters is reached, which is 2 in the theorem. The next theorem follows from the previous one.

Theorem 3 *The maximum number of parameters $r = \frac{1}{2}n(n+1)$ of Eq. (3) is admitted in projective Euclidean spaces.*

We will prove below that in non-projective Euclidean spaces, the number of independent parameters decreases. First, we will prove two lemmas, their proof is based on the following theorem (formulated by Mikeš and Moldobaev [16], and correctly proved by Peška [20, Theorem 13], see also [17, p. 302]).

Theorem 4 *Let in each coordinate system for the Riemannian curvature R (resp. the Weyl projective curvature W) tensor on A_n the following conditions hold: $R^H_{IJK} = 0$ or $R^H_{IIJ} = 0$ (resp. $W^H_{IJK} = 0$ or $W^H_{IIJ} = 0$), where H, I, J and K are fixed different indices. Then A_n is projectively flat.*

Theorem 4 generalizes the result obtained by Schouten and Struik [21].

Lemma 1 *In the non-projective Euclidean space A_n, $n \geq 4$, among the components of the tensor J there are $n-3$ components that are linearly dependent on their others components.*

Proof Obviously, the Eq. (3a) is a necessary and sufficient condition for the conformally Fedosov structure (2).

We find conditions for the integrability of the Eq. (2). To do this, we differentiate them with respect to x^l: $\nabla_l\nabla_k J^{ij} = \nabla_l a^i\delta^j_k - \nabla_l a^j\delta^i_k$, and then alternate with respect to the indices k and l:

$$\nabla_l \nabla_k J^{ij} - \nabla_k \nabla_l J^{ij} = \nabla_l a^i \delta_k^j - \nabla_l a^j \delta_k^i - \nabla_k a^i \delta_l^j + \nabla_k a^j \delta_l^i.$$

Based on the Ricci identities, which are of the form $\nabla_{[l} \nabla_{k]} J^{ij} = -J^{\alpha j} R^i_{\alpha kl} - J^{i\alpha} R^j_{\alpha kl}$, we obtain an expression that is the condition for the integrability of the Eq. (2)

$$J^{\alpha j} R^i_{\alpha kl} + J^{i\alpha} R^j_{\alpha kl} = -A^i_l \delta^j_k + A^j_l \delta^i_k + A^i_k \delta^j_l - A^j_k \delta^i_l, \tag{4}$$

where $A^i_l = \nabla_l a^i$. Since A_n is not projectively flat, then, based on Theorem 4, there exists a system of coordinates, in which

$$R^3_{112} \neq 0. \tag{5}$$

The formula (4) at $j = 3, k = 1, l = 2$ and $i \neq 1, 2$, will have the form

$$J^{\alpha 3} R^i_{\alpha 12} + J^{i\alpha} R^3_{\alpha 12} = 0. \tag{6}$$

The formula (6) can be written in the form

$$J^{i1} R^3_{112} + J^{\alpha 3} R^i_{\alpha 12} + \sum\nolimits_{\beta > 1} J^{i\beta} R^3_{\beta 12} = 0.$$

It can be seen that the components Y^{1i}, $i > 3$, depend linearly on the other components of the tensor Y^{ij}. This means that there are no less than $n - 3$ dependencies between the components of the tensor Y. □

Lemma 2 *In the non-projective Euclidean space A_n, $n \geq 4$, among the components of the tensor a^i there are $n - 3$ components that are linearly dependent on their others components and components of tensor J.*

Proof Next, we will covariantly differentiate the integrability conditions (4) with respect to x^m:

$$\begin{gathered}\nabla_m J^{\alpha j} R^i_{\alpha kl} + \nabla_m J^{i\alpha} R^j_{\alpha kl} + J^{\alpha j} \nabla_m R^i_{\alpha kl} + J^{i\alpha} \nabla_m R^j_{\alpha kl} = \\ -A^i_{lm} \delta^j_k + A^j_{lm} \delta^i_k + A^i_{km} \delta^j_l - A^j_{km} \delta^i_l,\end{gathered}$$

where $A^i_{lm} = \nabla_{lm} a^i$.

After excluding the derivative ∇J with the help of Eq. (2), after adjustments we get

$$\begin{gathered}a^\alpha R^i_{\alpha kl} \delta^j_m - a^j R^i_{mkl} - a^\alpha R^j_{\alpha kl} \delta^i_m + a^i R^j_{mkl} = \\ -A^i_{lm} \delta^j_k + A^j_{lm} \delta^i_k + A^i_{km} \delta^j_l - A^j_{km} \delta^i_l - J^{\alpha j} \nabla_m R^i_{\alpha kl} - J^{i\alpha} \nabla_m R^j_{\alpha kl}.\end{gathered} \tag{7}$$

Since A_n is not projectively flat, then, based on Theorem 4, there exists a system of coordinates in which (5): $R^3_{112} \neq 0$. The formula (7) at $j = 3, k = m = 1, l = 2$ and $i \neq 1, 2$, will have the form

$$a^3 R^i_{112} - a^i R^3_{112} = J^{\alpha 3} \nabla_1 R^i_{\alpha 12} + J^{i\alpha} \nabla_1 R^3_{\alpha 12}.$$

It can be seen that the components a^i, $i > 3$, depend linearly on the component a^3, and of the tensor Y^{ij}. □

Based on these Lemmas 1 and 2, it is evident that in non-projective Euclidean spaces, the number r of independent parameters decreases by $2(n-3)$, i.e. $r \leq \frac{1}{2}n(n+1) - 2(n-3) = \frac{1}{2}n(n-3) + 6$.

Theorem 5 *In the non-projective Euclidean space A_n, $n \geq 4$, general solution of Eq.* (3) *depends on $r \leq \frac{1}{2}n(n-3) + 6$ independent parameters.*

References

1. Almazbekov, Ch., Guseva, N.I. and Mikeš, J.: Conformal Fedosov structures and spaces. Preprint. 2023. Accepted in: Mat. zametki
2. Almazbekov, Ch., Mikeš, J. and Peška, P.: On the solutions of flat conformal Fedosov manifolds. Probl. Modern Math., 70th aniver. of A.A. Borubaev, June 15–19 (2021)
3. Banyaga, A.: Symplectic geometry and related structures. Cubo 6, 123–138 (2004)
4. Domashev, V.V. and Mikeš, J.: Theory of holomorphically projective mappings of Kählerian spaces. Math. Notes 23, 160–163 (1978)
5. Eastwood, M.G., Slovak, J.: Conformally Fedosov manifolds. Adv. Math. 349, 839–868 (2019)
6. Eisenhart L.P.: Continuous groups of transformations. Princeton Univ. Press (1933)
7. Hinterleitner, I. and Mikeš, J.: Geodesic mappings onto Riemannian manifolds and differentiability Geometry, Integrability and Quantization 18, 183–190 (2017)
8. Hinterleitner, I. and Mikeš, J.: Geodesic mappings of (pseudo-)Riemannian manifolds preserve class of differentiability. Miskolc Math. Notes 14:2, 575–582 (2013)
9. Gelfand, I.M., Retakh, V.S. and Shubin, M.A.: Fedosov manifolds. Adv. Math. 136, 104–140 (1998)
10. Kanetov, B., Mikeš, J. and Askulova, Z.: Conformally Fedosov manifolds and geodesic mappings. AIP Conf. Proc. 2334, No. 020014 (2021)
11. Kaushal, R., Kumar, R., Rani, R.: Harmonicity of slant conformal Riemannian maps. Math. Notes 113, No. 2, 243–254 (2023)
12. Kirichenko, V.F. and Pol'kina, E.A.: Contact Lie form and concircular geometry of locally conformally quasi-Sasakian manifolds. Math. Notes 99, No. 1, 52–62 (2016)
13. Kirichenko, V.F., Arsen'eva, O.E., Rustanov, A.R.: An example of a concircular vector field on a locally conformally Kähler manifold. Math. Notes 111, No. 4, 544–548 (2022)
14. Klepikov, P. N.: Conformally flat algebraic Ricci solitons on Lie groups. Math. Notes 104, No. 1, 53–62 (2018)
15. Lee, H.C.: A kind of even-dimensional differential geometry and its application to exterior calculus. Amer. J. Math. 65, 433–438 (1943)
16. Mikeš, J. and Moldobaev, Dz.: On some algebraic properties of tensors. In: Issled. po Neevklid. Geom., Kirghyz. Univ., Frunze, 60–64 (1982)
17. Mikeš, J. et al.: Differential geometry of special mappings. Palacky Univ. Press, Olomouc (2019)
18. Mikeš, J., Berezovski, V.: Geodesic mappings of affine-connected spaces onto Riemannian spaces Colloq. Math. Soc. J. Bolyai, 56. Diff. Geom. Eger Hungary, 491–494 (1989)
19. Mikeš, J., Rýparová, L. and Chudá, H.: On the theory of rotary mappings. Math. Notes 104, No. 4, 617–620 (2018)
20. Peška, P., Jukl, M. and Mikeš. J.: Tensor decompositions and their properties. Mathematics 2023, 11 (17), 3638
21. Schouten, J.A. and Struik, D.J.: Introduction into new methods in differential geometry. P. Noordhoff: Groningen, Germany, 1935

22. Sinyukov N.S.: Geodesic mappings of Riemannian spaces. Nauka, Moscow (1979)
23. Stepanov, S.E., Tsyganok, I.I.: Codazzi and Killing tensors on a complete Riemannian manifold. Math. Notes 109, No. 6, 932–939 (2021)
24. Vaisman, I.: Locally conformal symplectic manifolds. Internat. J. Math. Math. Sci. 8, 521–536 (1985)
25. Weyl, H.: Raum. Zeit. Materie. Vorlesungen über allgemeine Relativitätstheorie. 3. Auflage, Berlin, 1919.

Rotary Mappings of Equidistant Spaces

Lenka Vítková

Abstract The paper is devoted to study of rotary mappings equations of two-dimensional equidistant (pseudo-) Riemannian spaces. The general solution of these equations is found beyond these spaces under minimal requirements for the differentiability of the studied objects.

Keywords Equidistant space · Rotary mapping · (pseudo-) Riemannian space

1 Introduction

Special diffeomorphisms between (pseudo-) Riemannian manifold and manifold with affine or projective connection, for which any special curve maps onto a special curve, were studied in many works. For example, geodesic [2–5, 14, 23, 31], holomorphically-projective [6, 15, 16], and more special mappings [7, 24, 25, 28]. These problems can be found in the more developed form in monographs [17–19, 29].

Leiko studied interesting questions about rotary mappings in [8–13]. Newly found results have their application in theory about gravitational fields, for example, [10, 12]. He proved that certain surfaces of revolution whose metric is in special form admit rotary mapping [8]. Leiko continued this research with Vinnik cooperation, see [30].

New results in the theory of isoperimetric extremals of rotation and rotary diffeomorphisms were obtained by Mikeš, Stepanova and Sochor [22]. Generalization of these terms is in the work of Chudá, Mikeš and Sochor [1]. Global rotary mappings were constructed in the work as projections of spheres [21].

WWW home page: https://www.prf.upol.cz/kag/.

L. Vítková (✉)
Palacky University, 17. listopadu, 12, 77146 Olomouc, Czech Republic
e-mail: Lenka.Vitkova@upol.cz

© The Author(s), under exclusive license to Springer Nature Switzerland AG 2024
V. Rovenski et al. (eds.), *Differential Geometric Structures and Applications*,
Springer Proceedings in Mathematics & Statistics 440,
https://doi.org/10.1007/978-3-031-50586-7_11

In this paper, we deal with rotary mappings of equidistant two-dimensional (pseudo-) Riemannian manifolds, which are isometric with surfaces of revolution. Details about these isometries are given, for example, in [26]. Equidistant metrics of the conformally rotary correspondings are found explicitly. This paper is a continuation of our work [20, 27] and a generalization of Leiko's work for surfaces of revolution [8]. These results have local validity.

2 Isoperimetric Extremals of Rotation

Leiko [8] was the first one to introduce the term isoperimetric extremals of rotation on two-dimensional Riemannian spaces $\mathbb{V}_2$ and surfaces $\mathcal{S}_2$ with metric g. He proved that a curve ℓ is an isoperimetric extremal of rotation if and only if its Frenet curvature k and Gaussian curvature K are proportional $k = c \cdot K$, where c is a constant. For $c = 0$ we get a geodesic.

A more general approach to this can be found in work [22] by Mikeš, Stepanova and Sochor, also, in what follows.

Definition 1 A curve ℓ: $x = x(t)$ on surface or on two-dimensional (pseudo-) Riemannian space $\mathbb{V}_2$ is called the *isoperimetric extremal of rotation* if ℓ is extremal of functionals $\theta[\ell]$ and $s[\ell] = \text{const}$ with fixed ends.

Here $s[\ell] = \int_{t_0}^{t_1} |\lambda|\, dt$ and $\theta[\ell] = \int_{t_0}^{t_1} k(t)\, dt$, where $k(t)$ is the curvature and $|\lambda|$ is the length of the tangent vector λ of ℓ.

In [22], the authors found new simpler form of equations of isoperimetric extremal of rotation

$$\nabla_s \lambda = c \cdot K \cdot F\lambda,$$

where c is a constant, ∇ is covariant derivative on V_2, s is the arc length, K is the Gaussian curvature, and F is a tensor $\binom{1}{1}$ which satisfies the conditions

$$F^2 = -e \cdot Id, \quad g(X, FX) = 0, \quad \nabla F = 0, \tag{1}$$

for any tangent vector X.

For Riemannian manifold $\mathbb{V}_2$ is $e = +1$ and F is a *complex structure* and for (pseudo-) Riemannian manifold is $e = -1$ and F is a *product structure*. This tensor F is uniquely defined (with the respect to the sign) with using skew-symmetric and covariantly constant discriminant tensor ε, which is defined

$$F_j^h = g^{hi}\varepsilon_{ij}, \quad \varepsilon_{ij} = \sqrt{|g_{11}g_{22} - g_{12}^2|} \cdot \begin{pmatrix} 0 & 1 \\ -1 & 0 \end{pmatrix},$$

where $F_i^h, \varepsilon_{ij}, g_{ij}$ are components of F, ε, g, and $\|g^{ij}\| = \|g_{ij}\|^{-1}$.

3 Rotary Diffeomorphism

In [8], there was introduced the term of rotary diffeomorphism between two-dimensional Riemannian spaces $\mathbb{V}_2$ and surfaces $\mathcal{S}_2$ with metric g: A diffeomorphism between two-dimensional (pseudo-) Riemannian manifolds $\mathbb{V}_2$ and $\bar{\mathbb{V}}_2$ is called *rotary* if any geodesic on $\bar{\mathbb{V}}_2$ is mapped onto isoperimetric extremal of rotation on $\mathbb{V}_2$.

Chudá, Mikeš and Sochor ([1], [19, p. 434]) later generalized the definition itself to the following form:

Definition 2 A diffeomorphism $f\colon \mathbb{V}_2 \to \bar{\mathbb{A}}_2$ is called *rotary mapping* if any geodesic on manifold $\bar{\mathbb{A}}_2$ with affine connection $\bar{\nabla}$ is mapped onto isoperimetric extremal of rotation on two-dimensional (pseudo-) Riemanninan manifold $\mathbb{V}_2$.

Chudá, Mikeš and Sochor [1] later proved that (pseudo-) Riemannian manifold $\mathbb{V}_2$ admits rotary mapping onto $\bar{\mathbb{A}}_2$ if and only if in $\mathbb{V}_2$ holds equation

$$\nabla_j \theta^h = \theta^h(\theta_j + \partial_j \ln |K|) + \nu\, \delta^h_j, \tag{2}$$

where $\partial_1 = \partial/\partial x^i$, $\theta_i = g_{i\alpha}\theta^\alpha$, ν is a function on $\mathbb{V}_2$, and vector field θ^h is a special case of torse-forming field, which determined rotary mapping.

Connections ∇ and $\bar{\nabla}$ of spaces $\mathbb{V}_2$ and $\bar{\mathbb{A}}_2$ are related by formula

$$\Gamma^h_{ij} - \bar{\Gamma}^h_{ij} = \delta^h_i \varphi_j + \delta^h_i \varphi_j + \theta^h g_{ij}, \tag{3}$$

where Γ^h_{ij}, $\bar{\Gamma}^h_{ij}$ are components of ∇ and $\bar{\nabla}$, and φ_i is a components of linear form.

Let us remind you that in Leiko's work [8], he claims that the form ψ (in components $\psi_i = g_{ij}\theta^j$) is gradient ($\psi_i = \partial_i \Psi$), from which it follows that the vector field θ is concircular. In the work [20] by Mikeš, Rýparová and Chudá, we prove that this statement is wrong. A counterexample was constructed and a general solution to these equations was found.

4 Rotary Mappings on Surfaces of Revolution

Leiko [8] has also studied rotary mappings of surfaces of revolution $\mathcal{S}_2$ which is given by the parametrization $x = r\cos\varphi,\ y = r\sin\varphi,\ z = f(r)$. Here, he used the metric of the surface of revolution $\mathcal{S}_2$ in the following form

$$ds^2 = f(r)dr^2 + r^2 d\varphi^2. \tag{4}$$

Leiko analyzed the Eq. (2) and proved a theorem that a vector field θ exists in Riemannian space $\mathbb{V}_2$ with metric (4) if and only if $\mathbb{V}_2$ is isometric with surface of revolution $\mathcal{S}_2$ and the metrics of $\mathbb{V}_2$ has one from the following forms:

$$(\tilde{g}_{ij}) = \frac{f(r)}{A^2(B+\sqrt{f(r)})^2}\,\mathrm{diag}(f(r), r^2), \quad \text{or} \quad (\tilde{g}_{ij}) = B^2 f(r)\,\mathrm{diag}(f(r), r^2),$$

where A and B are const. Above mentioned was formulated in [8, Theorem 2].

Let us remind that the metric of the Riemannian space $\mathbb{V}_2$, that is induced by the surface of revolution on E_3

$$\mathcal{S}_2\colon x = r(u)\cos v,\, y = r(u)\sin v,\, z = z(r), \quad r'(u)^2 + z'(u)^2 = 1,$$

can be written in the form $ds^2 = du^2 + f(u)\,dv^2$, where $f = r(u)^2$.

Note that this metric of the surface of revolution $\mathcal{S}_2$ is more general than the metric in form (4), which was used by Leiko. The metric (5) also includes "gorge circle", which, in (4), are basically excluded.

5 Equidistant Spaces

We note that in (pseudo-) Riemannian space $\mathbb{V}_2$ existence of coordinate system

$$ds^2 = du^2 + f(u)\,dv^2, \tag{5}$$

where f ($\neq 0$) is a certain function of u, $u \equiv x^1$, and $v \equiv x^2$,
is connected with existence of anisotropic *concircular* vector field λ which is characterized by the equations

$$\nabla_i \lambda^h = \rho\, \delta_i^h,$$

where ρ is a certain function on $\mathbb{V}_n$.

Concircular vector fields have been studied in 1940's by K. Yano and in 1950's by N.S. Sinyukov, who called spaces with those vector fields *equidistant*, see [18] pp. 140–155. Existence of these vector fields on the surface is a criteria of local isometry of surface of revolution.

In $\mathbb{V}_2$ vector field λ generates Killing vector. This vector has the following form $\nu = F\lambda$, where F is from formula (1). From the other side Killing vector ν generates concircular vector field.

In case the metric ds^2 is indefinite the surface is given by the equations
$x = r(u)\cosh v$, $y = r(u)\sinh v$, $z = z(u)$, where (x, y, z) are coordinates in Minkowski space M_3, which metric has the form $ds^2 = dx^2 - dy^2 + dz^2$, therefore for our example $ds^2 = (r'^2 + z'^2)\,du^2 - r^2\,dv^2$. The issues of immersion on E_3 and M_3 of surfaces of revolution are considered in detail in the paper [26].

Further [27], we are going to prove that any surface of revolution admits rotary mapping. Moreover, any Riemannian space $\mathbb{V}_2$ that is isometric with such a surface of revolution $\mathcal{S}_2$, and also any pseudo-Riemannian space $\mathbb{V}_2$ whose metric has the form (5) admits rotary mapping onto space $\bar{\mathbb{A}}_2$.

6 Rotary Mappings on Equidistant Spaces

In this section we prove the following theorem.

Theorem 1 *Any equidistant space $\mathbb{V}_2$ with differentiable Gaussian curvature K admits a conformally rotary mapping onto $\tilde{\mathbb{V}}_2$ if and only if in certain coordinate (u, v) their metrics have the following form:*

$$(g_{ij}) = \mathrm{diag}(1, f(u)) \quad and \quad (\tilde{g}_{ij}) = \frac{c\, f}{(f' + \alpha\sqrt{|f|}\,)^2}\,\mathrm{diag}(1, f(u)),$$

where $c\ (\neq 0)$ and α are constants.

Remark 1 From this it follows that if one of the spaces is equidistant, then the other is equidistant.

Proof The space $\mathbb{V}_2$ admits a conformally rotary mapping onto $\tilde{\mathbb{V}}_2$ if and only if the mapping is conformal and at the same time rotary.

Spaces $\mathbb{V}_2$ and $\tilde{\mathbb{V}}_2$ are conformal if and only if for the metrics and connections of these spaces satisfy $\tilde{g} = e^{-2\psi}\, g$, and

$$\Gamma^h_{ij} - \tilde{\Gamma}^h_{ij} = -\delta^h_i \psi_j - \delta^h_j \psi_i + \psi^h g_{ij}, \tag{6}$$

where $\psi_i = \partial_i \psi$ and $\psi^i = g^{ij}\psi_j$. From (6) for $\varphi_i = \psi_i$ and $\psi^h = \theta^h$, using the formulas (2) and (3), we obtain that the conformal mapping is a rotary one. Formula (2) is equivalent to the following:

$$\nabla_i \psi_j \equiv \partial_i \psi_j + \psi_\alpha \Gamma^\alpha_{ij} = \psi_i(\psi_j + \partial_i \ln|K|) + \nu\, g_{ij}. \tag{7}$$

We choose the metric of the equidistant space $\mathbb{V}_2$ in the form (5). Components of the metric tensor g and its inverse tensor have the following form:

$$(g_{ij}) = \mathrm{diag}(1, f(u)) \quad \text{and} \quad (g^{ij}) = \mathrm{diag}(1, 1/f(u)).$$

Hence, we can calculate the Christoffel symbols $\Gamma^h_{ij} = g^{hk}\Gamma_{ijk}$, where $\Gamma_{ijk} = 1/2\,(\partial_i g_{jk} + \partial_j g_{ik} - \partial_k g_{ij})$:

$$\Gamma^2_{12} = \Gamma^2_{21} = \frac{f'(u)}{2\, f(u)} \quad \text{and} \quad \Gamma^1_{22} = -\frac{1}{2}\, f'(u),$$

and the others vanish. The well known Gaussian curvature K satisfies the formula $R_{1212} = K \cdot (g_{11}g_{22} - g_{12}{}^2)$, where $R_{hijk} = g_{h\alpha} R^\alpha_{ijk}$ are components of Riemann tensor of first type and

$$R^h_{ijk} = \partial_j \Gamma^h_{ik} - \partial_k \Gamma^h_{ij} + \Gamma^\alpha_{ik}\Gamma^h_{\alpha j} - \Gamma^\alpha_{ij}\Gamma^h_{\alpha k}.$$

We calculate the Gaussian curvature K of the surface $\mathbb{V}_2$ as

$$K = \frac{1}{4}\left(\frac{f'(u)}{f(u)}\right)^2 - \frac{1}{2}\,\frac{f''(u)}{f(u)}. \tag{8}$$

Let us suppose $\psi = \psi(u)$, thus from (7) we obtain the following relations:

$$\psi''(u) = \psi'(u)\cdot\left(\psi'(u) + \frac{K'}{K}\right) + \nu(u), \quad \text{and} \quad \frac{1}{2}\,\psi'(u)\,\frac{f'(u)}{f(u)} = \nu(u).$$

Now we merge these relations and obtain the following equation:

$$\psi'' = \psi'^2 + \psi'\cdot\left(\frac{K'}{K} + \frac{1}{2}\frac{f'}{f}\right),$$

and after dividing by ψ' we get

$$\frac{\psi''}{\psi'} = \psi' + \frac{K'}{K} + \frac{1}{2}\frac{f'}{f}.$$

Integrating the last equation, we obtain

$$\ln|\psi'| = \psi + \ln|K| + \ln\sqrt{|f|} + \ln|c_1|, \quad c_1 = \text{const}.$$

Evidently, $\psi' e^{-\psi} = c_1\, K\, \sqrt{|f|}$, and integrating this equation, we obtain

$$-\,e^{-\psi} = c_1 \int K\,\sqrt{|f|}\,du.$$

As Gaussian curvature K has a special form (8), we obtain

$$-\,e^{-\psi} = c_1\left(c_2 + \frac{f'(x^1)}{2\sqrt{f(x^1)}}\right),$$

where c_2 is a constant of integration. By the above, $e^{2\psi} = \dfrac{c\,f}{(f' + \alpha\sqrt{|f|}\,)^2}$, where c and α are constants. Since $\tilde{g} = e^{2\psi}\,g$, the theorem is proved. □

References

1. Chudá, H., Mikeš, J. and Sochor, M.: Rotary diffeomorphism onto manifolds with affine connection. Geometry, Integrability & Quantization 18, 130–137 (2017)
2. Dini, U.: On a problem in the general theory of the geographical representations of a surface on another. Anali di Mat. 3, 269–294 (1869)

3. Hinterleitner, I. and Mikeš, J.: Geodesic mappings of (pseudo-)Riemannian manifolds preserve class of differentiability. Miskolc Math. Notes 14, No. 2, 575–582 (2013)
4. Hinterleitner, I.: Geodesic mappings on compact Riemannian manifolds with conditions on sectional curvature. Publ. Inst. Math. 94, No. 108, 125–130 (2013)
5. Hinterleitner, I. and Mikeš, J.: Geodesic mappings and differentiability of metrics, affine and projective connections. Filomat 29, 1245–1249 (2015)
6. Hinterleitner, I. and Mikeš, J.: Fundamental equations of geodesic mappings and their generalizations. J. Math. Sci. 174, 537–554 (2011)
7. Kuzmina, I. and Mikeš, J.: On pseudoconformal models of fibrations determined by the algebra of antiquaternions and projectivization of them. Ann. Math. Inform. 42, 57–64 (2013)
8. Leiko, S.: Rotary diffeomorphisms on Euclidean spaces. Math. Notes 47, No. 3, 261–264 (1990)
9. Leiko, S.: Variational problems for rotation functionals, and spin-mappings of pseudo-Riemannian spaces. Sov. Math. 34, No. 10, 9–18 (1990)
10. Leiko, S.: Extremals of rotation functionals of curves in a pseudo-Riemannian space, and trajectories of spinning particles in gravitational fields. Russian Acad. Sci. Dokl. Math. 46, 84–87 (1993)
11. Leiko, S.: Isoperimetric extremals of a turn on surfaces in Euclidean space $\mathbb{E}^3$. Russ. Math. 40, No. 6, 22–29 (1996)
12. Leiko, S.: On the conformal, concircular, and spin mappings of gravitational fields. J. Math. Sci. 90, 1941–1944 (1998)
13. Leiko, S.: Isoperimetric problems for rotation functionals of the first and second orders in (pseudo) Riemannian manifolds. Russ. Math. 49, 45–51 (2005)
14. Mikeš, J.: Geodesic mappings of affine-connected and Riemannian spaces. J. Math. Sci. 78, 311–333 (1996)
15. Mikeš, J.: Holomorphically Projective mappings and their generalizations. J. Math. Sci. 89, 1334–1353 (1998)
16. Mikeš, J., Berezovski, V. E., Stepanova, E. and Chudá, H.: Geodesic mappings and their generalizations. J. Math. Sci. 217, No. 5, 607–623 (2016)
17. Mikeš, J., Vanžurová, A. and Hinterleitner, I.: Geodesic mappings and some generalizations. Palacky Univ. Press, Olomouc (2009)
18. Mikeš, J., et al.: Differential geometry of special mappings. Palacky Univ. Press, Olomouc (2015)
19. Mikeš, J., et al.: Differential geometry of special mappings. 2th ed. Palacky University Press, Olomouc (2019)
20. Mikeš, J., Rýparová, L. and Chudá, H.: On the theory of rotary mappings. Math. Notes 104, No. 4, 617–620 (2018)
21. Mikeš, J., Guseva, N. I., Peška, P. and Rýparová, L.: Rotary mappings and projections of a sphere. Math. Notes 110, No. 1, 152–155 (2021)
22. Mikeš, J., Sochor, M. and Stepanova, E.: On the existence of isoperimetric extremals of rotation and the fundamental equations of rotary diffeomorphism. Filomat 29, No. 3, 517–523 (2015)
23. Najdanović, M. S., Zlatanović, M. and Hinterleitner, I.: Conformal and geodesic mappings of generalized equidistant spaces. Publ. Inst. Math. 98, No. 112, 71–84 (2015)
24. Najdanović, M. S. and Velimirović, L. S.: On the Willmore energy of curves under second order infinitesimal bending. Miskolc Math. Notes 17, No. 2, 979–987 (2016)
25. Petrov, A.: Modeling of the paths of test particles in gravitation theory. Gravit. and the Theory of Relativity 4, No. 5, 7–21 (1968)
26. Peška P., Mikeš J., Ryparová L. and Chepurna O.: On general solutions of equidistant vector fields on two-dimensional (pseudo-) Riemannian spaces. Filomat 37, No. 25, 8569–8574 (2023)
27. Rýparová, L. and Mikeš, J.: Rotary mappings of surfaces of revolution. Math., Inf. Technol. and Appl. Sci., 2017. Univ. of Defence, Brno, 2017, 208–216 (2017)
28. Stepanov, S., Shandra, I. and Mikeš, J.: Harmonic and projective diffeomorphisms. J. Math. Sci. 207, 658–668 (2015)
29. Sinyukov N.S.: Geodesic mappings of Riemannian spaces. Nauka, Moscow (1979)

30. Vinnik, A., Leiko, S.: The property of reciprocity of rotary diffeomorphisms of two-dimensional Riemannian spaces. Differ. Geom. Mnogoobr. Figur 29, 13–16 (1998)
31. Zlatanović, M., Velimirović, L. and Stanković, M.: Necessary and sufficient conditions for equitorsion geodesic mapping. J. Math. Anal. Appl. 435, 578–592 (2016)

Smith-Gysin Sequence

J. I. Royo Prieto, M. Saralegi-Aranguren, and R. Wolak

Dedicated to Vladimir Rovenski on his 70th birthday

Abstract Given a smooth semifree action of S^3 on a manifold M, we have the Smith-Gysin sequence:

$$\cdots \to H^*(M) \to H^{*-3}(M/S^3, M^{S^3}) \oplus H^*(M^{S^3}) \to H^{*+1}(M/S^3, M^{S^3}) \to H^{*+1}(M) \to \cdots$$

In this paper, we construct a Smith-Gysin sequence that does not require the semifree condition. This sequence includes a new term, referred to as the *exotic term*, which depends on the subset M^{S^1}:

$$\cdots \to H^*(M) \to H^{*-3}(M/S^3, \Sigma/S^3) \oplus H^*(M^{S^3}) \oplus (H^{*-2}(M^{S^1}))^{-\mathbb{Z}_2} \\ \to H^{*+1}(M/S^3, M^{S^3}) \to H^{*+1}(M) \to \cdots$$

Partially supported by Ministerio de Ciencia e Innovación, Spain, grant PID2022-139631NB-100. The authors acknowledge that the research cooperation was funded by the program Excellence Initiative – Research University at the Jagiellonian University in Krakow within the framework of the research group Reeb-Reinhart 2022.

J. I. Royo Prieto
Matematika Saila. Zientzia eta Teknologia Fakultatea, University of the Basque Country UPV/EHU, Barrio Sarriena s/n, 48940 Leioa, Spain
e-mail: joseignacio.royo@ehu.eus

M. Saralegi-Aranguren
Laboratoire de Mathématiques de Lens, EA 2462. Université d'Artois. SP18, rue Jean Souvraz, 62307 Lens Cedex, France
e-mail: martin.saraleguiaranguren@univ-artois.fr

R. Wolak (✉)
Institute of Mathematics, Jagiellonian University, ul. Łojasiewicza 6, 30-348 Krakow, Poland
e-mail: robert.wolak@im.uj.edu.pl

© The Author(s), under exclusive license to Springer Nature Switzerland AG 2024

V. Rovenski et al. (eds.), *Differential Geometric Structures and Applications*,
Springer Proceedings in Mathematics & Statistics 440,
https://doi.org/10.1007/978-3-031-50586-7_12

Here, $\Sigma \subset M$ is the subset of points in M whose isotropy group is infinite. The group $\mathbb{Z}_2$ acts on M^{S^1} by $j \in S^3$.

Keywords Long exact sequence · S^3-actions

2010 Mathematics Subject Classification. 57S15 · 55N10

Let us consider a smooth action of S^3 on a manifold M. The cohomological relationship between the different actors involved in this action can be expressed through a long exact sequence known as the Gysin sequence. When the action is free, we have the following exact sequence:

$$\cdots \to H^*(M) \to H^{*-3}(M/S^3) \to H^{*+1}(M/S^3) \to H^{*+1}(M) \to \cdots,$$

see, for example, [4, Sect. 8.2]. The initial map in this sequence involves integrating along the fibers of the principal bundle $\pi : M \to M/S^3$, which is associated with the free action. The subsequent map involves multiplying by the Euler class $[e] \in H^4(M/S^3)$. Finally, the third map is simply the pull-back π^*. Suppose that the action is free outside the fixed point set M^{S^3} (*semi-free action*). We have

$$\cdots \to H^*(M) \to H^{*-3}(M/S^3, M^{S^3}) \to H^{*+1}(M/S^3) \to H^{*+1}(M) \to \cdots,$$

see, for example, [6, Corollary 4.8].

In the case of a generic action, an exotic term appears in the Gysin sequence:

$$\cdots \to H^*(M) \to H^{*-3}(M/S^3, \Sigma/S^3) \oplus (H^{*-2}(M^{S^1}))^{-\mathbb{Z}_2} \to H^{*+1}(M/S^3) \to H^{*+1}(M) \to \cdots,$$

cf. [5]. Here,

- $\Sigma \subset M$ is the subset of points in M whose isotropy group is infinite,
- the subset of fixed points under the action of the circle $S^1 \subset S^3$ is M^{S^1},
- the group $\mathbb{Z}_2$ acts on M^{S^1} by the quaternion number j of S^3 and
- $(-)^{-\mathbb{Z}_2}$ denotes the subspace of antisymmetric elements.

Notice that in this context, the orbit space M/S^3 is a pseudomanifold and not necessarily a manifold.

On the other hand, in the case of a semifree action, we also have the Smith-Gysin sequence

$$\cdots \to H^*(M) \to H^{*-3}(M/S^3, M^{S^3}) \oplus H^*(M^{S^3}) \to H^{*+1}(M/S^3, M^{S^3}) \to H^{*+1}(M) \to \cdots \quad (1)$$

see, for example, [1, Theorem 2.2] and [2, Exercise 12 on page 169].

The Gysin sequence and the Smith-Gysin sequence can be utilized to compute the cohomology of M in terms of *basic* objects, such as M/S^3 and M^{S^3}. In the case of the former, we employ the cohomology $H^*(M/S^3)$, while for the latter, we make use of the cohomology $H^*(M/S^3, M^{S^3})$. The latter is, a priori, easier to compute since the action of S^3 on (M, M^{S^3}) has no fixed points.[1]

In this paper, we extend the scope of the Smith-Gysin sequence to any smooth action of S^3 on a manifold M, obtaining the long exact sequence

$$\begin{array}{ccccc} \cdots \longrightarrow H^*(M) & \longrightarrow & H^{*-3}(M/S^3, \Sigma/S^3) \oplus \left(H^{*-2}\left(M^{S^1}\right)\right)^{-\mathbb{Z}_2} \oplus H^*\left(M^{S^3}\right) \\ & & \downarrow \\ \cdots \longleftarrow H^{*+1}(M) & \longleftarrow & H^{*+1}\left(M/S^3, M^{S^3}\right) \end{array}$$

In what follows, we consider a smooth action $\Phi\colon S^3 \times M \to M$, where M is a second-countable, Hausdorff, smooth manifold of dimension m and without boundary. We further assume that the action is effective. For definitions and properties related to actions of compact Lie groups, we refer the reader to [2]. The cohomology $H^*(-)$ refers to the singular cohomology with real coefficients.

1 Verona's Differential Forms

The action Φ induces the following invariant filtration[2] on M:

$$F = \{x \in M \mid \dim S^3_x = 3\} \subset \Sigma = \{x \in M \mid \dim S^3_x = 1 \text{ or } 3\} \subset M.$$

Notice that F is the fixed point set M^{S^3}.

The complex of controlled forms $\Omega^*_V(M)$ (or, Verona's forms), which is composed of differential forms defined on $M \setminus \Sigma$, computes the usual cohomology $H^*(M)$. In fact, a Verona's form can be represented as a triple $(\omega, \omega_0, \omega_1)$, where

$$\omega \in \Omega^*(M \backslash \Sigma), \qquad \omega_0 \in \Omega^*(\Sigma \backslash F), \qquad \omega_1 \in \Omega^*(F)$$

such that $\omega = \omega_0 = \omega_1$ near F, and $\omega = \omega_0$ near $\Sigma \backslash F$ (see [5, Definition 1.2] for the precise statement). In this work, we use the relative complex

$$\Omega^*_V(M, F) = \{\omega \in \Omega^*_V(M) \mid \omega_1 = 0\}.$$

Proposition 1 *Let $\Phi\colon S^3 \times M \to M$ be a smooth action. The short sequence*

$$0 \to \Omega^*_V(M, F) \hookrightarrow \Omega^*_V(M) \xrightarrow{\rho_F} \Omega^*(F) \to 0, \tag{2}$$

where $\rho_F(\omega) = \omega_1$, is exact. Moreover, we have $H^*(\Omega^\cdot_V(M, F)) = H^*(M, F)$.

[1] Recall that $H^*(M, M^{S^3}) = H^*(M \backslash M^{S^3}, T \backslash M^{S^3})$; here, T is a tubular neighborhood of the fixed point set M^{S^3}.

[2] We refer to [5, Sect. I] for the notions appearing in this presentation.

Proof The long sequence is exact if we prove that the map ρ_F is surjective. It suffices to proceed as in [5, Lemma 1.2]. On the other hand, Verona's forms are a particular case of intersection differential forms. In fact, we have $\Omega^*_V(M, F) = \Omega^*_{\overline{p}}(M)$, where $\overline{p}$ is the perversity taking the value $-\infty$ on the strata of F and 0 on the strata of $\Sigma \setminus F$. The result comes from [7, Proposition 3.2.3]. □

The exact sequence (2) induces the long exact sequence of the pair (M, F):

$$\cdots \to H^*(M, F) \to H^*(M) \to H^*(F) \to H^{*+1}(M, F) \to \cdots .$$

The orbit space M/S^3 is a pseudomanifold, not necessarily a manifold, but its cohomology can be also computed using differential forms. More specifically, the following complex computes the cohomology $H^*(M/S^3)$, cf. [8]:

$$\Omega^*_V(M/S^3) = \{\omega \in \Omega^*_V(M) \mid i_X\omega = i_X d\omega = 0 \ \ \forall X \in \mathfrak{X}_\Phi(M)\}.$$

Here, we denote by $\mathfrak{X}_\Phi(M)$ the family of vector fields of M tangent to the orbits of Φ. We also use the relative complex

$$\Omega^*_V(M/S^3, F) = \{\omega \in \Omega^*_V(M, F) \mid i_X\omega = i_X d\omega = 0 \ \ \forall X \in \mathfrak{X}_\Phi(M)\}.$$

Proceeding as in the previous proposition, we obtain the following result:

Proposition 2 *Let $\Phi\colon S^3 \times M \to M$ be a smooth action. The short sequence*

$$0 \to \Omega^*_V(M/S^3, F) \hookrightarrow \Omega^*_V(M/S^3) \xrightarrow{\rho'_F} \Omega^*(F) \to 0 \tag{3}$$

with $\rho'_F(\omega) = \omega_1$ is exact. Moreover, we get $H^*(\Omega^{\cdot}_V(M/S^3, F)) = H^*(M/S^3, F)$.

The exact sequence (3) induces the long exact sequence of the pair $(M/S^3, F)$:

$$\cdots \to H^*(M/S^3, F) \to H^*(M/S^3) \to H^*(F) \to H^{*+1}(M/S^3, F) \to \cdots .$$

2 Smith-Gysin Sequence

In order to obtain the Smith-Gysin sequence, we use the notion of a braid.

Definition 1 Let us consider six chain complexes A^*, B^*, C^*, D^*, E^* and F^*. A *braid* is a diagram of chain maps of the form

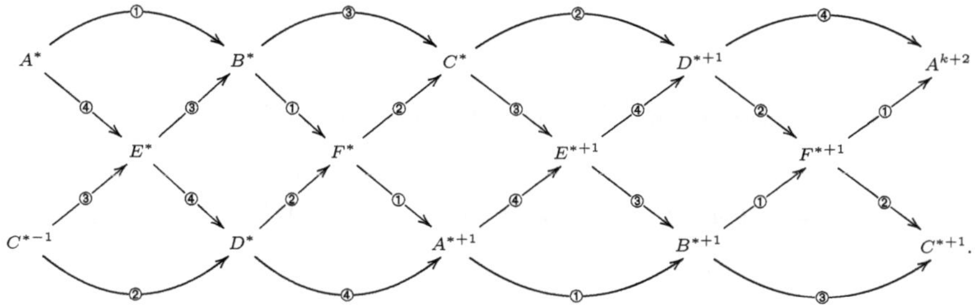

It is a *commutative braid* when all the triangles and diamonds are commutative. If the long sequences ①, ②, ③ and ④ are exact we say that the braid is *exact braid.*

An exact and commutative braid has the following property: the long sequence shown below is exact (see, for example, [3, pag. 39-41]).

$$\cdots \longrightarrow E^* \xrightarrow{(③,④)} B^* \oplus D^* \xrightarrow{①-②} F^* \xrightarrow{③\ ②} E^{*+1} \longrightarrow \cdots \tag{4}$$

The main result of this work is the following , where we use these conventions:

- the subset of fixed points under the action of the circle $S^1 \subset S^3$ is M^{S^1},
- the group $\mathbb{Z}_2$ acts on M^{S^1} by the quaternion number j of S^3,
- $(-)^{-\mathbb{Z}_2}$ denotes the subspace of antisymmetric elements.

Proposition 3 *Consider a smooth action* $\Phi \colon S^3 \times M \to M$*, and let* $\Sigma = \{x \in M \mid \dim S^3_x = 1 \text{ or } 2\}$*. The Smith-Gysin sequence associated to this action is the following long exact sequence:*

$$\begin{array}{c} \cdots \longrightarrow H^*(M) \longrightarrow H^{*-3}(M/S^3, \Sigma/S^3) \oplus \left(H^{*-2}\left(M^{S^1}\right)\right)^{-\mathbb{Z}_2} \oplus H^*\left(M^{S^3}\right) \\ \downarrow \\ \cdots \longleftarrow H^{*+1}(M) \longleftarrow H^{*+1}\left(M/S^3, M^{S^3}\right) \end{array}$$

We also have the long exact sequence

$$\begin{aligned} \cdots \to H^*\left(M/S^3, M^{S^3}\right) &\to H^*\left(M/S^3\right) \oplus H^*\left(M, M^{S^3}\right) \to H^*(M) \\ &\to H^{*+1}\left(M/S^3, M^{S^3}\right) \to \cdots . \end{aligned}$$

Proof The short exact sequences (2) and (3) can be arranged in the following commutative diagram of short exact sequences:

$$
\begin{array}{ccccccccc}
 & & 0 & & 0 & & & & \\
 & & \downarrow & & \downarrow & & & & \\
0 & \longrightarrow & \Omega^*_V(M/S^3, M^{S^3}) & \hookrightarrow & \Omega^*_V(M/S^3) & \longrightarrow & \Omega^*(M^{S^3}) & \longrightarrow 0 & ③ \\
 & & \downarrow & & \downarrow & & \| & & \\
0 & \longrightarrow & \Omega^*_V(M, M^{S^3}) & \longrightarrow & \Omega^*_V(M) & \longrightarrow & \Omega^*(M^{S^3}) & \longrightarrow 0 & ② \\
 & & \downarrow & & \downarrow & & & & \\
 & & \dfrac{\Omega^*_V(M, M^{S^3})}{\Omega^*_V(M/S^3, M^{S^3})} & \xrightarrow{\cong} & \dfrac{\Omega^*_V(M)}{\Omega^*_V(M/S^3)} & & & & \\
 & & \downarrow & & \downarrow & & & & \\
 & & 0 & & 0 & & & & \\
 & & & & & & & & \\
 & & ④ & & ① & & & &
\end{array}
$$

The cohomology of the complex $\dfrac{\Omega^*_V(M)}{\Omega^*_V(M/S^3)}$ has been computed in [5, Lemma 2.1]. The result is given by

$$H^{*-3}(M/S^3, \Sigma/S^3) \oplus \left(H^{*-2}(M^{S^1})\right)^{-\mathbb{Z}_2}.$$

This diagram leads to the following braid (cf. Propositions 1 and 2):

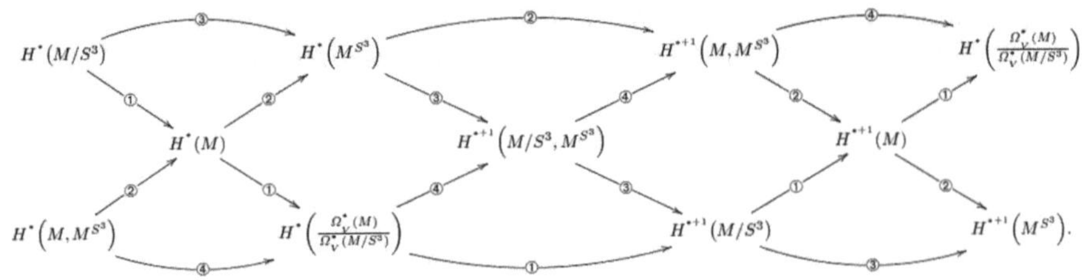

Furthermore, using property (4), we obtain the long exact sequences

$$\cdots \to H^*(M) \to H^{*-3}(M/S^3, \Sigma/S^3) \oplus \left(H^{*-2}(M^{S^1})\right)^{-\mathbb{Z}_2} \oplus H^*(M^{S^3})$$
$$\to H^{*+1}(M/S^3, M^{S^3}) \to H^{*+1}(M) \to \cdots,$$

the Smith-Gysin sequence, and

$$\cdots \to H^*(M/S^3, M^{S^3}) \to H^*(M/S^3) \oplus H^*(M, M^{S^3}) \to H^*(M)$$
$$\to H^{*+1}(M/S^3, M^{S^3}) \to \cdots.$$

We have constructed the two long exact sequences. □

Notes and Comments.

1. The exotic term can be directly computed by using the following formula:

$$\left(H^*\left(M^{S^1}\right)\right)^{-\mathbb{Z}_2} = H^*\left(M^{S^1}\right)/H^*\left(M^{S^1}/\mathbb{Z}_2\right),$$

cf. [5, 2.4 Remark (a)]. In this context, the $\mathbb{Z}_2$-action on M^{S^1} is defined by the quaternionic number $j \in S^3$.

2. The technique used to construct the Smith-Gysin sequence is the same as the one used in [2, page 161], where a smooth action[3] $\Phi\colon S^1 \times M \to M$ is given, and the following Smith-Gysin sequence is constructed:

$$\cdots \to H^*(M) \to H^{*-1}\left(M/S^1, M^{S^1}\right) \oplus H^*\left(M^{S^1}\right) \to H^{*+1}\left(M/S^1, M^{S^1}\right) \\ \to H^{*+1}(M) \to \cdots .$$

It is worth noting that the long exact sequence

$$\cdots \to H^*\left(M/S^1, M^{S^1}\right) \to H^*\left(M/S^1\right) \oplus H^*\left(M, M^{S^1}\right) \to H^*(M) \\ \to H^{*+1}\left(M/S^1, M^{S^1}\right) \to \cdots \tag{5}$$

can be also be obtained.

Recall (cf. [2, page 161]) that the Gysin sequence associated to the action Φ is

$$\cdots \to H^*(M) \to H^{*-1}\left(M/S^1, M^{S^1}\right) \to H^{*+1}\left(M/S^1\right) \to H^{*+1}(M) \to \cdots \tag{6}$$

3. We can obtain the Gysin sequences (1) and (6) from the Smith-Gysin sequences (1) and (5), respectively, through the following steps.

Assume that $G = S^k$ acts semifreely on M, where k is either 1 or 3, and let F be the set of fixed points. The quotient map $\pi : M \to B = M/S^k$ is continuous, and its restriction to $M \setminus F$ is a G-principal bundle over $B \setminus F$, where we have identified F with $\pi(F)$ for convenience. This restriction induces the pullback map $\pi^*\colon \Omega_v^*(B) \to \Omega_v^*(M)$ in the Verona complexes, which, in turn, induces a map in cohomology that we denote by $\pi^*\colon H^*(B) \to H^*(M)$, using isomorphism in cohomology. We have the following diagram:

[3] The result obtained by Bredon is more general than the one presented in this Remark, which only applies to smooth manifolds.

$$
\begin{array}{ccccccc}
0 & & 0 & & 0 & & 0 \\
\downarrow & & \downarrow & & \downarrow & & \downarrow \\
H^i(B,F) & \xrightarrow{\iota} & H^i(B) & \xrightarrow{\rho_B} & H^i(F) & \xrightarrow{\delta} & H^{i+1}(B,F) \\
\downarrow{\scriptstyle (1,\iota)} & & \downarrow{\scriptstyle (\pi^*,1_B)} & & \downarrow{\scriptstyle (0,1_F)} & & \downarrow{\scriptstyle (1,\iota)} \\
H^i(B,F) \oplus H^i(B) & \xrightarrow{(\pi^* \circ \iota, 1_B)} & H^i(M) \oplus H^i(B) & \xrightarrow{(\int, \pi^* \circ \rho_M)} & H^{i-k}(B,F) \oplus H^i(F) & \xrightarrow{(\wedge[e]+\delta, 0)} & H^{i+1}(B,F) \oplus H^{i+1}(B) \\
\downarrow{\scriptstyle \iota - 1_B} & & \downarrow{\scriptstyle 1_M - \pi^*} & & \downarrow{\scriptstyle proj_1} & & \downarrow{\scriptstyle \iota - 1_B} \\
H^i(B) & \xrightarrow{\pi*} & H^i(M) & \xrightarrow{\int} & H^{i-k}(B,F) & \xrightarrow{\wedge[e]} & H^{i+1}(B) \\
\downarrow & & \downarrow & & \downarrow & & \downarrow \\
0 & & 0 & & 0 & & 0
\end{array}
$$

The top horizontal sequence is the long exact sequence of the pair (B, F). The middle sequence is the direct sum of the Smith-Gysin sequence of [1] and the exact sequence

$$\cdots \to 0 \to H^i(B) \xrightarrow{1_B} H^i(B) \to 0 \to H^{i+1}(B) \xrightarrow{1_B} H^{i+1}(B) \to 0 \to \cdots,$$

which is exact as well. The above diagram is commutative, and we can verify its commutativity easily. By using the commutativity of the diagram, the exactness of the two horizontal sequences, and the vertical short sequences, we can perform a diagram chase that shows the exactness of the bottom long sequence (known as the Gysin sequence).

Example 1 We conclude this work with an example that illustrates the Smith-Gysin sequence from Proposition 3. Consider the smooth action $\Phi\colon S^3 \times S^{m+7} \to S^{m+7}$, where $m \in \mathbb{N}$, defined as follows. We use the identification $S^2 = S^3/S^1$ where the action of S^1 on S^3 is on the *right*. An element of S^2 is denoted by $\bar{k}$, where $k \in S^3$. The sphere S^{m+7} is the join

$$S^m * S^3 * S^2 = \{t_1 \cdot x + t_2 \cdot g + t_3 \cdot \bar{k}\}$$

(formal additions). Here, $x \in S^m$, $g \in S^3$, $\bar{k} \in S^2$, $t_1, t_2, t_3 \in [0, 1]$ and $t_1 + t_2 + t_3 = 1$. The action Φ is defined on the *left* as:

$$\ell \cdot (t_1 \cdot x + t_2 \cdot g + t_3 \cdot \bar{k}) = t_1 \cdot x + t_2 \cdot \ell \cdot g + t_3 \cdot \overline{\ell \cdot k}.$$

The action of S^3 on the first factor S^m is ineffective (no point is moved). It is a free action on the second factor S^3. The isotropy group of the action on the third factor has a dimension of 1 at each point. In this case, there are exactly two points, $\{N, S\} \subset S^2$, where the isotropy group is exactly S^1.

We compute the various terms of the Smith-Gysin sequence.

- Denote an element of the orbit space as $\langle - \rangle$, and let $e \in S^3$ denote the unity element. We have

$$\begin{aligned} M/S^3 &= \{\langle t_1 \cdot x + t_2 \cdot g + t_3 \cdot \bar{k} \rangle\} = \{\langle t_1 \cdot x + t_2 \cdot e + t_3 \cdot \overline{g^{-1}k} \rangle\} \\ &= \{\langle t_1 \cdot x + t_2 \cdot e + t_3 \cdot \bar{k} \rangle\}. \end{aligned}$$

  This gives,

$$M/S^3 = S^m * S^0 * S^2 = S^{m+4}.$$

- $M^{S^3} = S^m$.
- $H^*(M/S^3, M^{S^3}) = H^*(S^{m+4}, S^m) = \begin{cases} \mathbb{R} & \text{if } * = m+1 \text{ or } m+4 \\ 0, & \text{otherwise.} \end{cases}$
- $M^{S^1} = S^m * \{N, S\} = S^{m+1}$. The element $j \in S^3$ acts in this space by interchanging the points N and S. Then

$$\left(H^*\left(M^{S^1}\right)\right)^{-\mathbb{Z}_2} = \left(H^*\left(S^{m+1}\right)\right)^{-\mathbb{Z}_2} = H^{m+1}\left(S^{m+1}\right) = \mathbb{R}.$$

  We can also make use of the Note 1 on page 245 and the observation that $M^{S^1}/\mathbb{Z}_2$ is contractible.
- $\Sigma = S^m * S^2 = S^{m+3}$ and therefore $\Sigma/S^3 = S^m * \{point\}$ contractible. So,

$$H^*\left(M/S^3, \Sigma/S^3\right) = H^*\left(S^{m+4}, contractible\right) = H^{m+4}\left(S^{m+4}\right) = \mathbb{R}.$$

The Smith-Gysin sequence simplifies to the following isomorphisms: $H^0(M) = H^0(M/S^3)$, $H^m(M^{S^3}) = H^{m+1}(M/S^3, M^{S^3})$, $H^{m+7}(M) = H^{m+4}(M/S^3, \Sigma/S^3)$ and $\left(H^{m+1}\left(M^{S^1}\right)\right)^{-\mathbb{Z}_2} = H^{m+4}(M/S^3, M^{S^3})$.

In the case $m = 0$ we find the S^3-action on the sphere S^7, whose quotient space is the sphere S^4. It is important to note that this action does not give rise to the quaternionic Hopf fibration $\pi_{\mathbb{H}}: S^7 \to S^4$.

Acknowledgements The authors thank the referee for useful remarks to improve the presentation of the contents of this paper

References

1. Bredon, G. E.: Cohomology fibre spaces, the Smith-Gysin sequence, and orientation in generalized manifolds. Michigan Mathematical Journal, 10(4): 321–333, 1963
2. Bredon, G. E.: Introduction to compact transformation groups. Academic Press, New York-London, 1972. Pure and Applied Mathematics, Vol. 46

3. Eilenberg, S. and Moore, J. C.: Foundations of relative homological algebra. Mem. Amer. Math. Soc. No. 55:39, 1965
4. Greub, W., Halperin, S. and Vanstone R.: Connections, curvature, and cohomology. Vol. I: De Rham cohomology of manifolds and vector bundles Academic Press [A subsidiary of Harcourt Brace Jovanovich, Publishers], New York-London, 1973. Pure and Applied Mathematics, Vol. 47-I
5. Royo Prieto, J.I. and Saralegi-Aranguren, M.: The Gysin sequence for S^3-actions on manifolds. Publ. Math. Debrecen, 83(3): 275–289, 2013
6. Saralegi, M.: A Gysin sequence for semifree actions of S^3. Proc. Amer. Math. Soc., 118(4): 1335–1345, 1993
7. Saralegi-Aranguren, M.: de Rham intersection cohomology for general perversities. Illinois J. Math., 49(3): 737–758 (electronic), 2005
8. Verona, A.: Le théorème de de Rham pour les préstratifications abstraites. C. R. Acad. Sci. Paris Sér. A-B, 273: A886–A889, 1971

An Ay-Lê-Jost-Schwachhöfer Type Characterization of Quantitatively Weakly Sufficient Statistics

Kaori Yamaguchi and Hiraku Nozawa

Abstract We explain the notion of a δ-almost sufficient statistic, which is a quantitative weak version of sufficient statistics introduced in [11], based on the example of coin tosses. We review the basic notions in information geometry, e.g., parametrized measure models, statistics and Fisher quadratic forms. Then we state characterizations of sufficient statistics due to Ay-Jost-Lê-Schwachhöfer and an analogous characterizations of sufficient statistics due to [11].

Keywords Information geometry · Sufficient statistic · Statistical manifold · Fisher metric

1 Parametrized Measure Models and Its Integrability

Information geometry is a geometry arising from statistics. Recently, a rigorous formalization of the basic notions of information geometry have been established by Ay-Jost-Lê-Schwachhöfer [5], which we will follow. Based on an example of coin tosses, we will explain the quantitative version of sufficient statistics that we introduced in [11].

Let us begin with the definition of parametrized measure models. For a measurable space Ω, let $\mathcal{M}(\Omega)$ denote the space of all finite measures on Ω.

Definition 1 *(Parametrized measure models* [2, 5]*)* A parametrized measure model is a triple $(M, \Omega, \mathbf{p})$, where

K. Yamaguchi
Graduate School of Science and Engineering, Ritsumeikan University, Nojihigashi 1-1-1, Kusatsu, Shiga 525-8577, Japan
e-mail: ra0097vv@ed.ritsumei.ac.jp

H. Nozawa (✉)
Department of Mathematical Sciences, Faculty of Science and Engineering, Ritsumeikan University, Nojihigashi 1-1-1, Kusatsu, Shiga 525-8577, Japan
e-mail: hnozawa@fc.ritsumei.ac.jp

© The Author(s), under exclusive license to Springer Nature Switzerland AG 2024
V. Rovenski et al. (eds.), *Differential Geometric Structures and Applications*, Springer Proceedings in Mathematics & Statistics 440,
https://doi.org/10.1007/978-3-031-50586-7_13

- M is a Banach manifold,
- Ω is a measurable space and
- The map $\mathbf{p} : M \to \mathcal{M}(\Omega)$ is of class C^1.

A parametrized measure model $(M, \Omega, \mathbf{p})$ is a family of measures on Ω parametrized by a manifold M in a differentiable way. Let us consider an example of a model for n-times coin tosses for $n \in \mathbb{N}$.

Example 1 *(The model 1 for n-times coin tosses)* Let

- $M = (0, 1)$,
- $\Omega = \{0, 1\}^n$ with the counting measure and
- $\mathbf{p}(\xi)(\{\omega\}) = \xi^{a(\omega)}(1 - \xi)^{n-a(\omega)}$ for $\omega \in \Omega$, where $a(\omega)$ is the number of 1 that ω contains.

In this example, the element $\xi \in M$ represents the probability of the head of the coin, and the element $\omega \in \Omega$ represents the result of the n-time coin toss where 0 means the tail and 1 means the head. The element ξ in M parametrizes the measures $\mathbf{p}(\xi)$ on Ω. For example, $\omega = (0, 0, \dots, 0, 1)$ means that we get the tail $(n-1)$-times and then we get the head. Such model can be used, for example, to estimate the probability ξ of the head of the given coin.

We can also consider a simpler model of n-times coin tosses as follows:

Example 2 *(The model 2 for n-times coin tosses)* Let

- $M = (0, 1)$,
- $\Omega' = \{0, 1, ..., n\}$ with the counting measure μ_0 and
- $\mathbf{p}'(\xi)(\{k\}) = \sum_{k=0}^{n} C_k \xi^k (1 - \xi)^{n-k}$ $(\forall k \in \Omega')$.

The element $\xi \in M$ represents the probability of the head of the coin as in the last example, and the element $k \in \Omega'$ represents the number of times that we get the head in the n-time tosses.

In order to define the Fisher metrics following [5] later, we need certain regularity for parametrized measure model. First let us recall the power of measures. It is easy to see that the set $\mathcal{M}(\Omega)$ of finite measures on Ω forms a directed set with the relation $\leq$, where $\omega_1 \leq \omega_2$ if ω_1 is dominated by ω_2 for $\omega_1, \omega_2 \in \mathcal{M}(\Omega)$. For $r > 0$, consider the direct limit

$$\mathcal{S}^r(\Omega) := \lim_{\to} L^{1/r}(\Omega, \mu)$$

on $\mathcal{M}(\Omega)$. If $\mu_1 \leq \mu_2$, then the canonical injection $L^{1/r}(\Omega, \mu_1) \to L^{1/r}(\Omega, \mu_2)$ is an isometry of Banach spaces. Therefore $\mathcal{S}^r(\Omega)$ has a structure of a Banach space. For $\phi \in L^{1/r}(\Omega, \mu)$, the image of $L^{1/r}(\Omega, \mu) \to \mathcal{S}^r(\Omega)$ is denoted by $\phi\, \mu^r$. Thus, the r-th power μ^r of μ is defined. For $r > 0$, let

$$\mathcal{M}^r(\Omega) := \{\phi\, \mu^r \in \mathcal{S}^r(\Omega) \mid \mu \in \mathcal{M}(\Omega),\ \phi \geq 0\}.$$

Definition 2 For $k \in \mathbb{N}$, a parametrized measure model $(M, \Omega, \mathbf{p})$ is defined to be k-integrable if

$$\mathbf{p}^{1/k} : M \to \mathcal{M}^{1/k}(\Omega), \quad \xi \mapsto \mathbf{p}(\xi)^{1/k}$$

is a C^1-map.

If $\mathbf{p}(\xi)$ is of the form $\mathbf{p}(\xi) = p(\cdot, \xi)\mu_0$ for some $\mu_0 \in \mathcal{M}(\Omega)$ and a positive function $p(\cdot, \xi)$, then k-integrability implies that $\partial_v \log p \in L^k(\Omega, \mathbf{p}(\xi))$. The Fisher metric is defined based on this fact.

2 Sufficient Statistics and Its Characterization Due to Ay-Jost-Lê-Schwachhöfer

Statistics is a basic notion in statistics and information geometry.

Definition 3 *(Statistics)* For measurable spaces Ω and Ω', a measurable map $\Omega \to \Omega'$ is called a statistic.

Let us see how statistics relates different parametrized measure models.

Example 3 Let $\Omega = \{0, 1\}^n$ and $\Omega' = \{0, 1, ..., n\}$ equipped with respective counting measure. The mapping

$$a \colon \Omega \to \Omega',$$

which maps $\omega \in \Omega$ to the number of 1 contained in ω.

We can see that the above two models on n-times coin tosses (Examples 1 and 2) are closely related with this map a. Indeed, we have that, for all $\xi \in M$,

$$\mathbf{p}'(\xi) = a_* \mathbf{p}(\xi).$$

, We can see that the parametrized measure model $(M, \Omega', \mathbf{p}')$ is determined by $(M, \Omega, \mathbf{p})$ and a statistics a. This is a particular case of the parametrized measure model induced by a statistic.

Definition 4 Given a parametrized model $(M, \Omega, \mathbf{p})$ and a statistics $\kappa : \Omega \to \Omega'$, the parametrized measure model $(M, \Omega', \mathbf{p}')$ with $\mathbf{p}'(\xi) = \kappa_* \mathbf{p}(\xi)$ for $\forall \xi \in M$ is called the model induced by κ.

The model induced by a statistics can be regarded as the original model observed by the statistic. Intuitively, the induced model has less information than the original model. The notion of sufficiency of statistics arises from the following question:

Question 1 How are the difference between the quantity of the information between two models?

Fisher [7] considered and Neyman [10] formalized statistics that do not change the quantity of information in a sense as follows:

Definition 5 *(Fisher-Neyman sufficient statistics)* Given a parametrized model $(M, \Omega, \mathbf{p})$ and a statistics $\kappa : \Omega \to \Omega'$, the statistics κ is called Fisher-Neyman sufficient if there is a finite measure μ_0 on Ω such that

$$\mathbf{p}(\xi) = q(\kappa(\cdot); \xi)\mu_0'$$

for some $q(\cdot; \xi) \in L^1(\Omega', \mu_0')$, where $\mu_0' = \kappa_* \mu_0$.

If κ is Fisher-Neyman sufficient, then it is clear that the the difference between $\mathbf{p}(\xi)$ and $\mathbf{p}(\xi')$ for $\xi, \xi' \in M$ depends only on $\kappa(\xi)$ and $\kappa(\xi')$, which intuitively shows that κ does not change the difference of the quantity of information on $\mathbf{p}(\xi)$ and $\mathbf{p}(\xi')$.

Ay-Jost-Lê-Schwachhöfer formalized this sufficient property based on the Fisher metric.

Definition 6 *(Sufficient statistics)* Let $(M, \Omega, \mathbf{p})$ be a 2-integrable parametrized measure model. Assume that $\mathbf{p}(\xi)$ is dominated by μ_0, namely, $\mathbf{p}$ is of the form $\mathbf{p}(\xi) = p(\omega; \xi)\mu_0$ for a density function p and a finite measure μ_0 on Ω. Then the Fisher quadratic form on M is given by

$$\mathfrak{g}(v, w) = \int_\Omega (\partial_v \log p(\cdot; \xi))\,(\partial_w \log p(\cdot; \xi))\, d\mathbf{p}(\xi) \tag{1}$$

for tangent vectors v, w on M.

Note that the 2-integrability condition is required so that the integration in the right hand side of (1) is well-defined. In the case where M is of dimension one, the Fisher quadratic form was known as Fisher information, which is nothing but the L^2-norm of the infinitesimal change of the entropy.

Example 4 Let us compute the Fisher metrics of the above two models in Examples 1 and 2: Let $p(\cdot; \xi)$ and $p'(\cdot; \xi)$ be the density function of $\mathbf{p}(\xi)$ and $\mathbf{p}'(\kappa(\xi))$ with respect to μ_0 and $\kappa_*\mu_0$, respectively, i.e., $\mathbf{p}(\omega) = p(\cdot; \xi)\mu_0$ and $\mathbf{p}'(\kappa(\omega)) = p'(\cdot; \xi)\kappa_*\mu_0$. By $\sum_{k=0}^n {}_nC_k \xi^k (1-\xi)^{n-k} = 1$, $\sum_{k=0}^n k\,{}_nC_k \xi^k (1-\xi)^{n-k} = n\xi$ and $\sum_{k=0}^n k^2\,{}_nC_k \xi^k (1-\xi)^{n-k} = n(n-1)\xi^2 + n\xi$, we have

$$\begin{aligned}
\|\partial_\xi \log p(\xi)\|_2^2 &= \sum_{\omega \in \emptyset} \left(\frac{a(\omega)}{\xi} - \frac{n - a(\omega)}{1 - \xi} \right)^2 \xi^{a(\omega)} (1-\xi)^{n-a(\omega)} \\
&= \sum_{k=0}^n \sum_{\omega \in a^{-1}(k)} \left(\frac{a(\omega)}{\xi} - \frac{n - a(\omega)}{1 - \xi} \right)^2 \xi^{a(\omega)} (1-\xi)^{n-a(\omega)} \\
&= \sum_{k=0}^n {}_nC_k \left(\frac{k}{\xi} - \frac{n-k}{1-\xi} \right)^2 \xi^k (1-\xi)^{n-k}
\end{aligned}$$

$$
\begin{aligned}
&= \frac{1}{\xi^2(1-\xi)^2} \sum_{k=0}^{n} {}_nC_k (k - \xi n)^2 \xi^k (1-\xi)^{n-k} \\
&= \frac{1}{\xi^2(1-\xi)^2} \sum_{k=0}^{n} {}_nC_k \left(k^2 - 2\xi nk + \xi^2 n^2\right) \xi^k (1-\xi)^{n-k} \\
&= \frac{1}{\xi^2(1-\xi)^2} \left(n(n-1)\xi^2 + n\xi - 2n^2\xi^2 + n^2\xi^2\right) = \frac{n}{\xi(1-\xi)}
\end{aligned}
$$

$$
\begin{aligned}
\|\partial_\xi \log p'(\xi)\|_2^2 &= \sum_{k=0}^{n} {}_nC_k \left(\frac{k}{\xi} - \frac{n-k}{1-\xi}\right)^2 \\
&= \frac{1}{\xi^2(1-\xi)^2} \left(n(n-1)\xi^2 + n\xi - 2n^2\xi^2 + n^2\xi^2\right) = \frac{n}{\xi(1-\xi)}.
\end{aligned}
$$

Thus, we can see that they coincide, which intuitively shows that κ preserves the difference of the information between the measures.

The difference $\mathfrak{g}(v, v) - \mathfrak{g}'(v, v)$ is called the (second order) information loss, which is non-negative by the monotonicity theorem [1, 3, 4, 9].

Ay-Jost-Lê-Schmachhöfer redefined sufficient statistics as follows:

Definition 7 *(Sufficient statistics)* A statistics κ on a 2-integrable parametrized measure model $(M, \Omega, \mathbf{p})$ is defined to be sufficient if the information loss is 0 everywhere, namely, for every tangent vector v on M we have

$$\mathfrak{g}'(v, v) = \mathfrak{g}(v, v)\,.$$

By Example 4, the statistics a in Example 3 is sufficient in this sense.

We conclude this section with results of Ay-Jost-Lê-Schwachhöfer [5], which characterizes sufficient statistics by using the conditional probability. Together with Fisher-Neyman characterization [10], they showed that the sufficient statistics are closely related to Fisher-Neyman sufficient statistics.

Theorem 1 ([3], [5, Chap. 5], Fisher-Neyman characterization [10]) *Let $(M, \Omega, \mathbf{p})$ be a 2-integrable parametrized measure model of the form*

$$\mathbf{p}(\xi) = p(\omega; \xi)\mu_0$$

for a finite measure μ_0 on Ω such that p is positive. Let $\kappa : \Omega \to \Omega'$ be a statistics and let $p'(\omega; \xi)$ be the density function of $\kappa_*\mathbf{p}(\xi)$ with respect to $\kappa_*\mu_0$. Assume that M is connected. Then the following are equivalent:

(i) *κ is sufficient for $(M, \Omega, \mathbf{p})$.*
(ii) *We have $\partial_v \log p = \partial_v \log \kappa^* p'$ for every tangent vector v on M.*
(iii) *The map $\xi \mapsto \frac{p(\cdot;\xi)}{p'(\kappa(\cdot);\xi)}$ is a constant map.*

(iv) There exist a measurable function $s : \Omega' \times M \to \mathbb{R}$ and a function $t \in L^1(\Omega, \mu_0)$ such that we have

$$p(\omega; \xi) = s(\kappa(\omega); \xi)t(\omega), \qquad \mu_0\text{-}a.e.\ \omega \in \Omega \text{ and } \forall \xi \in M.$$

In order to define the Fisher quadratic form, we need to assume that the parametrized measure model are 2-integrable.

The condition (iv) can be regarded as a Fisher-Neyman sufficiency in a weak sense. It is easy to see from Definition 5 that the following two conditions are equivalent (see [5, Theorem 5.3]):

- κ is Fisher-Neyman sufficient.
- in addition to the condition (iv), we have $s(\cdot; \xi) \in L^1(\Omega', \kappa_* \mu_0)$ for all $\xi \in M$.

3 A Quantitatively Weak Version of Sufficient Statistics and Its Characterization

Let us recall a quantitatively weak version of the sufficient statistics, which we introduced in [11]:

Definition 8 Let $0 < \delta \leq 1$. A statistics κ is defined to be δ-almost sufficient if for every tangent vector v on M we have

$$\delta^2 \mathfrak{g}(v, v) \leq \mathfrak{g}'(v, v)\,.$$

It is clear that a 1-almost sufficient statistics is sufficient. By the monotonicity theorem $\mathfrak{g}'(v, v) \leq \mathfrak{g}(v, v)$, we have

$$\delta^2 \mathfrak{g}(v, v) \leq \mathfrak{g}'(v, v) \leq \mathfrak{g}(v, v),$$

which means that $\mathfrak{g}'$ is bi-Lipschitz equivalent to $\mathfrak{g}$.

First let us present some examples of δ-almost statistics.

Example 5 Let us consider $(M, \Omega, \mathbf{p})$ in Example 4 with $n = 2$. Namely,

- $M = (0, 1)$,
- $\Omega = \{0, 1\}^2$ with the counting measure,
- $\mathbf{p}(\xi)(\{\omega\}) = \xi^{a(\omega)}(1 - \xi)^{2-a(\omega)}$ for $\omega \in \Omega$, where $a(\omega)$ is the number of 1 that ω contains.

Let $\Omega' = \{0, 1\}$ and consider the statistics $\kappa : \Omega \to \Omega'$ given by $\kappa(\omega_1, \omega_2) = \omega_1$ for $\forall\, (\omega_1, \omega_2) \in \Omega$. By Example 4, we have

$$\|\partial_\xi \log p(\cdot;\xi)\|_2^2 = \frac{2}{\xi(1-\xi)}$$

On the other hand, we have

$$\|\partial_\xi \log \kappa^* p'(\cdot;\xi)\|_2^2 = \frac{1}{\xi(1-\xi)}.$$

Since $\frac{\|\partial_\xi \log \kappa^* p'\|_2^2}{\|\partial_\xi \log p\|_2^2} = \frac{1}{2}$, κ is $\frac{1}{\sqrt{2}}$-almost sufficient.

The following example of an almost sufficient statistics was proposed by Masayuki Asaoka.

Example 6 We consider a parametrized measure model with $M = (0, \frac{1}{2})$, $\Omega = \prod^2\{0, 1\}$ and $\mathbf{p}(\xi) = p(\cdot;\xi)\mu_0$, where μ_0 is the counting measure on Ω and density function p is given by

$$p((0,0);\xi) = \frac{1}{2} - \xi, \qquad p((0,1);\xi) = \xi,$$
$$p((1,0);\xi) = \xi^2, \qquad p((1,1);\xi) = \frac{1}{2} - \xi^2$$

for $0 < \xi < \frac{1}{2}$. This can be regarded as a model of toss of two coins such that the result of the second toss depends on the first one. Let $\Omega' = \{0, 1, 2\}$ and define a statistics $\kappa : \Omega \to \Omega'$ by $\kappa((x, y)) = x + y$. The density function p' of the induced model $\kappa_*\mathbf{p}$ with respect to $\kappa_*\mu_0$ is given by

$$\begin{aligned}
p'(0;\xi) &= p((0,0);\xi) = \frac{1}{2} - \xi, \\
p'(1;\xi) &= \frac{1}{2}\big(p((0,1);\xi) + p((1,0);\xi)\big) = \frac{\xi + \xi^2}{2}, \\
p'(2;\xi) &= p((1,1);\xi) = \frac{1}{2} - \xi^2.
\end{aligned}$$

Then we have

$$\begin{aligned}
\mathfrak{g}(\partial_\xi, \partial_\xi) &= \|\partial_\xi \log p(\cdot;\xi)\|^2 = \frac{1}{\frac{1}{2} - \xi} + \frac{5}{\xi} + \frac{4\xi^2}{\frac{1}{2} - \xi^2}, \\
\mathfrak{g}'(\partial_\xi, \partial_\xi) &= \|\partial_\xi \log \kappa^* p'(\cdot;\xi)\|^2 = \frac{1}{\frac{1}{2} - \xi} + \frac{(2\xi + 1)^2}{2\xi(\xi + 1)} + \frac{4\xi^2}{\frac{1}{2} - \xi^2}.
\end{aligned}$$

It is easy to see that

$$\frac{1}{10}\|\partial_\xi \log p(\cdot;\xi)\|^2 < \|\partial_\xi \log \kappa^* p'(\cdot;\xi)\|^2$$

for $0 < \xi < 1/2$ and $\frac{\|\partial_\xi \log \kappa^* p'(\cdot;\xi)\|^2}{\|\partial_\xi \log p(\cdot;\xi)\|^2} \to \frac{1}{10}$ as $\xi \to 0$ as $\xi \to 0$. It follows that κ is $\frac{1}{\sqrt{10}}$-almost sufficient, and it is not δ-almost sufficient for any $\delta > \frac{1}{\sqrt{10}}$.

We can characterize δ-almost sufficient statistics in terms of the conditional probability in a way similar to the characterization of sufficient statistics by Ay-Jost-Lê-Schwachhöfer (Theorem 1) as follows.

Theorem 2 ([11, Theorem 1.4]) *Let $(M, \Omega, \mathbf{p})$ be a 2-integrable parametrized measure model of the form*

$$\mathbf{p}(\xi) = p(\omega; \xi)\mu_0$$

for a finite measure μ_0 on Ω such that p is positive. Assume that M equipped with the Fisher quadratic form is a finite-dimensional Riemannian manifold of class C^2 such that every pair of points on M is connected by a geodesic and every tangent vector is tangent to a geodesic. Let $\kappa : \Omega \to \Omega'$ be a statistics and $0 < \delta \leq 1$. Then the following are equivalent:

1. *κ is δ-almost sufficient for $(M, \Omega, \mathbf{p})$.*
2. *We have*
$$\left\| \partial_v \log \frac{p(\cdot;\xi)}{\kappa^* p'(\cdot;\xi)} \right\| \leq \sqrt{1-\delta^2} \|\partial_v \log p(\cdot;\xi)\|$$
for every tangent vector v on M, where $\|\cdot\|$ denotes the L^2-norm.
3. *The map $M \to L^2(\Omega, \mu_0); \xi \mapsto \log \frac{p(\cdot;\xi)}{p'(\kappa(\cdot);\xi)}$ is $\sqrt{1-\delta^2}$-Lipschitz with respect to the distance defined by the Fisher quadratic form and the L^2-metric.*
4. *There exist measurable functions $s : \Omega' \times M \to \mathbb{R}$ and $t : \Omega \times M \to \mathbb{R}_{>0}$ such that $\log t(\cdot;\xi) \in L^2(\Omega, \mu_0)$ for $\forall \xi \in M$,*
   - *$p(\omega;\xi) = s(\kappa(\omega);\xi)t(\omega;\xi)$ for μ_0-a.e. $\omega \in \Omega, \forall \xi \in M$, and*
   - *the map $M \to L^2(\Omega, \mu_0); \xi \mapsto \log t(\cdot;\xi)$ is $\sqrt{1-\delta^2}$-Lipschitz with respect to the distance defined by the Fisher quadratic form and the L^2-metric.*

The proof of Theorem 2 is essentially the same as the proof of [11, Theorem 1.4]. The statement of Theorem 2 is different from the statement of [11, Theorem 1.4]. The Lipschitz condition in Theorem 2 is replaced to the local Lipschitz condition. With the assumption that every pair of points on M is connected by a geodesic and every tangent vector is tangent to a geodesic, we can apply the argument of the proof of [11, Theorem 1.4].

We need to assume that $\dim M < \infty$ and the Fisher quadratic form is of class C^2 so that the geodesics are well-defined. Note that the Fisher quadratic form of a 2-integrable parametrized measure model is merely continuous in general. In the general case, geodesics with given initial data may not exist (see e.g., [8]). For Riemannian manifolds of infinite dimension, it is known that there are many subtle problems even for their existence (see [6]).

References

1. Amari, S. and Nagaoka, H.: *Methods of Information Geometry*, Translated from the 1993 Japanese original by Daishi Harada. Transl. Math. Monogr. 191. American Mathematical Society, Providence, RI; Oxford University Press, Oxford, 2000
2. Ay, N., Jost, J., Lê, H.V. and Schwachhöfer, L.: *Parametrized measure models*, Bernoulli 24 (3) 1692–1725, (2015). https://doi.org/10.3150/16-BEJ910
3. Ay, N., Jost, J., Lê, H.V. and Schwachhöfer, L.: Information geometry and sufficient statistics, *Probab. Theory Relat. Fields*, 162 327–364, (2015). https://doi.org/10.1007/s00440-014-0574-8
4. Ay, N., Jost, J., Lê, H.V. and Schwachhöfer, L.: Invariant Geometric Structures on Statistical Models. In: Nielsen, F., Barbaresco, F. (eds.) Geometric Science of Information. GSI 2015. LN in Computer Science, vol. 9389. Springer, Cham
5. Ay, N., Jost, J., Lê, H.V. and Schwachhöfer, L.: *Information Geometry*, Ergeb. Math. Grenzgeb. (3) Folge. A Series of Modern Surveys in Mathematics, 64. Springer, Cham, 2017
6. Ekeland, I.: The Hopf-Rinow theorem in infinite dimension, *J. Diff. Geom.* 13, no. 2, 287–301, (1978). https://doi.org/10.4310/jdg/1214434494
7. Fisher, R. A.: On the mathematical foundations of theoretical statistics, *Philos. Trans. R. Soc. Lond. Ser. A Math. Phys. Eng. Sci.* containing Papers of a Mathematical or Physical Character, 222, 309–368. https://doi.org/10.1098/rsta.1922.0009
8. Hartmann, P. and Wintner, A.: On the Problems of Geodesics in the Small, *Amer. J. Math.* Vol. 73 132–148, (1951)
9. Lê, H.V.: The uniqueness of the Fisher metric as information metric, *Ann. Inst. Statist. Math.* 69, no. 4, 879–896, (2017)
10. Neyman, J.: Sur un teorema concernente le considette statistiche sufficienti, *Giorn. Istit. Ital. Att.* 6, 320–334, (1935)
11. Yamaguchi, K. and Nozawa, H.: On statistics which are almost sufficient from the viewpoint of the Fisher metrics, Preprint, arXiv:2305.04199 (2023)

Lower Bounds for High Derivatives of Smooth Functions With Given Zeros

Gil Goldman and Yosef Yomdin

Abstract Let $f : B^n \to \mathbb{R}$ be a $d+1$ times continuously differentiable function on the unit ball B^n, with $\max_{z \in B^n} |f(z)| = 1$. A well-known fact is that if f vanishes on a set $Z \subset B^n$ with a non-empty interior, then for each $k = 1, \ldots, d+1$ the norm of the k-th derivative $\|f^{(k)}\|$ is at least $M = M(n, k) > 0$. A natural question to ask is: *what happens for other sets Z? In particular, for finite, but sufficiently dense sets?* This question was partially answered in [16] and [20–22]. This study is naturally related to a certain special settings of the Whitney's smooth extension problem. Our goal in this paper is threefold: first, to provide an overview of the relevant questions and existing results in the general Whitney's problem. Second, we provide an overview of our specific setting and some available results. Third, we provide some new results in our direction, which extend the recent result of [21], where an answer to the above question is given via the topological information on Z.

Keywords Whitney smooth extension · Smooth rigidity · Nested ovals

Mathematics Subject Classification (2010). 26D05 · 26D10 · 58K05

1 Introduction

We continue the study started (chronologically) in ([4, 5, 16, 20–22]), of certain special settings of the Whitney's smooth extension problem (see [2, 6, 7, 11–13]).

The paper is organized as follows: in Sect. 2 we give (an extremely sketchy) overview of the Whitney's smooth extension problem. In Sect. 3 we provide (a slightly more extended) overview of our results in [16, 20–22] and [4, 5]), devoted

G. Goldman (✉)
Department of Applied Mathematics, Tel Aviv University, Tel Aviv 69978, Israel
e-mail: ggoldman@tauex.tau.ac.il

Y. Yomdin
Department of Mathematics, Weizmann Institute of Science, Rehovot 76100, Israel
e-mail: yosef.yomdin@weizmann.ac.il

© The Author(s), under exclusive license to Springer Nature Switzerland AG 2024
V. Rovenski et al. (eds.), *Differential Geometric Structures and Applications*, Springer Proceedings in Mathematics & Statistics 440,
https://doi.org/10.1007/978-3-031-50586-7_14

to what we call "smooth rigidity", which is, essentially, a special case of the general Whitney's problem: we want to smoothly extend a zero function from a given set Z. Finally, in Sect. 4 we present new results of this paper, which extend the results of [21], bounding smooth rigidity in topological terms.

2 Whitney's Smooth Extension Problem

Let's recall shortly what this problem is about, and what is its current status.

Let $E \subset B^n \subset \mathbb{R}^n$ be a closed subset of the unit ball B^n in Euclidean n-space, and let $\bar{f}$ be a real function defined on E. Can $\bar{f}$ be extended to a C^d-smooth function f on $\mathbb{R}^n$, and, if the extension is possible, what is the minimal C^d-norm of f? (The C^d-norm of f is defined in Sect. 3 below). Whitney posing this problem, and partially answering it in [11–13], provided two main results:

1. In dimension $n = 1$, the necessary and sufficient extendability condition is (roughly) that for all the $d+1$–tuples of points $X = \{x_1, \ldots, x_{d+1}\} \in E$ the divided finite differences of $\bar{f}$ on X are uniformly bounded ([12]).
2. In dimension $n \geq 2$, Whitney considers "more informative" initial data: at each $x \in E$ the Taylor polynomial of degree d of f is given. This kind of data is usually called "Whitney fields" or "Jet fields". In [13] necessary and sufficient conditions are given for smooth extendability of the Whitney fields.

We shall not touch Whitney fields anymore in this paper, instead we complete the discussion with the following two important remarks:

A. The recent solution of the general Whitney problem in [6] was, essentially, based on a reconstruction of the Whitney field of $\bar{f}$ from its point-wise value data.
B. Whitney fields are important in many areas of mathematics. Let us just mention the profound work of Kolmogorov on the complexity of functional classes (see [8], and the references therein, and the huge research area opened by these works). For applications of Whitney fields in numerical analysis, see, e.g., [14].

Recent exciting developments in the general Whitney problem (see [2, 6, 7] and references therein), provide essentially a complete answer to the general Whitney extension problem in any dimension. In particular, the results of [2, 6, 7], provide an important information on this problem, which was earlier available only in dimension one. The "finiteness principle" achieved in this recent work, claims that, as in classical Whitney's results in dimension one ([12]), it is enough to check only finite subsets of Z with cardinality bounded only in terms of n and d. For example, by [2], for $n = 2,\ d = 2$, i.e., for 2-smooth functions of two variables, it is enough to check *the extendability of $\bar{f}$ from all the six-points subsets of E*. The explicit criteria of the extendability from the six-points subsets are also given in [2].

In general, an algorithmic way of providing an extension and estimating its minimum norm, is presented in [2, 6, 7] and in other related publications.

However, a possibility of an explicit answer, as in dimension one, through a kind of *multi-dimensional divided finite differences*, remains an open problem.

3 Smooth Rigidity

Let us now turn to our study in [4, 16] and [20–22], where our goal was to consider some quite special cases of the general Whitney problem, but to provide more explicit answers. Specifically, let us describe in more detail the setting of the recent study in [20–22], which we use below. Let $Z \subset B^n \subset \mathbb{R}^n$ be a closed subset of the unit ball B^n. In [20–22], we look for C^{d+1}-smooth functions $f : B^n \to \mathbb{R}$, vanishing on Z. Such C^{d+1}-smooth (and even C^∞) functions f always exist, since any closed set Z is the set of zeroes of a certain smooth function.

We normalize the extensions f requiring $\max_{B^n} |f| = 1$. Put the pointwise norm $\|f^{(d+1)}(x)\|$ to be the sum of the absolute values of $d+1$-st order derivatives of f at $x \in B^n$, and let the global norm $\|f^{(d+1)}\|$ be the maximum of $\|f^{(d+1)}(x)\|$ for all $x \in B^n$. We ask the question about the minimum possible norm of the last derivative $\|f^{(d+1)}\|$, which we call *the d-rigidity* $\mathcal{RG}_d(Z)$ *of* Z. In other words, for each normalized C^{d+1}-smooth function $f : B^n \to \mathbb{R}$ vanishing on Z, we have

$$\|f^{(d+1)}\| \geq \mathcal{RG}_d(Z),$$

and $\mathcal{RG}_d(Z)$ is the maximum number with this property.

Bounding from below $\mathcal{RG}_d(Z)$ is, essentially, very close to the answering the Whitney problem for smooth extensions of the zero function $\bar{f} \equiv 0$ on Z.

Papers [20–22], provide certain bounds on $\mathcal{RG}_d(Z)$, in terms of the fractal geometry (or of the topology) of Z. Before stating these results more accurately, and in order to compare them with the general results, available today in Whitney's extension theory, let's make the following remark.

The results of [6, 7], in principle, provide an algorithmic way to estimate our quantities $\mathcal{RG}_d(Z)$ for any closed $Z \subset B^n$ by considering finite subsets of Z of bounded cardinality. However, our goal in this paper, as in our previous papers related to Smooth Rigidity, is somewhat different: *we seek an explicit answer in terms of simple and directly computable geometric or topological properties of* Z.

Let us now move on to specific results. As a "model" example, consider the case of dimension $n = 1$. Here we have the following important fact, which can be proven using standard interpolation (or finite difference) formulas:

Proposition 1 *For any* $Z \subset B^1$ *we have the following:*
$\mathcal{RG}_d(Z) \geq \frac{(d+1)!}{2^{d+1}}$ *if* Z *consists of at least* $d+1$ *different points, and*
$\mathcal{RG}_d(Z) = 0$ *if* Z *consists of at most* d *different points.*

Another simple observation is the following.

Proposition 2 *For any* $Z \subset B^n$ *with non-empty interior we get the following inequality:* $\mathcal{RG}_d(Z) \geq \frac{(d+1)!}{2^{d+1}}$.

This fact easily follows from Proposition 1. We just restrict any function f to a certain straight line ℓ, passing through z_0 with $|f(z_0)| = 1$, and through an interior point of Z. Let us mention here also an old result of [16], related to smooth rigidity. Informally it can be stated as follows: if the set of zeros $Y(f)$ of a smooth function f on B^n does not look like a union of smooth hypersutfaces, of a total area bounded by a constant, depending only on n, d, then the norm of $f^{(d+1)}$ is not smaller than a certain positive constant, depending only on n, d.

Many of the "near-polynomiality" results of [17, 23] can be naturally interpreted in terms of smooth rigidity. We plan to present some new results in this direction separately.

3.1 Rigidity and Remez Constant of Z

Here we present the results of [20]. We need a definition and some properties of the Remez (or Lebesgue, or norming, ...) constant (see, e.g., [3, 10, 18]).

Definition 1 For a set $Z \subset B^n \subset \mathbb{R}^n$ the Remez constant $\mathcal{R}_d(Z)$ is the minimal K for which the inequality

$$\sup_{x \in B^n} |P(x)| \le K \sup_{x \in Z} |P(x)|$$

is valid for any real polynomial $P(x) = P(x_1, \ldots, x_n)$ of degree d.

Clearly, we always have $\mathcal{R}_d(Z) \ge 1$. For some Z the Remez constant $\mathcal{R}_d(Z)$ may be equal to ∞. In fact, $\mathcal{R}_d(Z)$ is infinite if and only if Z is contained in the set of zeroes

$$Y_P = \{x \in \mathbb{R}^n \mid P(x) = 0\}$$

of a certain polynomial P of degree d. Sometimes it is convenient to use the inverse Remez constant $\hat{\mathcal{R}}_d(Z) := \frac{1}{\mathcal{R}_d(Z)}$.

In [19, 20] we showed that the rigidity $\mathcal{RG}_d(Z)$ and the Remez constant $\hat{\mathcal{R}}_d(Z)$ are closely related:

Theorem 1 ([20]) *For any $Z \subset B^n$, we get $\frac{(d+1)!}{2}\hat{\mathcal{R}}_d(Z) \le \mathcal{RG}_d(Z)$.*

The proof is heavily based on the result of [19], published in the Proceedings of the current series of conferences a few years ago.

3.2 Rigidity and Test Curves

As it was mentioned above, *for Z with a non-empty interior we always have $\mathcal{RG}_d(Z) \ge \frac{(d+1)!}{2^{d+1}}$, independently of the size and the geometry of Z.*

The following question was discussed in [20, 21]: *Can this last property be extended to other Z, beyond those with a non-empty interior? In particular, is it true for sufficiently dense finite sets Z?*

A partial answer was given in [22]:

Theorem 2 *If the box dimension $dim_e(Z)$ is greater than $n - \frac{1}{d+1}$, then*

$$\mathcal{RG}_d(Z) \geq M = M(n, d) > 0,$$

where the positive constant M depends only on n and d.

The box (or Minkowski, or entropy ...) dimension of a set X is, informally, the power β in the expression $M(\mathbf{e}, X) \sim \left(\frac{1}{\mathbf{e}}\right)^{\beta}$, where $M(\mathbf{e}, X)$ is the covering number of X by the $\mathbf{e}$-balls.

In particular, the result of Theorem 2 provides examples of discrete, but sufficiently dense, sets Z for which $\mathcal{R}_d(Z)$ behaves in the same way as for sets with a non-empty interior.

Let us describe our basic approach to the proof of this result in [22]. If we could find a straight line ℓ in $\mathbb{R}^n$, passing through the point z_0, where the absolute value $|f(z)|$ is equal to one, and through some $d+1$ distinct points in Z, we could immediately get the required lower bound $\mathcal{RG}_d(Z) \geq \frac{(d+1)!}{2^{d+1}}$ via the basic properties of d-rigidity, mentioned above (Proposition 1).

However, for a generic finite set Z any straight line ℓ meets Z at one or two points at most. Instead we replace ℓ by a smooth curve ω, and try to mimic the calculations for ℓ. This requires analysis of the high order chain-rule expressions, on one side, and construction of curves ω with small high-order derivatives, passing through some $d+1$ distinct points in Z, on the other side.

We show, using a kind of "discrete integral geometry", that already for finite or discrete sets Z, which are dense enough (in particular, when $\dim_e(Z) > n - \frac{1}{d+1}$), the required curves exist.

We call the smooth curves ω as above "the test curves". We expect such curves to play an important role in investigation of the Smooth Rigidity, the multidimensional divided finite differences (D.F.D's), and, ultimately, the general Whitney smooth extension problem.

3.3 *Polynomial Test Curves of Higher Degree*

Another application of test curves is given in [5]. Here, we assume that the set Z of zeroes of f sits on a polynomial (or near-polynomial) curve ω of degree s, and provide certain specific smooth rigidity results. Our main results in [5], Theorem 3 and Corollaries 1 and 2 below, can be considered as generalizations of Proposition 1 from straight lines to curves of higher degree.

Consider polynomial parametric curves ω of degree s in $\mathbb{R}^n$. The curves ω are given in the coordinate form by $\omega(t) = (\omega_1(t), \ldots, \omega_n(t))$, with $t \in [-1, 1]$, and with $\omega_i(t)$ being polynomials in t of the degree at most s. Denote, as usual, by $[\eta]$ the integer part of η.

Theorem 3 *Let $f : B^n \to \mathbb{R}$ and ω be as above, with $\omega([-1, 1]) \subset B^n$. Put $g(t) = f(\omega(t))$. Then for each $t \in [-1, 1]$ we have*

$$\sum_{|\alpha|=[\frac{d+1}{s}]+1}^{d+1} \|f^{(\alpha)}(\omega(t))\| \geq C(n, d, s)\|g^{(d+1)}(t)\|.$$

with the positive constant $C(n, d, s)$, which is explicitly given in [5].

In particular, we have

$$\sum_{|\alpha|=[\frac{d+1}{s}]+1}^{d+1} \|f^{(\alpha)}\| \geq C(n, d, s)\|g^{(d+1)}\|.$$

Thus, for ω a curve of degree $s > 1$ we cannot translate the derivatives of $f(\omega)$ directly to the derivatives of f of the same order, as in the case of ω being the straight line. But still we can bound from below the sum of the norms of the derivatives of f of orders from $[\frac{d+1}{s}] + 1$ to $d + 1$ in terms of the derivatives of $g = f(\omega)$. As an immediate corollary we obtain the following.

Corollary 1 *In the assumptions as above, if the curve ω crosses the zero set Z of f at at least $d + 1$ points, and passes through a certain point z_0 with $|f(z_0)| \geq \gamma > 0$, then*

$$\sum_{|\alpha|=[\frac{d+1}{s}]+1}^{d+1} ||f^{(\alpha)}|| \geq C(n, d, s)\frac{\gamma(d+1)!}{2^{d+1}}.$$

Another corollary is as follows.

Corollary 2 *Let $f : B^n \to \mathbb{R}$ be an infinitely differentiable function on B^n, and let ω be as above, with $\omega([-1, 1]) \subset B^n$ not contained in the zero set $Z(f)$. Assume that $|Z \bigcap \omega([-1, 1])| = \infty$, i.e., there are infinitely many zeros of f on the curve ω. Then there is an infinite number of the derivatives orders m for which*

$$||f^{(m)}|| \geq C(d, m, f, \omega)\frac{(m+1)!}{2^{m+1}},$$

with the positive constants $C(d, m, f, \omega)$, which are explicitly given in [5].

3.4 *Rigidity Via the Topology of Z*

The following result is obtained in [21]: let B^n be the unit n-dimensional ball. For a given integer d, let $Z \subset B^n$ be a smooth compact hypersurface with $N = (d-1)^n + 1$ connected components Z_j *with disjoint interiors* U_j. Let μ_j be the n-volume of the interior U_j of Z_j, and put $\mu = \min\{\mu_j,\ j = 1, \ldots, N\}$.

Theorem 4 ([21]) *Under the conditions above we obtain the following:*
1. (Remez-type inequality). For each polynomial P of degree d on $\mathbb{R}^n$ we have

$$\frac{\max_{B^n} |P|}{\max_Z |P|} \leq \left(\frac{4n}{\mu}\right)^d.$$

2. (Smooth rigidity). As a consequence, via [20], we provide an explicit lower bound for the $(d+1)$-st derivatives of any smooth function f, which vanishes on Z, while being of order 1 on B^n:

$$||f^{(d+1)}|| \geq \frac{1}{(d+1)!}\left(\frac{4n}{\mu}\right)^d.$$

3.5 Rigidity Via Singular Points and Values

In a recent paper [4] we bounded the higher derivatives of f from below using the geometry of critical points and values of f. While this setting does not exactly fit the rigidity definition, given above, it certainly remains in the general Whitney's problem framework. Basically, to conclude that a certain geometric configuration *of the critical points of f* implies lower bounds on the high-order derivatives of f, we apply in [4] the results of [20, 21], to the first order partial derivatives of f, instead of f itself.

In turn, to conclude that a certain geometric configuration *of the critical values of f* implies lower bounds on the higher derivatives of f, we try to "read backward" the (pretty old) results of [15, 23]. It turns out to be a non-trivial problem. We consider the results of [4] as quite instructive, and plan to present further results in this direction separately.

4 New Results

Now we finally come to the new results of this paper. These results form a direct continuation and extension of the "topological" results of [21], presented in Sect. 3.4.

Our topological assumption in [21] was that all the interiors of the components of the hypersurface Z of zeroes of f are disjoint. This assumption strongly simplifies the considerations, but also strongly restricts the applicability of the result. In this paper *we drop this assumption, investigating arbitrary nesting configurations of the components of Z*. We do it here only on the plane, by two reasons: first, the presentation becomes truly elementary, and, second, pretty delicate topological considerations (in the line of the Jordan-Brower separation theorem in higher dimension)

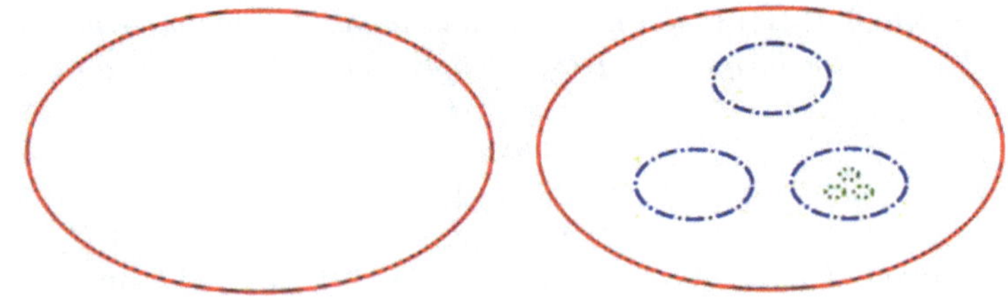

Fig. 1 This is an example of a nested configuration

can be avoided. We strongly believe that our results below *can be verbally extended to higher dimensions*, and plan to present them separately.

Let $Z = \bigcup_{i=1}^{s} Z_j$ be the union of s disjoint smooth ovals Z_i in the plane $\mathbb{R}^2$. By "oval" we mean a closed curve without self-intersections in $\mathbb{R}^2$. For each two ovals Z_{i_1} and Z_{i_2} the important for us relations between them are the following:

1. Z_{i_1} is entirely in the interior of Z_{i_2} (or wise versa).
2. the interiors of Z_{i_1} and of Z_{i_2} are disjoint.

Here we strongly depend on the Jordan separation theorem, claiming that a continuous closed curve S embedded in $\mathbb{R}^2$ divides $\mathbb{R}^2$ into two components—the exterior and the interior of S. However, with this theorem in dispose (which, of course, is the main topic, requiring a careful presentation in higher dimensions), we can provide a complete topological description of any collection of plane ovals.

First we give a verbal topological description of Z: it consists of a certain number s_0 of disjoint ovals $O_l,\ l = 1, \dots, s_0$. In turn, each oval O_l contains inside it "depth one" disjoint ovals $O_{l,l,1},\ l_1 = 1, \dots, s_{l,1}$. We continue inductively, with the "depth" of the nested configuration of the ovals, see Fig. 1. We denote the total number of the ovals in Z as N.

Finally, we describe the compact domains W_j, whose boundaries are formed by the ovals Z_i. We start with the "depth one" ovals $O_l,\ l = 1, \dots, s_0$, and for each O_l we consider the domain W_l, which is bounded by the oval O_l from the outside, and by the depth two ovals $O_{l,l_1},\ l_1 = 1, \dots, s_{l,1}$ from the inside. Then we continue inductively. This construction defines for each Z as above a unique collection $\mathcal{U}(Z)$ of the compact domains W_j, with non-intersecting interiors, whose boundaries are formed by the ovals Z_i, see Fig. 2.

The fact important for our final result is the following:

Lemma 1 *For each configuration Z of plane ovals, the total number of the compact domains $W_j \in \mathcal{U}(Z)$, as above, is equal to the total number N of the ovals in Z.*

Proof We apply induction with respect to the nesting depth of Z. For all ovals in Z disjoint, we are in situation of [21], and the conclusion is immediate. Here we start the induction. As an example to the opposite situation, consider the "totally nested case".

Now assume that the result is proved for a nesting depth l. To pass to a configuration of the nesting depth $l + 1$, we add inside some ovals of depth l some new non-nested

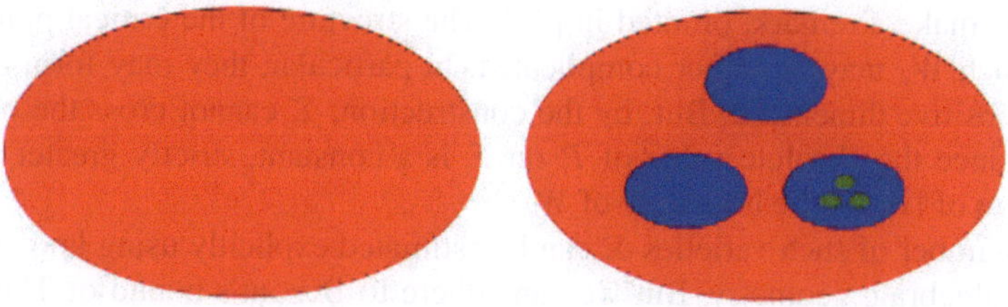

Fig. 2 This is a visual representation of the domains W_j, corresponding to the ovals configuration in Fig. 1. Notice the number of the domain is the same as the number of the ovals, confirming Lemma 1 below in this special case

ovals (of the depth $l+1$). Next, in each oval of the depth l, we add its interior W', up to the new ovals, as one of new domains W_j, and we add also all the interiors W'' of new ovals of the nesting depth $l+1$ (separately).

Clearly, if the number of the newly added ovals of the depth $l+1$ was q, the total number N of the ovals increased by q. The total number of the new domains W', W'' is $q+1$, but one of these domains existed before the reconstruction. We conclude that the total number of domains and ovals remains the same. □

Let μ_j be the n-volume of the domain W_j. For Z as above we put

$$\mu(Z) = \min\{\mu_j,\ W_j \in \mathcal{U}(Z)\}.$$

Now we are ready to state and prove our main result:

Theorem 5 *For a given integer d, let* $Z = \bigcup_{i=1}^{s} Z_j$ *be the union of the disjoint* $s \geq (d-1)^2 + 1$ *smooth ovals* Z_i *in the unit ball* B^2 *in the plane* $\mathbb{R}^2$*. Then for each polynomial P of degree d on* $\mathbb{R}^2$ *we have the Remez-type inequality*

$$\frac{\max_{B^n} |P|}{\max_Z |P|} \leq \left(\frac{8}{\mu(Z)}\right)^d.$$

As a consequence, via [20], we have an explicit lower bound for the $(d+1)$*-st derivatives of any smooth function f, which vanishes on Z, while being of order* 1 *on* B^2 *(smooth rigidity):* $||f^{(d+1)}|| \geq \frac{1}{(d+1)!}\left(\frac{8}{\mu(Z)}\right)^d$.

Proof We apply exactly the same arguments as in [21] to the domains W_j. Thus, for the detailed proof we refer to [21]. The idea is that if the polynomial P with $\max_{B^2} |P|$ is too small on Z, i.e., at the boundaries of the domains W_j, it must still be "large" inside each W_j. Otherwise, it will be too small on the entire ball B^2 due to the classical Brudny-Ganzburg-Remez inequality [1], applied separately to each W_j. Hence, being "small" on Z but "large" inside each W_j, P must have a critical point inside each W_j. As a result, we get more than $(d-1)^2$ critical points of P, which contradicts to Bezout's theorem. □

Let us make a remark, omitted in [21]. The structure of the critical points of P inside each W_j may be rather complicated. In particular, they may form a variety Σ of a positive dimension. But, by the construction, Σ cannot cross the boundary of W_j, since the absolute value of P on Σ is a constant, strictly greater than the maximum of $|P|$ on the boundary of W_j.

The number of such varieties Σ can be estimated explicitly using known results of semi-algebraic geometry. But we can adhere to Bezout's bound of Theorem 5, using the following standard trick: we make a small perturbation to $\tilde{P} = P + \xi$, adding to it an arbitrarily small linear form ξ, such that all critical points of $\tilde{P}$ are non-degenerate. Since the perturbation ξ is small, all new non-degenerate critical points of $\tilde{P}$ remain inside the corresponding domains W_j. Now the usual Bezout's bound can be applied.

We conclude our discussion with the following open question: real algebraic geometry provides various restrictions on the topology of the sets of zeros of real polynomials of a given degree d. If for a given smooth f any such restriction is violated, then f cannot be a polynomial of degree d, hence the $(d+1)$-st derivative of f is non-zero. The problem is *to provide an effective lower bound on this derivative.*

As an example, a plane real curve S of degree 6 can have at most 11 ovals; the possible mutual positions of these ovals are given in the solution of the first part of Hilbert's 16-th problem. In particular, if the zero set Z of f contains 12 or more ovals, then the 7-st derivative of f is non-zero. *But we are not aware of any approach to providing an effective lower bound.*

In contrast, if there are $5^2 + 1 = 26$ ovals of the zero set Z of f, then Theorem 5 provides an effective lower bound for the 7-st derivative of f.

Let us assume that zeros of f are transversal, with an explicit lower bound γ for the transversality measure. Then we expect that the results of [9] would provide effective lower bounds for higher derivatives of f, in case where the topology of $Z(f)$ violates the polynomial restrictions. However, these bounds blow up as $\gamma \to 0$. Note that there are no transversality assumptions in Theorem 5.

References

1. Brudnyi, Y. and Ganzburg, M.: On an extremal problem for polynomials of n-variables. Math. USSR Izv. 37, 344–356 (1973)
2. Brudnyi, Y. and Shvartsman, P.: Generalizations of Whitney's extension theorem, IMRN 1994, no. 3, 129–139
3. Brudnyi, A. Yomdin, Y.: Norming sets, and related Remez-type inequalities, J. Aust. Math. Soc. 100 (2016) 163–181
4. Goldman, G. and Yomdin, Y.: Higher derivatives of functions with given critical points and values, arXiv:2308.14722
5. Goldman, G. and Yomdin, Y.: Higher derivatives of functions with zeros on algebraic curves, arXiv:2309.03975
6. Fefferman, C.: Whitney extension problem for C^m, Ann. Math. 164 (2006), 313–359
7. Fefferman, C. and Klartag, B.: Fitting a C^m-smooth function to data. Part I, *Ann. Math.* (2) 169 (2009), 315-346. Part II, *Rev. Mat. Iberoam.* 25 (2009), 49–273

8. Kolmogorov, A.N. and Tikhomirov, V.M.: ε-entropy and ε-capacity of sets in functional spaces, Usp. Math. Nauk, 14 (1959), 3–86
9. Lerario, A. and Stecconi, M.: What is the degree of a smooth hypersurface? arXiv:2010.14553v1
10. Remez, E.J.: Sur une propriete des polynomes de Tchebycheff, *Comm. Inst. Sci. Kharkov*, '13 (1936) 93–95
11. Whitney, H.: Analytic extensions of differentiable functions defined in closed sets, Trans. Amer. Math. Soc., Vol. 36 (1934), 63–89
12. Whitney, H.: Differentiable Functions Defined in Closed Sets. I, Trans. Amer. Math. Soc., Vol. 36, No. 2 (1934), 369–387
13. Whitney, H.: Functions differentiable on the boundaries of regions, Ann. of Math. 35 (1934), 482–485
14. Wiener, Z. and Yomdin, Y.: Numerical From Formal Numerical Solutions of Elliptic PDE's to the True Ones, Mathematics of Computation, Vol. 69, No. 229 (Jan., 2000), 197–235
15. Yomdin, Y.: The Geometry of Critical and Near-Critical Values of Differentiable Mappings. Mathematische Annalen, 264, (1983) 495–516
16. Yomdin, Y.: The set of zeroes of an "almost polynomial" function, *Proc. AMS*, Vol. 90, No. 4 (1984), 538–542
17. Yomdin, Y.: Global bounds for the Betti numbers of regular fibers of differentiable mappings. Topology 24 (1985), no. 2, 145–152
18. Yomdin, Y.: Remez-Type Inequality for Discrete Sets, *Isr. J. of Math.*, Vol 186 (2011), 45–60
19. Yomdin, Y.: Remez-Type Inequality for smooth functions, in: "Geometry and its Applications", V. Rovenski, P. Walczak, Editors, Springer 2014, see https://link.springer.com/book/10.1007/978-3-319-04675-4
20. Yomdin, Y.: Smooth rigidity and Remez-type inequalities, Anal. Math. Phys. 11, 89 (2021)
21. Yomdin, Y.: Smooth rigidity and Remez inequalities via Topology of level sets, Journal of Singularities, Vol. 25 (2022), 443–455
22. Yomdin, Y.: Higher derivatives of functions vanishing on a given set, arXiv:2108.02459v1
23. Yomdin, Y. and Comte, G.: Tame geometry with application in smooth analysis, Springer, LNM 1843, 2004

Interactions Between Differential Geometry and Production Theory

Alina-Daniela Vîlcu and Gabriel-Eduard Vîlcu

Abstract The use of differential geometry in the theory of production is an interesting topic that has attracted the attention of many researchers in the last two decades, both from the fields of mathematics and economics. The purpose of this work is to provide a comprehensive survey on the properties of the homogeneous and quasi-homogeneous production models, most of these properties being of geometric nature and obtained using a differential geometric treatment. Some open problems are also discussed at the end of the article.

Keywords Production function · Hypersurface · Gauss-Kronecker curvature · Mean curvature

1 Introduction

The notion of production function is one of the key concepts of mainstream neo-classical theories, both in microeconomics and macroeconomics, being used in the mathematical modeling of the relationship between physical output of a production process and the inputs that have been used in obtaining it (like labor, capital, land and raw materials). In the economic literature, it is generally considered that the

A.-D. Vîlcu
Department of Information Technology, Mathematics and Physics, Petroleum-Gas University of Ploieşti, Bd. Bucureşti 39, Ploieşti 100680, Romania
e-mail: daniela.vilcu@upg-ploiesti.ro

G.-E. Vîlcu (✉)
Department of Mathematics and Informatics, National University of Science and Technology Politehnica Bucharest, Splaiul Independenţei 313, Bucharest 060042, Romania

Gheorghe Mihoc-Caius Iacob Institute of Mathematical Statistics and Applied Mathematics of the Romanian Academy, Calea 13 Septembrie 13, Bucharest 050711, Romania

Research Center in Geometry, Topology and Algebra, University of Bucharest, Str. Academiei 14, Bucharest 70109, Romania
e-mail: gabriel.vilcu@upb.ro

© The Author(s), under exclusive license to Springer Nature Switzerland AG 2024
V. Rovenski et al. (eds.), *Differential Geometric Structures and Applications*,
Springer Proceedings in Mathematics & Statistics 440,
https://doi.org/10.1007/978-3-031-50586-7_15

English economist Philip Wicksteed [64] introduced the concept of production function in 1894. However, according to some very well documented studies [34, 42], it seems that the German mathematical economist Johann Heinrich von Thünen was the first person to algebraically formulate the relationship between output and inputs, in 1840, as a mapping $f : \mathbb{R}^n_+ \to \mathbb{R}_+$, $f = f(x_1, \ldots, x_n)$, where f is the quantity of output, n is the number of the inputs and $x_1, \ldots, x_n$ are the factor inputs, such as labor, capital, land, raw materials etc. In order that these functions to model as well as possible the economic reality, in the classical theory they are required to have several properties. We recall the following most important of them with appropriate economic interpretations [28, 46, 53]:

1. f vanishes in the absence of an input, that means the factor inputs are necessary.
2. $\frac{\partial f}{\partial x_i} > 0$, for all $i \in \{1, ..., n\}$, which indicates that the production function is strictly increasing with respect to any factor of production.
3. $\frac{\partial^2 f}{\partial^2 x_i} < 0$, $\forall i = \overline{1, n}$, which means that the production has decreasing efficiency with respect to any factor of production.
4. $f(x + y) \geq f(x) + f(y)$, for all $x, y \in \mathbb{R}^n_+$, that signifies the production has non-decreasing global efficiency.

The simplest production model used in economics is the famous *Cobb-Douglas* (CD) production function, a model introduced in 1928 by the mathematician C.W. Cobb and the labor economist P.H. Douglas [19], in order to describe the distribution of the national income of the United States, in the form

$$f = CK^{\alpha}L^{\beta},$$

where f stands for total production, K for capital input, L for labor input and C is a positive constant which signifies the total factor productivity. We note that in the original definition it is required that $\alpha + \beta = 1$, but this condition has been later relaxed; usually, C, α and β are estimated from empirical data. A generalized CD production function depending on n-inputs is given by

$$f(x_1, ..., x_n) = A \cdot \prod_{i=1}^{n} x_i^{\alpha_i}, \tag{1}$$

where $A, \alpha_1, \ldots, \alpha_n > 0$. The advantages of the Cobb-Douglas production function in its generalized form were highlighted by many authors (see, e.g., K.V.B. Murthy [43]).

The Cobb-Douglas production model was generalized by K.J. Arrow, H.B. Chenery, B.S. Minhas and R.M. Solow [6], they introducing the so-called *Constant Elasticity of Substitution* (CES) production function. This model was extended to the n-inputs by H. Uzawa [57] and D. McFadden [35], they defining independently a new production function, usually called *generalized CES* production function, *Armington aggregator* or *ACMS* (Arrow-Chenery-Minhas-Solow) production function, by

$$f(x_1, ..., x_n) = A\left(\sum_{i=1}^{n} \alpha_i^{\rho} x_i^{\rho}\right)^{\frac{\gamma}{\rho}}, \tag{2}$$

where $A > 0$, $\rho < 1$, $\rho \neq 0$, $\gamma > 0$ and $\alpha_i > 0$, for all $i \in \{1, ..., n\}$. We note that CES production functions are of great interest in the field of economics because of their invariant characteristic, namely that the elasticity of substitution between the parameters is constant on their domains [38].

We also remark that generalized CES production functions include as special cases many other famous production models, like multinomial production functions or Leontief production functions (see [42] for other examples and definitions). We recall that the *multinomial* production function is obtained by taking $\rho \to 1$ in (2). Moreover, if $\gamma = 1$, then a multinomial production model is said to be a *linear* production model, also called a *perfect substitutes* production model. Notice that when $n = 2$, the multinomial production function is called *binomial.*

We recall next the main indicators of production. Let f be a production function with n inputs $x_1, x_2, ..., x_n$, $n \geq 2$ and we denote the partial derivatives $\frac{\partial f}{\partial x_i}$, $\frac{\partial^2 f}{\partial x_i \partial x_j}$, ..., etc. by $f_{x_i}, f_{x_i x_j}, \ldots$, etc. The *elasticity of production* with respect to a certain factor of production x_i is defined as

$$E_i = \frac{x_i}{f} f_{x_i},$$

and the *marginal rate of technical substitution* of input x_j for input x_i is given by

$$\text{MRS}_{ij} = \frac{f_{x_j}}{f_{x_i}}.$$

It is known that the elasticity of production with respect to x_i measures the change of production (in percent) when we have a one percent change in the input x_i, holding unchanged the rest of inputs. On the other hand, the marginal rate of technical substitution between the inputs x_i and x_j represents the amount by which the quantity of the input x_j must be reduced when one extra unit of the input x_i is used, so that to have the same level of output.

A production function is said to satisfy the *proportional marginal rate of substitution* (PMRS) property if and only if

$$\text{MRS}_{ij} = \frac{x_i}{x_j}$$

for all $1 \leq i \neq j \leq n$.

On the other hand, the function H_{ij} defined by

$$H_{ij}(x_1, \ldots, x_n) = \frac{\frac{1}{x_i f_{x_i}} + \frac{1}{x_j f_{x_j}}}{-\frac{f_{x_i x_i}}{f_{x_i}^2} + \frac{2 f_{x_i x_j}}{f_{x_i} f_{x_j}} - \frac{f_{x_j x_j}}{f_{x_j}^2}},$$

for all $(x_1, \ldots, x_n) \in \mathbb{R}^n_+$, is called the *Hicks elasticity of substitution* of the ith production factor with respect to the jth production factor, where $i, j \in \{1, \ldots, n\}$,

$i \neq j$. It measures the elasticity of the ratio of two inputs with respect to their marginal rate of substitution. Moreover, the function A_{ij} defined as

$$A_{ij}(x_1, \ldots, x_n) = -\frac{x_1 f_{x_1} + \cdots + x_n f_{x_n}}{x_i x_j} \frac{\Delta_{ij}}{\Delta},$$

for all $(x_1, \ldots, x_n) \in \mathbb{R}^n_+$, where Δ is the determinant of the bordered matrix

$$A(f) = \begin{pmatrix} 0 & f_{x_1} & \cdots & f_{x_n} \\ f_{x_1} & f_{x_1x_1} & \cdots & f_{x_1x_n} \\ \vdots & \vdots & \vdots & \vdots \\ f_{x_n} & f_{x_1x_n} & \cdots & f_{x_nx_n} \end{pmatrix}$$

and Δ_{ij} is the co-factor of the element $f_{x_ix_j}$ in the determinant Δ ($\Delta \neq 0$ is assumed), is said to be the *Allen elasticity of substitution* of the ith production factor with respect to the jth production factor, where $i, j \in \{1, \ldots, n\}, i \neq j$. Moreover, $A(f)$ is said to be the *Allen's matrix* of f and we call Δ the *Allen determinant* (see, e.g., [7]). This elasticity was introduced independently by Allen and Uzawa because the Hicks elasticity of substitution fails when the inputs x_i and x_j are non-separable with respect to the rest of inputs. We note that in case of two inputs, Hicks elasticity of substitution and Allen elasticity of substitution coincide. So $H_{ij} = A_{ij}$ and in this case the indicator is simply called the *elasticity of substitution* between the two factors of production. Recall that if

$$H_{ij}(x_1, \ldots, x_n) = \sigma \tag{3}$$

holds for all combinations of inputs $(x_1, \ldots, x_n) \in \mathbb{R}^n_+$ and all $i, j \in \{1, \ldots, n\}$, $i \neq j$, where σ is a non-zero real constant, then f is said to have *CES property*.

As it was pointed in [63], the classical treatment of the production functions makes use of the projections of production functions on a plane, but this approach leads to limited conclusions. Fortunately, this problem can be solved by identifying a production function f with the graph of f, i.e., the nonparametric hypersurface of the $(n+1)$-dimensional Euclidean space $\mathbb{E}^{n+1}$, defined by [63]

$$L(x_1, ..., x_n) = (x_1, ..., x_n, f(x_1, ..., x_n)), \quad (x_1, \ldots, x_n) \in \mathbb{R}^n_+, \tag{4}$$

and applying a differential geometric treatment. Therefore, using this approach, several basic properties of production models can be interpreted in terms of the geometry of their graphs [15, 60]. This hypersurface is called the *production hypersurface* [63]. Obviously, in the particular case of two inputs, we have a surface, called *production surface*, and an analysis of some basic production models from the point of view of space differential geometry has been realized in [29] and [66]. Notice that in the analysis of production functions, a fundamental role is played by isoquants. An *isoquant* represents the set of all possible combinations of inputs $(x_1, ..., x_n)$ that produce a specified level of output f^*. This is a basic concept in economic analysis, because isoquants assist firms to adjust the inputs in order to maximize outputs, and thus profits. If we have only two inputs, then the isoquants are curves on the production surface and in the general case, they are $(n-1)$-dimensional submanifolds of the corresponding hypersurface of the production function.

It is well-known that the study of the shape and the properties of the production possibility frontier is a subject of great interest in economic analysis. Some conditions under which the production possibility frontier is a surface lying in the edge of a cone, cylinder or plane were derived in [30]. Moreover, in [1], the authors investigate the condition for the production surface to contain a flat portion. In the following, we present the main results on the geometry of homogeneous and quasi-homogeneous production models, these being two of the most important functions used in the mathematical modeling of production.

2 Preliminaries and Basic Aspects

If M is a hypersurface of the Euclidean space $\mathbb{E}^{n+1}$, then it is known that the Gauss map $\nu : M \to S^n$ maps M to the unit hypersphere S^n of $\mathbb{E}^{n+1}$. With the help of the differential $d\nu$ of ν it can be defined a linear operator on the tangent space T_pM, denoted by S_p and known as the *shape operator*, by

$$g(S_p v, w) = g(d\nu(v), w), \quad v, w \in T_pM,$$

where g is the metric tensor on M induced from the Euclidean metric on $\mathbb{E}^{n+1}$. The determinant of the shape operator S_p, denoted by $K(p)$, is called the *Gauss-Kronecker curvature*. When $n = 2$, the Gauss-Kronecker curvature is simply called the Gauss curvature, which is intrinsic due to famous Gauss's *Theorema Egregium*. The trace of the shape operator S_p is called the *mean curvature* of the hypersurface. In contrast to the Gauss-Kronecker curvature, the mean curvature is extrinsic, depending on the immersion of the hypersurface. A hypersurface is said to be *minimal* if its mean curvature vanishes identically.

On another hand, the *Riemann curvature tensor* R of M is given by

$$R(u, v)w = \nabla_u \nabla_v w - \nabla_v \nabla_u w - \nabla_{[u,v]} w,$$

where ∇ is the Levi-Civita connection of g. It is well known that R measures non-commutativity of the covariant derivative, and as such is the integrability obstruction for the existence of an isometry with Euclidean space [18]. A Riemannian manifold is said to be *flat* if its Riemann curvature tensor vanishes identically.

Next we recall the following result that provides us simple formulas for the Gauss-Kronecker curvature, mean curvature, sectional curvature and Riemann curvature tensor of a graph hypersurface.

Lemma 1 [16] *Let L be a hypersurface of* $\mathbb{E}^{n+1}$ *defined by* (4) *and*

$$w = \sqrt{1 + \sum_{i=1}^{n} f_{x_i}^2}. \tag{5}$$

Then we have the following.

(i) *The Gauss-Kronecker curvature K is given by*

$$K = \frac{\det(f_{x_i x_j})}{w^{n+2}}. \tag{6}$$

(ii) *The mean curvature H is given by*

$$H = \frac{1}{n} \sum_{i=1}^{n} \frac{\partial}{\partial x_i}\Big(\frac{f_{x_i}}{w}\Big).$$

(iii) *The sectional curvature* K_{ij} *of the plane section spanned by* $\frac{\partial}{\partial x_i}, \frac{\partial}{\partial x_j}$ *is*

$$K_{ij} = \frac{f_{x_i x_i} f_{x_j x_j} - f_{x_i x_j}^2}{w^2(1 + f_{x_i}^2 + f_{x_j}^2)}.$$

(iv) *The Riemann curvature tensor R and the metric tensor g satisfy*

$$g\Big(R\Big(\frac{\partial}{\partial x_i}, \frac{\partial}{\partial x_j}\Big)\frac{\partial}{\partial x_k}, \frac{\partial}{\partial x_\ell}\Big) = \frac{f_{x_i x_\ell} f_{x_j x_k} - f_{x_i x_k} f_{x_j x_\ell}}{w^4}.$$

Remark 1 A smooth surface $\mathcal{S}$ of $\mathbb{E}^3$ with vanishing Gauss curvature is said to be *developable* [55]. This name is motivated by the fact that such a surface is known to be locally isometric to a planar domain [51]. As the Gauss curvature of $\mathcal{S}$ is $K = k_1 k_2$, where k_1 and k_2 denote the principal curvatures of $\mathcal{S}$, it follows that the points of a developable surface are either planar or parabolic. It is known that developable surfaces have been classified as the plane, the cone, the cylinder, or the tangent surface of a space curve, provided that the points of the surface are all planar or all are parabolic (see [36, 49, 58]). However, there exists many examples of developable surfaces containing points of both types. A first strong result regarding such developable surfaces was reported (without proof) only in 1956 by Pogorelov [44]. Therefore, if $\mathcal{S} \subset \mathbb{E}^3$ is a complete surface exhibiting vanishing Gauss curvature, then $\mathcal{S}$ is a generalized cylinder. Several proofs for this nice result can be found in [21, 33, 41, 50]. If the assumption of completeness is removed, then the situation is even more complicated (see, e.g., [41] and [65, p. 355]). However, some remarkable results can be stated even in this case in some particular setting. For example, considering developable graph surfaces, i.e., those surfaces defined as graphs of some functions $z = f(x, y)$, for $(x, y) \in D$, where $D \subset \mathbb{R}^2$ is an open subset, it follows from Lemma 1 that the developable surfaces are nothing but the solutions of the famous homogeneous Monge-Ampère equation (see [20, p. 10])

$$f_{xx} f_{yy} - f_{xy}^2 = 0. \tag{7}$$

Note that if $f_{xx} \neq 0$ everywhere, then Eq. (7) was completely solved by Ushakov in [56]. His result allows us to explicitly describe the developable surfaces in $\mathbb{E}^3$ having no planar points.

3 Classification Results for Homogeneous Production Models

There are some special classes of production functions that are often analyzed in both macroeconomics and microeconomics. The most important is given by the class of homogeneous production functions. A production function f defined on a set D for which $(\lambda x_1, \ldots, \lambda x_n) \in D$ whenever $\lambda > 0$ and $(x_1, \ldots, x_n) \in D$, is said to be *homogeneous of degree p* if there exists a real number p such that

$$f(\lambda x_1, \ldots, \lambda x_n) = \lambda^p f(x_1, \ldots, x_n)$$

for all $(x_1, \ldots, x_n) \in D$ and all $\lambda > 0$. This means that if the inputs are multiplied by same factor, then the output is multiplied by some power of this factor. If $p = 1$ then the function is said to have a *constant return to scale*, if $p > 1$ then we have an *increased return to scale* and if $p < 1$ then we say that the function has a *decreased return to scale*. We remark that both generalized Cobb-Douglas and CES production functions are homogeneous.

Remark 2 Suppose f is a generalized CD production function given by (1) and let L be the corresponding hypersurface defined by (4). Then from (1) it follows that for all $i, j \in \{1, ..., n\}, i \neq j$, we have

$$f_{x_i} = \tfrac{\alpha_i}{x_i} f(x_1, ..., x_n), \tag{8}$$

$$f_{x_i x_i} = \tfrac{\alpha_i(\alpha_i - 1)}{x_i^2} f(x_1, ..., x_n), \tag{9}$$

$$f_{x_i x_j} = \tfrac{\alpha_i \alpha_j}{x_i x_j} f(x_1, ..., x_n). \tag{10}$$

Using now (8) and (5), we obtain

$$w = \sqrt{1 + f^2(x_1, ..., x_n) \sum\nolimits_{i=1}^{n} \left(\frac{\alpha_i}{x_i}\right)^2}. \tag{11}$$

On the other hand, using (9) and (10), we derive after a direct computation that

$$\det(f_{x_ix_j}) = \frac{(-1)^n \left(1 - \sum_{i=1}^n \alpha_i\right) f^n(x_1, \ldots, x_n) \prod_{i=1}^n \alpha_i}{\prod_{i=1}^n x_i^2}. \tag{12}$$

Substituting now (11) and (12) in (8), we get

$$K(x_1, \ldots, x_n) = \frac{(-1)^n\left(1 - \sum_{i=1}^n \alpha_i\right) f^n(x_1, \ldots, x_n) \prod_{i=1}^n \alpha_i}{\left[1 + f^2(x_1, \ldots, x_n) \sum_{i=1}^n \left(\frac{\alpha_i}{x_i}\right)^2\right]^{\frac{n}{2}+1} \prod_{i=1}^n x_i^2}. \tag{13}$$

Now, we can easily conclude that a generalized Cobb-Douglas production function has constant return to scale if and only if

$$\varepsilon = \sum_{i=1}^n \alpha_i = 1$$

and from (13) we deduce the equivalence with the vanishing of Gauss-Kronecker curvature for the corresponding Cobb-Douglas production hypersurface. Hence, we have the next result, originally established in [63].

Theorem 1 *([63]) A Cobb-Douglas production hypersurface has vanishing Gauss-Kronecker curvature if and only if the corresponding generalized Cobb-Douglas production function has constant return to scale.*

Remark 3 In 1907, Tzitzéica discovered a new class of surfaces in $\mathbb{E}^3$ of great importance due to their nice applications in several areas of mathematics and physics (see e.g., [9, 23, 45, 52]). These surfaces, nowadays known under the name of Tzitzéica surfaces (see [2]), are characterized by a nice geometric property: the ratio of the Gauss curvature at a point x and the fourth power of the distance from a fixed point O to the tangent plane at x is constant [54]. The generalization of this concept to higher dimensions is called a Tzitzéica hypersurface. This is nothing but a hypersurface satisfying

$$K(x) = k \cdot d^{n+2}(x), \tag{14}$$

where $K(x)$ stands for the Gauss-Kronecker curvature at x, $d(x)$ is the distance between the origin and the hyperplane tangent to the hypersurface at x, while $k \in \mathbb{R}$ is a constant. The simplest Tzitzeica hypersurface has the equation

$$x_1 \cdot \ldots \cdot x_{n+1} = 1.$$

In the following, we are going to investigate if there is any connection between Tzitzéica hypersurfaces and Cobb-Douglas hypersurfaces.

Since the tangent hyperplane to a hypersurface defined through a map $L : U \to \mathbb{R}^{n+1}$, $L = L(x_1, \ldots, x_n)$, U being an open subset of $\mathbb{R}^n$, contains the vectors $\frac{\partial L}{\partial x_i}$, for all $i = 1, \ldots, n$, it is clear from (6) that the tangent hyperplane H to a Cobb-Douglas hypersurface at a point $x = (x_1, ..., x_n)$ has the equation:

$$(H): \, det \begin{bmatrix} X^1 - x_1 & X^2 - x_2 & \ldots & X^n - x_n & X^{n+1} - f(x_1, \ldots, x_n) \\ 1 & 0 & \ldots & 0 & \frac{\alpha_1}{x_1} f(x_1, \ldots, x_n) \\ 0 & 1 & \ldots & 0 & \frac{\alpha_2}{x_2} f(x_1, \ldots, x_n) \\ \vdots & \vdots & \vdots & \vdots & \vdots \\ 0 & 0 & \ldots & 1 & \frac{\alpha_n}{x_n} f(x_1, \ldots, x_n) \end{bmatrix} = 0,$$

from which we derive

$$\sum_{i=1}^{n} (X^i - x_i) \frac{\alpha_i f(x_1, \ldots, x_n)}{x_i} = X^{n+1} - f(x_1, \ldots, x_n).$$

Therefore, applying a classical formula in $(n+1)$-dimensional Euclidean geometry, we deduce that the distance between the origin $O = (0, \ldots, 0)$ and the hyperplane tangent to the hypersurface at $x = (x_1, \ldots, x_n)$ is

$$d(x) = \frac{\left|1 - \sum_{i=1}^{n} \alpha_i\right| f(x_1, \ldots, x_n)}{\left[1 + f^2(x_1, \ldots, x_n) \sum_{i=1}^{n} \left(\frac{\alpha_i}{x_i}\right)^2\right]^{\frac{1}{2}}}. \tag{15}$$

From (13), (14) and (15) we deduce that the equation

$$x_{n+1} = A \prod_{i=1}^{n} x_i^{\alpha_i},$$

where $A > 0$, defines a Tzitzéica hypersurface if and only if $\alpha_1 = ... = \alpha_n = -1$ or $\sum_{i=1}^{n} \alpha_i = 1$. Hence, we can identify two examples of Tzitzéica hypersurfaces. Firstly, we have obtained a new class of Tzitzéica hypersurfaces, given by

$$x_{n+1} = A \prod_{i=1}^{n} x_i^{\alpha_i},$$

where $\sum_{i=1}^{n} \alpha_i = 1$ and secondly, we can identify the hypersurface given by $x_{n+1} = \frac{A}{x_1 \cdot \ldots \cdot x_n}$, that was originally discovered by Tzitzéica for $n = 2$ and later generalized by Calabi [11]. Taking into account now that for a generalized Cobb-Douglas production function the values of $\alpha_1, \ldots, \alpha_n$ are positive constants (see e.g., [37]), in fact they being the output elasticities of the factor inputs determined by available technology (see also [12]), we remark that the case $\alpha_1 = ... = \alpha_n = -1$ has no economical meaning. Therefore we can state the following result.

Theorem 2 *[63] A Cobb-Douglas production hypersurface is a Tzitzéica hypersurface if and only if the corresponding generalized Cobb-Douglas production function has constant return to scale.*

Remark 4 In [59], Theorem 1 was generalized to the case of CES production functions with n-inputs. Suppose f is a generalized CES production function. Then from (2) we obtain for all $i, j \in \{1, ..., n\}, i \neq j$:

$$f_{x_i} = A\gamma\alpha_i^\rho x_i^{\rho-1}\left(\sum_{i=1}^n \alpha_i^\rho x_i^\rho\right)^{\frac{\gamma}{\rho}-1}, \tag{16}$$

$$f_{x_ix_i} = A\gamma\alpha_i^\rho x_i^{\rho-2}\left(\sum_{i=1}^n \alpha_i^\rho x_i^\rho\right)^{\frac{\gamma}{\rho}-2}\left((\rho-1)\sum_{k\neq i}\alpha_k^\rho x_k^\rho + (\gamma-1)\alpha_i^\rho x_i^\rho\right), \tag{17}$$

$$f_{x_ix_j} = A\gamma(\gamma-\rho)\alpha_i^\rho\alpha_j^\rho x_i^{\rho-1}x_j^{\rho-1}\left(\sum_{i=1}^n \alpha_i^\rho x_i^\rho\right)^{\frac{\gamma}{\rho}-2}. \tag{18}$$

Using now (5) and (16), we obtain

$$w = \sqrt{1 + A^2\gamma^2\left(\sum_{i=1}^n \alpha_i^\rho x_i^\rho\right)^{2\frac{\gamma}{\rho}-2}\left(\sum_{i=1}^n \alpha_i^{2\rho} x_i^{2\rho-2}\right)}. \tag{19}$$

On the other hand, using (17) and (18), we derive after some tedious computations that

$$\det(f_{x_ix_j}) = (-1)^n(\gamma-1)(\rho-1)^{n-1}A^n\gamma^n\left(\sum_{i=1}^n \alpha_i^\rho x_i^\rho\right)^{n\frac{\gamma}{\rho}-n}\prod_{i=1}^n \alpha_i^\rho x_i^{\rho-2}. \tag{20}$$

Substituting now (19) and (20) in (16), we get

$$K(x_1,\ldots,x_n) = \frac{(-1)^n(\gamma-1)(\rho-1)^{n-1}A^n\gamma^n\left(\sum_{i=1}^n \alpha_i^\rho x_i^\rho\right)^{n\frac{\gamma}{\rho}-n}\prod_{i=1}^n \alpha_i^\rho x_i^{\rho-2}}{\left[1 + A^2\gamma^2\left(\sum_{i=1}^n \alpha_i^\rho x_i^\rho\right)^{2\frac{\gamma}{\rho}-2}\sum_{i=1}^n \alpha_i^{2\rho} x_i^{2\rho-2}\right]^{\frac{n}{2}+1}}, \tag{21}$$

and now, it follows easily from (21) with similar arguments as in Theorem 1, that the generalized CES production function has constant return to scale if and only if $\gamma = 1$ and from (21) we deduce the equivalence with the vanishing of Gauss-Kronecker curvature. Hence, we have the following result stated in [59].

Theorem 3 *[59] A CES production hypersurface exhibits vanishing Gauss-Kronecker curvature if and only if the corresponding generalized CES production function has constant return to scale.*

Remark 5 From (13) and (21) it follows that we also have a correspondence between the concepts of decreasing/increasing return to scale for generalized Cobb-Douglas and CES production functions and the sign of the Gauss-Kronecker curvature for the corresponding production hypersurfaces, depending on the parity of the number of production factors. We would like to point out that in the original proofs of Theorems 1 and 3 given in [59, 63], the coefficient $(-1)^n$ in (13) and (21) was lost. Fortunately, this does not affect the validity of assertions in Theorems 1 and 3, as this coefficient disappears when the condition $K = 0$ is imposed, but it is important if we want to establish the correspondence between the concepts of decreasing/increasing return to scale for generalized Cobb-Douglas and CES production functions and the positivity/negativity of the Gauss-Kronecker curvature for the corresponding production hypersurfaces. Therefore, [63, Theorem 3.1. (ii)+(iii)] and [59, Theorem 3.1. (ii)] are valid only if the number of inputs is even. Thus, a generalized CD/CES production function has decreasing/increasing return to scale if and only if the corresponding CD production hypersurface has positive/negative Gauss-Kronecker curvature, provided that the number of inputs is even. On the other hand, if the number of inputs is odd, then a generalized CD/CES production function has increasing/decreasing return to scale if and only if the corresponding CD/CES production hypersurface has negative/positive Gauss-Kronecker curvature. Because similar results are also valid for all classical production functions used in economics, a natural question arises, namely if a general result of this type can be stated for all homogeneous production functions. A first answer was given by B.-Y. Chen in [13]: the decreasing/increasing return to scale property cannot be determined by the Gauss-Kronecker curvature of production hypersurface; even for two-factor homogeneous production functions, as follows from [13, Example 4.1]. Moreover, an interesting result on the homogeneous production functions defining flat hypersurfaces follows from the below theorem which generalizes a classical result in differential geometry concerning homogeneous functions of degree two due to F. Brickell [10].

Theorem 4 *[18] Let f be a twice differentiable, r-homogeneous, non-constant, real valued function of n variables $(x_1, \ldots, x_n)$ on an open domain $D \subset \mathbb{R}^n$, $n \geq 2$. Then the hypersurface of $\mathbb{E}^{n+1}$ defined by*

$$z = f(x_1, \ldots, x_n), \quad (x_1, \ldots, x_n) \in D,$$

is flat if and only if either f is linearly homogeneous, i.e., $r = 1$, or

$$f = (c_1 x_1 + c_2 x_2 + \cdots + c_n x_n)^r, \quad r \neq 1,$$

for some real constants $c_1, \ldots, c_n$.

An immediate consequence of Theorem 4 is the following.

Corollary 1 *[18] Let f be a r-homogeneous, differentiable of class C^2, non-constant, real valued function of two variables (x, y) defined on an open domain $D \subset \mathbb{R}^2$. Then the surface defined by $z = f(x, y)$, $(x, y) \in D$, is developable if and only if either f is linearly homogeneous, i.e., $r = 1$, or $f(x, y) = (c_1 x + c_2 y)^r$, $r \neq 1$, where c_1 and c_2 are real constants.*

Now, from Theorem 4 we obtain immediately the complete classification of homogeneous production functions with an arbitrary number of inputs whose production hypersurfaces are flat, as follows.

Theorem 5 *[18] A homogeneous production function with an arbitrary number of inputs defines a flat hypersurface if and only if either it has constant return to scale or it is a multinomial production function.*

In particular, we deduce the classification of homogeneous production functions with two inputs whose production surfaces are developable, i.e., with vanishing Gauss curvature.

Theorem 6 *[18] A homogeneous production function with two inputs defines a production surface with vanishing Gauss curvature if and only if either it has constant return to scale or it is a binomial production function.*

Remark 6 Theorem 4 is false if the flatness of the hypersurface were replaced by vanishing of Gauss-Kronecker curvature as it was shown in [18, Remark 4.1]. Indeed, if we consider the r-homogeneous function

$$f(x, y, z) = (x + y + \sqrt{yz})^r, \quad (x, y, z) \in \mathbb{R}^3_+, \quad r > 1, \tag{22}$$

then the hypersurface in $\mathbb{E}^4$ defined by (22) is non-flat, but it has vanishing Gauss-Kronecker curvature.

4 Classification Results for Quasi-Homogeneous Production Models

By weakening the property of homogeneity of production functions to quasi-homogeneity, one arrives at some more general production models, known as *quasi-homogeneous* (in short QH) models. It is worth mentioning that this broader property for production models was first proposed in [25] and the importance of QH models has been further highlighted in various works (see, e.g., [24, Sect. 6.2], [31, Chap. 12], as well as [26, 40, 47]). Recently, such models with 2 inputs have been investigating in [32], the authors deriving their analytical expression in case of unit elasticity of substitution. We recall that a production function $f : \mathbb{R}^n_+ \to \mathbb{R}_+$ is called a *quasi-homogeneous* production function having quasi-degree q and weight vector $g = (g_1, \ldots, g_n) \in \mathbb{R}^n$, where $g_1^2 + \cdots + g_n^2 \neq 0$, if it satisfies

$$f(\lambda^{g_1} x_1, \ldots, \lambda^{g_n} x_n) = \lambda^q f(x_1, \ldots, x_n) \tag{23}$$

for all points $(x_1, \ldots, x_n) \in \mathbb{R}^n_+$ and all $\lambda > 0$. It is obvious that in the particular case when the weight vector is $(1, \ldots, 1)$, a QH function having quasi-degree q reduces to a q-homogeneous function. However, it is clear that the class of QH functions is considerably larger than that of homogeneous functions.

We would like to emphasize now that the quasi-homogeneity property (23) mathematically models a precise economic situation encountered in a production process when a multiplication of the inputs with different powers of an identical factor leads to a multiplication of the output by a power of the same factor. This situation can occur when it is not possible to identically multiply all the factors of production due to the lack of one or more physical inputs. Recently, it was proved in [8] that the quasi-homogeneity assumption is more useful than the homogeneity assumption in analyzing the Japan's economy. This is mainly due to the fact that assuming the hypothesis of quasi-homogeneity instead of that of homogeneity, new production functions with CES property can be determined. Recall that according to a very important result in mathematical economics, there are only two homogeneous production models satisfying CES property, namely Cobb-Douglas and ACMS production functions (see, e.g., [14, 38]). This result was generalized in [5] to the setting of quasi-homogeneous production models, the authors proving that in this case there are five production functions exhibiting the CES property, as follows.

Theorem 7 *[5] Suppose that f is a twice differentiable quasi-homogeneous production function having quasi-degree $q \neq 0$ and weights $(g_1, \dots, g_n)$. Then:*

(i) *f exhibits unitary elasticity of substitution, that is f satisfies (3) for $\sigma = 1$, if and only if f reduces to a CD production function expressed as*

$$f(x_1, \dots, x_n) = A x_1^{\alpha_1} x_2^{\alpha_2} \cdot \dots \cdot x_n^{\alpha_n},$$

where A and $\alpha_i \neq 0$ are real constants such that $A > 0$, $\alpha_i \neq 0$, $i = 1, \dots, n$, and $\sum_{i=1}^{n} \alpha_i g_i = q$.

(ii) *If $n = 2$, then f satisfies CES property for a non-zero real constant $\sigma \neq 1$ if and only if one of the next situations occurs:*

a. *f reduces to a production model expressed by*

$$f(x_1, x_2) = \left(\frac{a_1^{\frac{\sigma-1}{\sigma}} x_1^{\frac{\sigma-1}{\sigma}}}{a_2^{\frac{\sigma-1}{\sigma}} x_2^{\frac{\sigma-1}{\sigma}} + 1} \right)^{\frac{\sigma q}{(\sigma-1)g_1}},$$

where a_1, a_2 are positive constants, provided that $g_2 = 0$.

b. *f reduces to a production model given by*

$$f(x_1, x_2) = \left(\frac{a_2^{\frac{\sigma-1}{\sigma}} x_2^{\frac{\sigma-1}{\sigma}}}{a_1^{\frac{\sigma-1}{\sigma}} x_1^{\frac{\sigma-1}{\sigma}} + 1} \right)^{\frac{\sigma q}{(\sigma-1)g_2}},$$

where a_1, a_2 are positive constants, provided that $g_1 = 0$.

c. *f reduces to a two-input ACMS production model expressed by*

$$f(x_1, x_2) = \left(a_1^{\frac{\sigma-1}{\sigma}} x_1^{\frac{\sigma-1}{\sigma}} + a_2^{\frac{\sigma-1}{\sigma}} x_2^{\frac{\sigma-1}{\sigma}} \right)^{\frac{\sigma q}{(\sigma-1)g_1}},$$

where a_1, a_2 are positive constants, provided that $g_1 = g_2$.

d. *f reduces to a production model expressed by*

$$f(x_1, x_2) = A x_2^{\frac{q}{g_2}} e^{V\left(\frac{x_1^{g_2}}{x_2^{g_1}}\right)},$$

where A is a positive constant and V is an antiderivative of the function v of variable $u = \frac{x_1^{g_2}}{x_2^{g_1}}$ *defined implicitly by the identity*

$$\left[1 - \frac{q}{g_2(g_1 - g_2)} \cdot \frac{1}{uv(u)}\right]^{g_1 - g_2} = Bu^{\frac{\sigma-1}{\sigma}}\left[1 - \frac{q}{g_1 g_2} \cdot \frac{1}{uv(u)}\right]^{g_1},$$

for some positive constant B, provided that $g_1 g_2 \neq 0$ *and* $g_1 \neq g_2$.

Note that the classification of quasi-homogeneous production functions having a constant elasticity with respect to a single production factor, with respect to all production factors, as well as with PMRS property, was established in [62].

Theorem 8 *[62] Let* $f : \mathbb{R}_+^n \to \mathbb{R}_+$ *be a twice differentiable production function of n inputs* $(x_1, x_2, \ldots, x_n)$, $n \geq 2$. *If f is a quasi-homogeneous production function of quasi-degree q with weight vector* $(g_1, \ldots, g_n)$, *then:*

(i) *The production has constant elasticity* k_i *with respect to an input* x_i *if and only if f reduces to the one of the following:*

(a) *a function of the form*

$$f(x_1, x_2, \ldots, x_n) = x_i^{\frac{q}{g_i}} F(x_1, \ldots, x_{i-1}, x_{i+1}, \ldots, x_n),$$

where F is a twice differentiable real valued function of $n-1$ *variables, provided that the weights are*

$$g_1 = 0, \ldots, g_{i-1} = 0, g_i = \frac{q}{k_i} \neq 0, g_{i+1} = 0, \ldots, g_n = 0;$$

or

(b) *a function of the form*

$$f(x_1, x_2, \ldots, x_n) = x_i^{k_i} x_j^{\frac{q-k_i g_i}{g_j}} F(u_1, \ldots, u_{n-2}),$$

where j is any element settled from the set $\{1, \ldots, n\} \setminus \{i\}$ *for which* $g_j \neq 0$ *and F is a twice differentiable real valued function of* $n-2$ *variables*

$$\{u_1, \ldots, u_{n-2}\} = \left\{\frac{x_k^{g_j}}{x_j^{g_k}} | k \in \{1, \ldots, n\} \setminus \{i, j\}\right\}.$$

(ii) *The production has constant elasticity* k_i *with respect to all inputs* x_i, $i \in \{1, 2, \ldots, n\}$, *if and only if*

$$g_1 k_1 + g_2 k_2 + \cdots + g_n k_n = q$$

and f reduces to the following CD production function:

$$f(x_1, x_2, \ldots, x_n) = Cx_1^{k_1} x_2^{k_2} \ldots x_n^{k_n},$$

where C is a positive real constant.

(iii) *f satisfies the PMRS property if and only f reduces to the one of the following:*

(a) *a quasi-homogeneous function of quasi-degree 0 defined by the next homothetic CD production function*

$$f(x_1, x_2, \ldots, x_n) = F\left(x_1^{g_j} x_2^{g_j} \ldots x_n^{g_j}\right),$$

where F is a twice differentiable real valued function of one variable and j is any element settled from the set $\{1, \ldots, n\}$ for which $g_j > 0$, provided that $\sum_{i=1}^{n} g_i = 0$; or

(b) *a quasi-homogeneous function of quasi-degree $q \neq 0$ defined by the next CD production function*

$$f(x_1, x_2, \ldots, x_n) = Cx_1^{\frac{q}{\sum_{i=1}^{n} g_i}} x_2^{\frac{q}{\sum_{i=1}^{n} g_i}} \ldots x_n^{\frac{q}{\sum_{i=1}^{n} g_i}},$$

where C is a positive real constant, provided that $\sum_{i=1}^{n} g_i \neq 0$.

Particularizing the above Theorem to the case of the main two inputs used in economics, namely the capital and the labor, we obtain the following.

Corollary 2 *[62] Let $f : \mathbb{R}_+^2 \to \mathbb{R}_+$ be a twice differentiable production function with two inputs (K, L), where K denotes the capital and L stands for labor. If f is a quasi-homogeneous production function of quasi-degree q with weight vector (g_K, g_L), then:*

(i) *The output elasticity with respect to capital is a constant k if and only if f reduces to the one of the following:*

(a) *a function having the form:*

$$f(K, L) = K^{\frac{q}{g_K}} F(L),$$

where F is a twice differentiable real valued function of one real variable, provided that the weights are

$$g_K = \frac{q}{k} \neq 0, \quad g_L = 0;$$

or

(b) *a CD production function given by:*

$$f(K, L) = CK^k L^{\frac{q-kg_K}{g_L}},$$

where C is a positive real constant, provided that $g_L \neq 0$.

(ii) *The output elasticity with respect to labor is a constant k if and only if f reduces to the one of the following:*

(a) *a function having the form:*

$$f(K, L) = L^{\frac{q}{g_L}} F(K),$$

where F is a twice differentiable real valued function of one real variable, provided that the weights are

$$g_K = 0, \quad g_L = \frac{q}{k} \neq 0;$$

or

(b) *a CD production function given by:*

$$f(K, L) = CK^{\frac{q-kg_L}{g_K}} L^k,$$

where C is a positive real constant, provided that $g_K \neq 0$.

(iii) *The output elasticities of labor and capital are both constant, k and l, respectively, if and only if*

$$g_K k + g_L l = q$$

and f reduces to the following CD production function:

$$f(K, L) = CK^k L^l,$$

where C is a positive real constant.

(iv) *f satisfies the PMRS property if and only if f reduces to the one of the following:*

(a) *a quasi-homogeneous function of quasi-degree 0 defined by the next homothetic CD production function*

$$f(K, L) = F\left(K^{g_K} L^{-g_L}\right),$$

where F is a twice differentiable real valued function of one real variable, provided that $g_K + g_L = 0$; or

(b) *a quasi-homogeneous function of quasi-degree $q \neq 0$ defined by the next CD production function*

$$f(K, L) = CK^{\frac{q}{g_K+g_L}} L^{\frac{q}{g_K+g_L}},$$

where C is a positive real constant, provided that $g_K + g_L \neq 0$.

In [17], it was obtained the next result regarding the classification of developable surfaces which are defined as graphs of quasi-homogeneous functions.

Theorem 9 *[17] Suppose $\Omega \subset \mathbb{R}^2$ is an open domain and let $f : \Omega \to \mathbb{R}$ be a non-constant weighted-homogeneous C^2 function of quasi-degree q and weights (a, b). Then the graph surface S defined by $z = f(x, y)$, $(x, y) \in \Omega$, is developable if and only if one of the following situations occur.*

(i) *f is a linear homogeneous function, that is $a = b = q$.*
(ii) *f is a $\frac{q}{a}$-homogeneous function of the form*

$$f(x, y) = (Ax + By)^{\frac{q}{a}},$$

for some real constants A and B, provided that $a = b \neq q$.
(iii) *f is given by*

$$f(x, y) = Cy^{\frac{q}{b}}(Dx - 1)^{1-\frac{q}{b}}, \tag{24}$$

for some non-zero constants C and D, provided that $a = 0$ and $q \neq b$.
(iv) *f is given by*

$$f(x, y) = Ex^{\frac{q}{a}}(Fy - 1)^{1-\frac{q}{a}}, \tag{25}$$

for some non-zero constants E and F, provided that $b = 0$ and $q \neq a$.
(v) *f has the form*

$$f(x, y) = y^{\frac{q}{b}} e^{V\left(\frac{x^b}{y^a}\right)}, \tag{26}$$

where V is a primitive of the real function v of one real variable $u = \frac{x^b}{y^a}$, with $v(u) = \frac{w(u)}{u}$ and w given by the implicit equation

$$\left[1 - \frac{q}{b(a-b)w}\right]^{q+a} \cdot \left[1 - \frac{q-b}{b(a-b)w}\right]^{b-q-a} = \left(\frac{G}{u}\right)^{\frac{a}{a-b}}, \tag{27}$$

for some non-zero real constant G, provided that $ab \neq 0$ and $a \neq b$.

Obviously, Theorem 9 generalizes Corollary 1. Moreover, translating Theorem 9 in language of production functions, we obtain the following classification result for developable quasi-homogeneous production functions, that generalizes Theorem 6.

Theorem 10 *[17] Let f be a quasi-homogeneous production function with two inputs x and y, having the quasi-degree q and weights (a, b). Then the production surface of f is developable if and only if one of the following cases occurs.*

(i) *f reduces to a constant return to scale production model, i.e., $a = b = q$.*
(ii) *$a = b \neq q$ and f reduces to a binomial production model with degree of homogeneity $\frac{q}{a}$.*
(iii) *$a = 0$, $q \neq b$ and f reduces to a two-input production model having the form* (24).
(iv) *$b = 0$, $q \neq a$ and f reduces to a two-input production model having the form* (25).
(v) *$ab \neq 0$, $a \neq b$ and f reduces to a two-input production model having the form* (26), *where V is a primitive of the real function v of real variable $u = \frac{x^b}{y^a}$, with $v(u) = \frac{w(u)}{u}$ and w given by the implicit Eq. (27).*

5 Further Developments

Next, we will list some open problems in the field.

Problem 1 Generalize Theorem 7(ii) for $n \geq 3$, i.e., classify quasi-homogeneous production models with at least 3 inputs satisfying the CES property for a constant $\sigma \neq 1$.

Problem 2 Generalize Theorem 9 for $n \geq 3$, i.e., classify quasi-homogeneous production models with at least 3 inputs, whose production hypersurfaces exhibit vanishing Gauss-Kronecker curvature.

Problem 3 Classify homothetic production functions (see [27]) with PMRS property.

Problem 4 Classify quasi-homogeneous production functions with constant Allen elasticity of substitution.

Problem 5 Classify quasi-sum and quasi-product production functions (see [3, 22, 39, 61]) with constant Allen elasticity of substitution.

Problem 6 Classify semi-homogenous production models (see [48]), whose corresponding production hypersurfaces are minimal or flat.

Problem 7 Classify ray-homothetic production models (see [26]) that have the PMRS property.

Problem 8 Classify ray-homothetic production models, whose corresponding production hypersurfaces are minimal or flat.

Problem 9 Classify almost ray-homothetic production models (see [4]), whose corresponding production hypersurfaces are minimal or flat.

Problem 10 Investigate the curvature properties of isoquants for homogeneous, quasi-homogeneous, homothetic, quasi-sum and quasi-product production models.

References

1. Abe, K., Okamoto, H., and Tawada, M.: A note on the production possibility frontier with pure public intermediate goods. Canadian J. Econ. 19(2), 351–356 (1986). https://doi.org/10.2307/135290
2. Agnew, A., Bobe, A., Boskoff, W., Suceavă, B.: Gheorghe Ţiţeica and the origins of affine differential geometry. Historia Mathematica 36, No. 2, 161–170 (2009). https://doi.org/10.1016/j.hm.2008.09.001
3. Alodan, H., Chen, B.-Y., Deshmukh, S., Vîlcu, G.-E.: On some geometric properties of quasi-product production models. J. Math. Anal. Appl. 474(1), 693–711 (2019). https://doi.org/10.1016/j.jmaa.2019.01.072
4. Al-Ayat, R., Färe, R.: Almost ray-homothetic production correspondences. Z. Nationalökonomie 39(1/2), 143–152 (1979) https://doi.org/10.1007/BF01289298
5. Alodan, H., Chen, B.-Y., Deshmukh, S., Vîlcu, G.-E.: Solution of the system of nonlinear PDEs characterizing CES property under quasi-homogeneity conditions. Adv. Differ. Equ. 2021, 257 (2021). https://doi.org/10.1186/s13662-021-03417-6
6. Arrow, K.J., Chenery, H.B., Minhas, B.S., Solow, R.M.: Capital-labor substitution and economic efficiency. Rev. Econ. Stat. 43(3), 225–250 (1961). https://doi.org/10.2307/1927286
7. Aydin, M.E., Ergüt, M.: Composite functions with Allen determinants and their applications to production models in economics. Tamkang J. Math. 45(4), 427–435 (2014). https://doi.org/10.5556/j.tkjm.45.2014.1569
8. Bayanjargal, D., Yerkyebulan, B., Battsukh, T.: A new class of production function. Theor. Econ. Lett. 10, 356–365 (2020). https://doi.org/10.4236/tel.2020.102024
9. Borisov, A.B., Zykov, S.A. , Pavlov, M.V.: Tzitzéica equation and proliferation of nonlinear integrable equations. Theor. Math. Phys. 131, No. 1, 550–557 (2002). https://doi.org/10.1023/A:1015162021255
10. Brickell, F.: A theorem on homogeneous functions. J. London Math. Soc. 42, 325–329 (1967). https://doi.org/10.1112/jlms/s1-42.1.325
11. Calabi, E.: Complete affine hyperspheres, I. Sympos. Math. 10, 19–38 (1972)
12. Charnes, A., Cooper, W.W., Schinnar, A.P.: A theorem on homogeneous functions and extended Cobb-Douglas forms. Proc. Natl. Acad. Sci. USA 73, No. 10, 3747–3748 (1976). https://doi.org/10.1073/pnas.73.10.3747
13. Chen, B.-Y.: On some geometric properties of h-homogeneous production functions in microeconomics. Kragujevac J. Math. 35, 343–357 (2011)
14. Chen, B.-Y.: Classification of h-homogeneous production functions with constant elasticity of substitution. Tamkang J. Math. 43(2), 321–328 (2012). https://doi.org/10.5556/j.tkjm.43.2012.1145
15. Chen, B.-Y.: Solutions to homogeneous Monge-Ampère equations of homothetic functions and their applications to production models in economics. J. Math. Anal. Appl. 411, 223–229 (2014). https://doi.org/10.1016/j.jmaa.2013.09.029
16. Chen, B.-Y.: On some geometric properties of quasi-sum production models. J. Math. Anal. Appl. 392(2), 192–199 (2012). https://doi.org/10.1016/j.jmaa.2012.03.011
17. Chen, B.-Y., Vîlcu, A.-D., Vîlcu, G.-E.: Classification of graph surfaces induced by weighted-homogeneous functions exhibiting vanishing Gaussian curvature. Mediterr. J. Math. 19, 162 (2022). https://doi.org/10.1007/s00009-022-02106-2
18. Chen, B.-Y., Vîlcu, G.-E.: Geometric classifications of homogeneous production functions. Appl. Math. Comput. 225, 345–351 (2013). https://doi.org/10.1016/j.amc.2013.09.052
19. Cobb C.W., Douglas, P.H.: A theory of production. Am. Econ. Rev. 18, 139–165 (1928). https://www.jstor.org/stable/1811556
20. Courant, R., Hilbert, D.: Methods of Mathematical Physics, Vol. II: Partial Differential Equations. Interscience Publishers, New York–London (1962)
21. do Carmo, M.: Differential geometry of curves and surfaces. Prentice-Hall, Inc., Englewood Cliffs, N.J. (1976)

22. Du, Y., Fu, Y., Wang, X.: On the minimality of quasi-sum production models in microeconomics. Math. Met. Appl. Sc. 46(12), 7607–7630 (2022) https://doi.org/10.1002/mma.8265
23. Dunajski, M., Plansangkate, P.: Strominger-Yau-Zaslow geometry, affine spheres and Painlevé. III. Commun. Math. Phys. 290, No. 3, 997–1024 (2009). https://doi.org/10.1007/s00220-009-0861-x
24. Eichhorn, W.: Theorie Der Homogenen Produktionsfunktion. Springer-Verlag: Berlin-Heidelberg-New York (1970)
25. Eichhorn, W., Oettli, W.: Mehrproduktunternehmungen mit linearen expansionswegen. Oper. Res. Verfahren 6, 101–117 (1969)
26. Färe, R., Shephard, R.W.: Ray-homothetic production functions. Econometrica 45, 133–146 (1977). https://doi.org/10.2307/1913291
27. Försund, F.R.: The homothetic production function. Swedish J. Econ. 77(2), 234–244 1975). https://doi.org/10.2307/3438919
28. Fuss, M., McFadden, D., Mundlak, Y.: A survey of functional forms in the economic analysis of production. In: Fuss, M. and McFadden, D., editors, Production Economics: A Dual Approach to Theory and Applications, volume 1, vol. 1, McMaster University Archive for the History of Economic Thought (1978) https://econpapers.repec.org/bookchap/hayhetcha/fuss1978-4.htm
29. Ioan, C.A., Ioan, G.: A generalization of a class of production functions. Appl. Econ. Letters 18(18), 1777–1784 (2011). https://doi.org/10.1080/13504851.2011.564117
30. Inoue, T.: On the shape of the production possibility frontier with more commodities than primary factors. Int. Econ. Rev. 25(2), 409–424 (1984). https://doi.org/10.2307/2526206
31. Jensen, B.: The Dynamic Systems of Basic Economic Growth Models. Mathematics and Its Applications, Springer: Dordrecht, Netherlands (1994)
32. Khatskevich, G.A., Pranevich, A.F.: On quasi-homogeneous production functions with constant elasticity of factors substitution. J. Belarus. State Univ. Econ. 1, 46–50 (2017). https://elibrary.ru/item.asp?id=36293496
33. Hartman, P., Nirenberg, L.: On spherical image maps whose Jacobians do not change sign. Amer. J. Math. 81, 901–920 (1959). https://doi.org/10.2307/2372995
34. Humphrey, T.M.: Algebraic production functions and their uses before Cobb-Douglas. Federal Reserve Bank of Richmond Economic Quarterly 83(1), 51–83 (1997). https://www.richmondfed.org/publications/research/economic_quarterly/1997/winter/humphrey
35. McFadden, D.: Constant elasticity of substitution production functions. Rev. Econ. Stud. 30(2), 73–83 (1963). https://EconPapers.repec.org/RePEc:oup:restud:v:30:y:1963:i:2:p:73-83
36. Klingenberg, W.: A course in differential geometry. Graduate Texts in Mathematics, Vol. 51, Springer-Verlag, New York-Heidelberg (1978)
37. Liu, S.-T.: A geometric programming approach to profit maximization. Appl. Math. Comput. 182, No. 2, 1093–1097 (2006). https://doi.org/10.1016/j.amc.2006.04.061
38. Losonczi, L.: Production functions having the CES property, Acta Math. Acad. Paedagog. Nyházi. (N.S.) 26(1), 113–125 (2010). https://eudml.org/doc/231704
39. Luo, Y., Wang, X.: Some extrinsic geometric characterizations of quasi-product production functions in microeconomics. J. Math. Anal. Appl. (2023), in press. https://doi.org/10.1016/j.jmaa.2023.127675
40. Mak, K.-T.: General homothetic production correspondences. In: Applications of Modern Production Theory: Efficiency and Productivity. Dogramaci A., Färe R., Eds.; Springer: Dordrecht, Netherlands (1988)
41. Massey, W.: Surfaces of Gaussian curvature zero in Euclidean 3-space. Tôhoku Math. J. (2) 14, 73–79 (1962). https://doi.org/10.2748/tmj/1178244205
42. Mishra, S.K.: A brief history of production functions. The IUP Journal of Managerial Economics 8(4), 6–34 (2010). https://ideas.repec.org/a/icf/icfjme/v08y2010i3p6-34.html
43. Murthy, K.V.B.: Arguing a case for the Cobb-Douglas production function. Rev. Commer. Stud. 20–21(1), 75–94 (2002). https://ideas.repec.org/p/wpa/wuwpem/0409012.html
44. Pogorelov, A.V.: Extensions of the theorem of Gauss on spherical representation to the case of surfaces of bounded extrinsic curvature. Dokl. Akad. Nauk. SSSR (N.S.) 111, 945–947 (1956) (Russian)

45. Schief, W.K.: Isothermic surfaces in spaces of arbitrary dimension: integrability, discretization, and Bäcklund transformations–a discrete Calapso equation. Stud. Appl. Math. 106, No. 1, 85–137 (2001). https://doi.org/10.1111/1467-9590.00162
46. Shepard, R.W.: Theory of cost and production functions. Princeton University Press (1970)
47. Shepard, R.W.: Some remarks on the theory of homogeneous production functions. Z. Nationalökonomie 31, 251–256 (1971) https://doi.org/10.1007/BF01287235
48. Shephard, R.W.: Semi-homogeneous production functions and scaling of production. In Production Theory; Eichhorn, W., Henn, R., Eds.; Springer: New York, NY, USA; pp. 253–285 (1974)
49. Spivak, M.: A comprehensive introduction to differential geometry, vol. III, Publish or Perish, Inc., Berkeley (1979)
50. Stoker, J.J.: Developable surfaces in the large. Comm. Pure Appl. Math. 14, 627–635 (1961). https://doi.org/10.1002/cpa.3160140333
51. Stoker, J.J.: Differential geometry. Wiley-Interscience, New York (1969)
52. Tamizhmani, K.M., Satsuma, J., Grammaticos, B., Ramani, A.: Nonlinear integrodifferential equations as discrete systems. Inverse Probl. 15, No. 3, 787–791 (1999). https://doi.org/10.1088/0266-5611/15/3/310
53. Thompson, A.: Economics of the firm: theory and practice. Prentice Hall (1985)
54. Tzitzéica, G.: Sur une nouvelle classe de surfaces. C. R. Acad. Sci. Paris 144, 1257–1259 (1907)
55. Toponogov, V.A.: Differential geometry of curves and surfaces, A concise guide. With the editorial assistance of Vladimir Y. Rovenski. Birkhäuser, Boston, Inc., Boston, MA (2006)
56. Ushakov, V.: The explicit general solution of trivial Monge-Ampère equation. Comment. Math. Helv. 75(1), 125–133 (2000). https://doi.org/10.1007/s000140050115
57. Uzawa, H.: Production functions with constant elasticities of substitution. Rev. Econ. Stud. 29(4), 291–299 (1962). https://EconPapers.repec.org/RePEc:oup:restud:v:29:y:1962:i:4:p:291-299
58. Vaisman, I.: A first course in differential geometry. Monographs and Textbooks in Pure and Applied Mathematics, 80, Dekker, New York (1984)
59. Vîlcu, A.-D., Vîlcu, G.-E.: On some geometric properties of the generalized CES production functions. Appl. Math. Comput. 218(1), 124–129 (2011). https://doi.org/10.1016/j.amc.2011.05.061
60. Vîlcu, A.-D., Vîlcu, G.-E.: A survey on the geometry of production models in economics. Arab J. Math. Sc. 23(1), 18–31 (2017). https://doi.org/10.1016/j.ajmsc.2016.08.003
61. Vîlcu, A.-D., Vîlcu, G.-E.: Some characterizations of the quasi-sum production models with proportional marginal rate of substitution. C. R. Math. Acad. Sci. Paris 353, 1129–1133 (2015). https://doi.org/10.1016/j.crma.2015.09.019
62. Vîlcu, A.-D., Vîlcu, G.-E.: On quasi-homogeneous production functions. Symmetry 11(8), 976 (2019). https://doi.org/10.3390/sym11080976
63. Vîlcu, G.-E.: A geometric perspective on the generalized Cobb-Douglas production functions. Appl. Math. Lett. 24(5), 777–783 (2011)
64. Wicksteed, P.H: An Essay on the Coordination of the Laws of Distribution. Macmillan & Co. (1894)
65. Wintner, A.: On Frenet's equations. Amer. J. Math. 78, 349–356 (1956). https://doi.org/10.2307/2372520
66. Zakhirov, M.: Econometric and geometric analysis of Cobb-Douglas and CES production functionss. ROMAI Journal, 1:237–242 (2005). https://rj.romai.ro/arhiva/2005/caim05v1nr1.pdf

A Lagrangian Program Detecting the Weighted Fermat-Steiner-Fréchet Multitree for a Fréchet N-multisimplex in Euclidean N-space

Anastasios N. Zachos

Dedicated to Vladimir Rovenski on his 70th birthday.

Abstract We introduce the Fermat-Steiner-Fréchet (FSFR) problem for a given $\frac{1}{2}N(N+1)$-tuple of positive real numbers determining the edge lengths of an N-simplex in $\mathbb{R}^N$ in order to study its solution called the "FSFR multitree", which consist of a union of Fermat-Steiner (FS) trees for all derived pairwise incongruent N-simplexes in the sense of Blumenthal, Herzog for $N=3$ and Dekster-Wilker for $N \geq 3$. We obtain a method to determine the FSFR multitree in $\mathbb{R}^N$ based on the theory of Lagrange multipliers, whose equality constraints depend on $N-1$ independent solutions of the inverse weighted Fermat problem for an N-simplex in $\mathbb{R}^N$. A fundamental application of the Lagrangian program for the FSFR problem in $\mathbb{R}^N$ is the detection of the FS tree with global minimum length having $N-1$ equally weighted FS points among $\frac{[\frac{1}{2}N(N+1)]!}{(N+1)!}$ incongruent N-simplexes determined by an $\frac{1}{2}N(N+1)$-tuple of consecutive natural numbers controlled by Dekster-Wilker, Blumenthal-Herzog conditions and enriched with the fundamental evolutionary processes of Nature (Minimum communication networks, minimum mass transfer, maximum volume of incongruent simplexes). Furthermore, we obtain the unique solution of the inverse weighted Fermat problem, referring to the unique set of $(N+1)$ weights, which correspond to the vertices of an N-boundary simplex in $\mathbb{R}^N$. Additionally, we give a negative answer for an intermediate weighted Fermat-Steiner-Fréchet multitree having one node (weighted Fermat point) for m boundary closed polytopes ($m \geq N+2$), which is determined by m prescribed rays meeting at a fixed weighted Fermat point, by deriving a linear dependence for the m variable weights (Plasticity of an Intermediate FSFR multitree for m boundary closed polytopes). By entering in the plasticity of an intermediate FSFR multitree for m

A. N. Zachos (✉)
Department of Mathematics, University of Patras, GR-26500 Rion, Greece
e-mail: azachos@gmail.com

© The Author(s), under exclusive license to Springer Nature Switzerland AG 2024

V. Rovenski et al. (eds.), *Differential Geometric Structures and Applications*, Springer Proceedings in Mathematics & Statistics 440,
https://doi.org/10.1007/978-3-031-50586-7_16

boundary closed polytopes a two-way mass transport from k vertices to the unique weighted Fermat point and from this point to the $m-k$ vertices, and reversely, we derive the equations of "mutation" of intermediate FSFR multitrees in $\mathbb{R}^N$.

Keywords Lagrange mutlipliers · Convex program · Fermat-Steiner tree · Fermat-Steiner-Fréchet problem · Fermat-Steiner-Fréchet multitree · Fermat-Steiner minimal tree · Incongruent simplexes · Plasticity of multitrees

1 Introduction

The problem considered in the present paper is a result of synthesis of two problems: the weighted Fermat-Steiner (WFS) problem on the shortest networks for boundary N-simplexes in $\mathbb{R}^N$ and Fréchet's problem on seeking the necessary and sufficient conditions for a given $\frac{1}{2}N(N+1)$-tuple of positive real numbers determining the edge lengths of an N-simplex in $\mathbb{R}^N$ for $N \geq 3$. The WFS problem for a boundary N-simplex is a problem on finding a shortest network of total weighted length connecting $N+1$ non-collinear and non-coplanar points possessing each of them a positive real number (weight) in $\mathbb{R}^N$. In [29], we found the solution of the WFS problem for a boundary tetrahedron ($N=3$). The solution is a tree T having two WFS points (nodes) inside the tetrahedron. Each node is a weighted Fermat-Steiner (WFS point) or Torricelli point. We call T a WFS tree, which is the union of one line segment between the two WFS points and of four line segments joining each WFS point with two fixed neighboring vertices. In [21], Rubinstein, Thomas and Weng solved the unweighted FS problem for tetrahedra in $\mathbb{R}^3$. In [18], Jarnik and Kossler introduced a generalization of the Fermat problem on optimal networks of a finite set of points in $\mathbb{R}^2$. In [4], Courant and Robbins restated and named the problem as the Steiner problem. In [15, 17], Ivanov and Tuzhilin studied properties of weighted minimal binary trees, which is an important generalization of WFS trees in $\mathbb{R}^2$. The Fréchet (FR) problem in $\mathbb{R}^3$ states that [8]: For a given sextuple of positive real numbers $a_{ij} = a_{ji}$ for $1 \leq i \neq j \leq 4$ what are the necessary and sufficient conditions that them be the edge lengths of the tetrahedron $A_1A_2A_3A_4$ in $\mathbb{R}^3$?

In [1, Theorem 2.1], Blumenthal proved that a sextuple of positive real numbers $(a+nd)^{1/2}$, $n = 0, 1, 2, 3, 4, 5$, $0 < 4d \leq a$, forms a completely tetrahedral sextuple, which yields thirty incongruent tetrahedra in $\mathbb{R}^3$. In [9, Theorem 3], Herzog proved a theorem characterizing complete tetrahedral sextuples, which states that: Six positive real numbers $\{a, b, c, d, e, f\}$ satisfying the conditions $a > b > c > d > e > f > 0$ and $e + f \geq a$ form a complete tetrahedral sextuple if and only if the Caley-Menger determinant $D(a, b; c, f; e) \geq 0$ (the edges separated by semicolons are pairs of opposite edges of the tetraedron) that is if and only if the tetrahedron in which the faces are formed by the triples (a, c, d), (a, e, f), (b, c, e) and (b, d, f) is realizable in $\mathbb{R}^3$. In [9, Remarks 4, 5, 6], Herzog gave the following three important applications of his theorem:

(i) Six consecutive positive integers $x+5, x+4, x+3, x+2, x+1, x$ form a complete tetrahedral sextuple if and only if x is greater than or equal to the positive root ρ of the equation $D(x+5, x+4; x+3, x; x+2, x+1) = 0$: $6.09 < \rho < 6.10$ (Herzog sextuple of consecutive positive integers).
(ii) The sextuple of positive numbers $\{1, 1-\delta; 1-2\delta, \epsilon; 1-3\delta, 1-4\delta\}$ for $\delta = \delta(\epsilon) > 0$ and $0 < \epsilon < 1$ such that $1-4\delta > \epsilon$ forms a complete tetrahedral sextuple.
(iii) Blumenthal's sextuple $\{(a+nd)^{1/2}\}$, $n = 0, 1, 2, 3, 4, 5$, is a complete tetrahedral sextuple if and only if $\frac{a}{d}$ is greater than or equal to a cubic irrationality r : $1.91 < r < 1.92$.

In [5], Dekster and Wilker derived the condition $\frac{x}{x+5} \geq \lambda(3) \equiv \frac{1}{\sqrt{2}}$, which shows that $x \geq 13$ and leaves six runs from Herzog result sextuple of consecutive integers that works for $x \geq 7$.

We introduce the Fermat-Steiner-Fréchet (FSFR) problem for a given $\frac{1}{2}N(N+1)$-tuple of positive real numbers determining N-simplexes in $\mathbb{R}^N$.

Problem 1 (The weighted Fermat-Steiner-Fréchet–WFSFR–problem in $\mathbb{R}^N$) Given a $2N$-tuple of positive real numbers (weights) such that $N-1$ of them are equal and a $\frac{1}{2}N(N+1)$ of positive numbers determining the edge lengths of N-simplexes in $\mathbb{R}^N$, find the shortest networks (weighted Fermat-Steiner trees) of total weighted length for all derived incongruent N-simplexes in $\mathbb{R}^N$.

We call a WFSFR multitree the solution of the WFSFR problem in $\mathbb{R}^N$, which is the union of corresponding weighted Fermat-Steiner (WFS) trees for all derived incongruent N-simplexes in $\mathbb{R}^N$, and a Fréchet N-multisimplex the class of incongruent N-simplexes in $\mathbb{R}^N$ derived by the same given $\frac{1}{2}N(N+1)$-tuple of positive real numbers.

In [27], we solve the weighted Fermat-Fréchet (WFFR) problem for $N = 3$ in $\mathbb{R}^3$ and we find the condition to locate the corresponding weighted Fermat (WF) trees, by applying a generalization of the cosine law in $\mathbb{R}^3$, which depend only on edge lengths and the Caley-Menger determinant representing tetrahedral volume with respect to edge lengths.

In this paper, we obtain a Lagrangian program to detect a WFSFR multitree for a boundary N-Fréchet multisimplex in $\mathbb{R}^N$, which gives:

(i) The plasticity of a WFSFR multitree having one node (intermediate WFSFR multitree) for boundary m-closed polytopes in $\mathbb{R}^N$.
(ii) A characterization of the most natural of natural numbers obtained by a $\frac{1}{2}N(N+1)$-tuple of consecutive natural numbers.
(iii) The Bessel (random) plasticity of WFSFR multitree having one node (intermediate WFSFR multitrees) for boundary m-closed polytopes in $\mathbb{R}^N$.
(iv) An ϵ-approximation method for the weights that corresponds to the vertices of a Fréchet N-multisimplex in $\mathbb{R}^N$.

In Sect. 2, we mention some fundamental known results concerning the existence and uniqueness of solution for the WFS problem and the WF problem in $\mathbb{R}^N$ and the uniqueness of solution of the inverse weighted Fermat (INVWF) problem for a boundary tetrahedron in $\mathbb{R}^3$. In Sect. 3, we prove the uniqueness of the solution of the INVWF problem for N-simplexes in $\mathbb{R}^N$ and the plasticity solutions of a WFFR multitree with one node (WF point) for m-boundary closed polytopes in $\mathbb{R}^N$. In Sect. 4, we apply the uniqueness of the INVWF problem for N-simplexes in $\mathbb{R}^N$ to a Lagrangian program, in order to detect the WFSFR multitree for an N-Fréchet multisimplex for $N = 4$ and $N > 4$, respectively. In Sect. 5, we extend the Lagrangian program for intermediate weighted Fermat-Steiner-Fréchet (I.WFSFR) multitrees for an N-Fréchet multisimplex in $\mathbb{R}^N$, whose # of nodes is less than $N - 1$. In Sect. 6, we give the "mutation" equations of an I.WFSFR multitree having one node for m-boundary closed polytopes in $\mathbb{R}^N$, which is a generalization of the plasticity solutions given in Sect. 3. In Sect. 7, we construct the weights of a Fréchet N-multisimplex in $\mathbb{R}^N$, by introducing a method of ϵ-approximation of variable weighted multitrees in $\mathbb{R}^N$. In Sect. 8, we introduce the Bessel plasticity equations of an I.WFSFR multitree having one node for m-boundary closed polytopes in $\mathbb{R}^N$ such that one of the weights follows a Bessel motion.

2 Prelimiminaries: The Weighted Fermat-Steiner Problem and the Inverse Weighted Fermat Problem in $\mathbb{R}^N$

In this section, we state the weighted Fermat-Steiner (WFS) problem for an N-simplex in $\mathbb{R}^N$, which is a generalization of the weighted Fermat (WF) problem in $\mathbb{R}^N$. We mention some fundamental results established by Ivanov-Tuzhilin concerning the existence and uniqueness of the FS tree solutions and the characterization of solutions with respect to the weighted Fermat (WF) problem given by Sturm for the unweighted case and extended by Kupitz-Martini for the weighted case. Finally, we state the INVWF problem for m points in $\mathbb{R}^N$.

Denote by $A_{0,1}, A_{0,2}, \ldots A_{0,N-1}$ $N-1$ points inside the N-simplex $A_1 A_2 \ldots A_{N+1}$ in $\mathbb{R}^N$, by b_i a positive real number(weight), which corresponds to each vertex A_i, by $a_{(0,i),(0,j)}$ the length of the line segment $A_{0,i}A_{0,j}$ and by a_{k0} the length of the line segment $A_k A_0$, for $i, j = 1, 2, \ldots, N-1$ and $k = 1, 2, \ldots, N+1$.

Problem 2 (The WFS problem for $A_1 A_2 \ldots A_{N+1}$ in $\mathbb{R}^N$) Find $A_{0,i}$ in $\mathbb{R}^N$ for $i = 1, 2 \ldots, N-1$ with equal given weight B_{ST} such that

$$f(\{A_{0,i}\}) = \sum_{i=1,\ j_i\in\{1,2,\ldots,N-1\}}^{N+1} b_i\, a_{i,(0,j_i)} + b_{ST} \sum_{i=1,\ j_i\in\{1,2,\ldots,N-1\}}^{N-1} a_{(0,i),(0,j_i)} \to \min . \quad (1)$$

In [13, Corollary 1.1, Proposition 1.2], [14, Theorem 1.1, Corollary 1.2, Chap. 3], [15, 17], Ivanov and Tuzhilin studied the WFS problem and many variations starting from an embedding of a Steiner topological graph (one dimensional cell complex) on a manifold W. The following definitions of an immersed parametric network are given in [14, Definitions, p. 56, p. 58]: A parametric network Γ of type G in the manifold W is an arbitrary mapping $\phi : G \to W$. A network $\Gamma = \{\phi : G \to W\}$ is called piecewise smooth if each of its edges is a piecewise curve. A piecewise smooth curved is called immersed if all its edges are nondegenerate. We note that a description of weighted minimum networks in $\mathbb{R}^N$ is given in [16, Sect. 3]. Local minimal networks of type G can be realized as weighted minimal networks type G with constant weight function (positive real number) defined on the edge set of G.

Lemma 1 (Weighted Local Minimality criterion, Corollary 1.2, Chap. 3 in [14]; [15]) *Let Γ be an immersed weighted parametric network (Weighted Fermat-Steiner network). The network is local minimal if and only if all the edges of Γ are geodesic segments for any mobile vertex (Fermat-Steiner point) $A_{0,i}$ of Γ the linear combination of weighted unit vectors of the directions of edges of Γ going out of $A_{0,i}$ (with coefficients equal to the weights of these edges) vanishes.*

We mention an important theorem of Ivanov-Tuzhilin ([14]), which deals with the uniqueness for networks with boundaries in $\mathbb{R}^N$. A strong local minimal network is a parametric network Γ, if for any point x of the parametric graph of the reduced network (without degenerate edges) corresponding to Γ, there exists a strong local network Γ^x_{sloc}, such that any small deformation of the network Γ^x_{sloc}, that preserves its boundary does not decrease the length of Γ^x_{sloc} (see in [14, Definitions, p. 91]).

Lemma 2 (Uniqueness theorem of Ivanov-Tuzhilin for networks with boundaries in $\mathbb{R}^N$, see Theorem 3.1 in Chap. 2; Proof of Theorem 3.1 in Chap. 3 in [14]) *Let $M \subset \mathbb{R}^N$ be an arbitrary non-empty finite set of points from $\mathbb{R}^N$; denote by G an acyclic topological graph and let β be a boundary mapping that maps a subset of vertices of graph G onto the set M. Assume that among strong minimal networks in class $\mathcal{M}_G(\beta)$ there exists an embedded network Γ which does not possess mobile vertices $A_{0,i}$ of degree two. Then all other strong minimal networks in this class coincide with Γ (up to a parameterization); that is Γ is unique.*

By substituting $A_0 \equiv A_{0,i}$, for $i = 1, 2, \ldots N-1$ in (1), we derive the WF problem $A_1 A_2 \ldots A_{N+1}$ in $\mathbb{R}^N$.

Problem 3 (The WF problem for $A_1 A_2 ... A_{N+1}$ in $\mathbb{R}^N$) Find A_0 in $\mathbb{R}^N$ such that

$$f(\{A_{0,i}\}) = \sum_{i=1}^{N+1} b_i a_{0i} \to \min .$$

In [24], Sturm gave a complete characterization of the solutions of the unweighted Fermat problem for m given points in $\mathbb{R}^N$, and Kupitz and Martini extended this characterization in the weighted case, in [2, Theorem 18.37, pp. 250].

Theorem 1 (see [24] and Theorem 18.37, pp. 250 in [2]).

(I) The weighted Fermat point A_0 with respect exists and is unique.
(II) If for each point $A_i \in \{A_1, A_2, \ldots, A_m\}$

$$\|\sum_{j=1, j\neq i}^{m} b_j \vec{u}(A_j, A_i)\| > b_i,$$

for $i, j = 1, 2, \ldots, m$, *then*

(a) the weighted Fermat point A_0 *does not belong to* $\{A_1, A_2, \ldots, A_m\}$,

(b) weighted floating Fermat tree solution:

$$\sum_{i=1}^{m} b_i \vec{u}(A_0, A_i) = 0. \tag{2}$$

(III) if there is a point $A_i \in \{A_1, A_2, \ldots, A_m\}$ *satisfying*

$$\|\sum_{j=1, j\neq i}^{m} b_j \vec{u}(A_j, A_i)\| \leq b_i \quad (i, j = 1, 2, \ldots, m),$$

then the weighted Fermat point $A_0 \equiv A_i$ *(weighted absorbed Fermat tree solution).*

The $(m-1)$-INVWF for m points (non-collinear and non-coplanar) in $\mathbb{R}^N$ states that:

Problem 4 (The $(m-1)$-INVWF problem in $\mathbb{R}^N$) Given $A_0 \notin \{A_1, A_2, \ldots, A_m\}$ and $A_0, A_i \in \mathbb{R}^N$, does there exist a unique set of positive weights b_{i0} such that

$$b_{10} + b_{20} + \ldots + b_{m0} = c = const, \tag{3}$$

for which A_0

$$f(A_0) = \sum_{i=1}^{m} b_{i0} a_{i0} \to \min?$$

The solution of the 2-INVWF problem in $\mathbb{R}^2$ ($b_4 = b_5 = \ldots = b_m = 0$) is given in [11].

Proposition 1 *Solution of the 2-INVWF problem in* $\mathbb{R}^2$ *[11] is the following:*

$$b_i = \frac{C}{1 + \frac{\sin \alpha_{i0j}}{\sin \alpha_{j0k}} + \frac{\sin \alpha_{i0k}}{\sin \alpha_{j0k}}},$$

where $i, j, k = 1, 2, 3$ *and* $i \neq j \neq k$.

The weights b_1, b_2, b_3 depend on exactly two given angles $\alpha_{102}, \alpha_{203}$, because $\alpha_{103} = 2\pi - \alpha_{102} - \alpha_{203}$.

Problem 5 (The 3-INVWF problem in $\mathbb{R}^3$, see [26]) Given a point A_0, which belongs to the interior of $A_1A_2A_3A_4$ in $\mathbb{R}^3$, does there exist a unique set of positive weights b_{i0} such that

$$b_{10} + b_{20} + b_{30} + b_{40} = C > 0,$$

$$f(A_0) = \sum_{i=1}^{4} b_{i0}\, a_{i0} \to \min?$$

The unique solution of the 3-INWF problem for boundary tetrahedra has been established in [26], by taking into account (4)–(5) of Lemma 5.

Lemma 3 (Solution of the inverse weighted Fermat problem for tetrahedra in $\mathbb{R}^3$ [26, (3.12), (3.13), p. 120])

$$(\frac{b_{j0}}{b_{i0}})^2 = \frac{\sin^2\alpha_{kom} - \cos^2\alpha_{moi} - \cos^2\alpha_{koi} + 2\cos\alpha_{moi}\cos\alpha_{koi}\cos\alpha_{kom}}{\sin^2\alpha_{kom} - \cos^2\alpha_{moj} - \cos^2\alpha_{koj} + 2\cos\alpha_{moj}\cos\alpha_{koj}\cos\alpha_{kom}}$$

for $i, j, k = 1, 2, 3, 4$ *and* $i \neq j \neq k \neq m \neq i$.

Lemma 4 *[28, Proposition 1] The ratios* $\frac{b_{j0}}{b_{i0}}$ *depend on exactly five given angles* $\alpha_{102}, \alpha_{103}, \alpha_{104}, \alpha_{203}$ *and* α_{204}.

Thus, the solution of the 3-INVWF problem in $\mathbb{R}^3$ depends on exactly five given angles $\alpha_{102}, \alpha_{103}, \alpha_{104}, \alpha_{203}, \alpha_{204}$.

We mention the definitions of a tree topology, the degree of a vertex, a Fermat (F) tree topology, a Fermat-Steiner (FS) tree topology, an intermediate Fermat-Steiner (I.FS) tree topology, in order to study the solution of the WFS problem with respect to fixed tree topology for $A_1A_2 \ldots A_{N+1}$ in $\mathbb{R}^N$.

Definition 1 (see [10]) A tree topology is a connection matrix specifying which pairs of points from the list $A_1, A_2, \ldots, A_N, A_{0,1}, A_{0,2}, \ldots, A_{0,N-2}$ have a connecting line segment (edge).

Definition 2 (see [3, 14]) The degree of a vertex corresponds to the number of connections of the vertex with line segments.

Definition 3 (see [10]) (i) An F tree topology of a boundary N-simplex $A_1A_2 \ldots A_{N+1}$ in $\mathbb{R}^N$ is a tree topology, such that each boundary vertex A_i has degree one and the WF point $A_{0,1}$ has degree $N+1$. (ii) An FS tree topology of a boundary N-simplex $A_1A_2 \ldots A_{N+1}$ in $\mathbb{R}^N$ is a tree topology such that each boundary vertex A_i has degree one and each WF point $A_{0,i}$ has degree three for $i = 1, 2, \ldots, N-1$. (iii) An I.FS tree topology of a boundary N-simplex $A_1A_2 \ldots A_{N+1}$ in $\mathbb{R}^N$ is a tree topology, such that each boundary vertex A_i has degree one, and the number of WF points $A_{0,i}$ is less than $N-1$ having degree $< N+1$.

Definition 4 (i) A tree of weighted minimum length with a F tree topology is called a WF tree. (ii) A tree of weighted minimum length with an FS tree topology is called a WFS tree. (iii) A tree of weighted minimum length with an I.FS tree topology is called an I.WFS tree.

In [20], Melzak constructed an algorithm of circles to find FS tree topologies for an Nconvex polygon in $\mathbb{R}^2$. Unfortunately, Melzak cannot be extended in $\mathbb{R}^3$. In [21], Rubinstein, Thomas and Weng succeeded in solving numerically the unweighted problem, by applying a fixed point iteration method to a system of two equations with two variable line segments. In [29, Theorem 4], we extended Rubinstein, Thomas and Weng method to solve the weighted Fermat-Steiner problem for tetrahedra in $\mathbb{R}^3$.

We note that extended Rubinstein Thomas method takes into account the coordinates (x_i, y_i, z_i) of each vertex A_i, for $i = 1, 2, 3, 4$. Therefore, we need to find a method to use the six edge lengths of the tetrahedron and some variable edge lengths, in order to consider the WFSFR problem for Fréchet tetrahedra in $\mathbb{R}^3$.

We denote by $D(S)$ the Cayley-Menger determinant:

$$D(S) = \det \begin{pmatrix} 0 & a_{12}^2 & a_{13}^2 & a_{14}^2 & 1 \\ a_{12}^2 & 0 & a_{23}^2 & a_{24}^2 & 1 \\ a_{13}^2 & a_{23}^2 & 0 & a_{34}^2 & 1 \\ a_{14}^2 & a_{24}^2 & a_{34}^2 & 0 & 1 \\ 1 & 1 & 1 & 1 & 0 \end{pmatrix}.$$

We consider the 3-INVWF problem in $\mathbb{R}^3$. Let A_0 be a weighted Fermat point inside $A_1A_2A_3A_4$ in $\mathbb{R}^3$. In [26], the following relation is proved.

Lemma 5 (see formula (2.25) in [26]) *If b_{i0} is the weight corresponding to the vertex $A_i : \| \sum_{j=1,i\neq j}^{4} b_{j0}\vec{a}_{ij} \| > b_{i0}$ for $i, j = 1, 2, 3, 4$ holds, then*

$$\frac{b_{30}}{a_{03}\,\mathrm{Vol}(A_0A_1A_2A_4)} = \frac{b_{40}}{a_{04}\,\mathrm{Vol}(A_0A_1A_2A_3)} = C, \tag{4}$$

$$\frac{b_{30}}{a_{03}\,\mathrm{Vol}(A_0A_1A_2A_4)} = \frac{b_{10}}{a_{01}\,\mathrm{Vol}(A_0A_2A_3A_4)} = C,$$

$$\frac{b_{30}}{a_{03}\,\mathrm{Vol}(A_0A_1A_2A_4)} = \frac{b_{20}}{a_{02}\,\mathrm{Vol}(A_0A_1A_3A_4)} = C,$$

$$\frac{b_{10}}{a_{01}\,\mathrm{Vol}(A_0A_2A_3A_4)} = \frac{b_{20}}{a_{02}\,\mathrm{Vol}(A_0A_1A_3A_4)} = C, \tag{5}$$

where $C = \frac{\sum_{i=1}^{4} b_{i0}/a_{0i}}{\mathrm{Vol}(A_1A_2A_3A_4)}$.

The volumes of $A_0A_iA_jA_k$ $\mathrm{Vol}(A_0A_iA_jA_k)$ for $i, j, k = 1, 2, 3, 4$, can be computed via the Cayley-Menger determinant ([25, pp. 249–255]):

$$288\,\mathrm{Vol}(A_0A_iA_jA_k)^2 = D(\{a_{0i}, a_{0j}, a_{0k}, a_{ij}, a_{ik}, a_{jk}\}).$$

3 Plasticity of Weighted Fermat-Fréchet Multitrees for Boundary Closed Polytopes in $\mathbb{R}^N$

In this section, we obtain the unique solution with respect to the N-INVWF problem in $\mathbb{R}^N$, which depends on exactly $\frac{1}{2}N(N+1) - 1$ angles and the non-unique solution (dynamic plasticity) of the $(N+1)$-INVWF problem in $\mathbb{R}^N$. The dynamic plasticity of the $(N+1)$-INVWF problem in $\mathbb{R}^N$ leads to the plasticity of weighted Fermat-Fréchet (WFFR) multitrees for boundary closed polytopes in $\mathbb{R}^N$.

Theorem 2 (Solution of the N-INVWF problem in $\mathbb{R}^N$) *The weight B_i is uniquely determined by:*

$$B_i = \frac{C}{1+\left|\frac{\sin\alpha_{i,0k_1k_2\ldots k_{N-1}}}{\sin\alpha_{k_N,0k_1k_2\ldots k_{N-1}}}\right|+\left|\frac{\sin\alpha_{i,0k_1k_2\ldots k_{N-2}k_n}}{\sin\alpha_{k_{N-1},0k_1k_2\ldots k_{N-2}k_N}}\right|+\ldots+\left|\frac{\sin\alpha_{i,0k_2\ldots k_{N-1}k_N}}{\sin\alpha_{k_1,0k_2\ldots k_{N-1}k_N}}\right|} \tag{6}$$

for $i, k_1, k_2, \ldots, k_N = 1, 2, \ldots, N+1$ *and* $k_1 \neq k_2 \neq \ldots \neq k_N$.

Proof We consider $N+1$ unit vectors $\vec{u}(A_0, A_i) \in \mathbb{R}^N$,
for $i = 1, 2, \ldots, N+1$, which meet at the weighted Fermat point A_0 :

$$\vec{u}(A_0, A_1) = (1, 0, \ldots, 0),$$
$$\vec{u}(A_0, A_2) = (\cos\alpha_{102}, \sin\alpha_{102}, 0, \ldots, 0),$$
$$\vec{u}(A_0, A_3) = (\cos a_{3,102}\cos\omega_{3,102}, \cos a_{3,102}\sin\omega_{3,102}, \sin a_{3,102}, 0, \ldots 0),$$
$$\vec{u}(A_0, A_4) = (\cos a_{4,1023}\cos\omega_{4,1023}\cos z_{4,1023}, \cos a_{4,1023}\cos\omega_{4,1023}\sin z_{4,1023},$$
$$\cos a_{4,1023}\sin\omega_{4,1023}, \sin a_{4,1023}, 0, \ldots, 0),$$

...

$$\vec{u}(A_0, A_N) = (\cos a_{N,1023\ldots N-1}\cos\omega_{(1)N,1023\ldots N-1}\cdots\cos\omega_{(N-2)N,1023\ldots N-1},$$
$$\cos a_{N,1023\ldots N-1}\cos\omega_{(1)N,1023\ldots N-1}\cdots\sin\omega_{(N-2)N,1023\ldots N-1},$$
$$\ldots, \sin a_{N,1023\ldots N-1}), \tag{7}$$

$$\vec{u}(A_0, A_{N+1})$$
$$= (\cos a_{N+1,1023\ldots N-1}\cos\omega_{(1)(N+1),1023\ldots N-1}\cdots\cos\omega_{(N-2)(N+1),1023\ldots N-1},$$
$$\cos a_{N+1,1023\ldots N-1}\cos\omega_{(1)(N+1),1023\ldots N-1}\cdots\sin\omega_{(N-2)(N+1),1023\ldots N-1},$$
$$\ldots, \sin a_{N+1,1023\ldots N-1}). \tag{8}$$

Starting from $N=4$, and by using induction for the angles $\omega_{(i)(N),1023\ldots N-1}$, $\omega_{(i)(N+1),1023\ldots N-1}$, $\ldots\omega_{4,1023}$, $\omega_{3,102}$, $z_{4,1023}$ are eliminated, $\cos^2 a_{N,1023\ldots N-1}$, depend on $\frac{1}{2}N(N-1)$ angles α_{102}, α_{103}, $\alpha_{10(N-2)}$, α_{10N}, α_{203}, $\ldots$ $\alpha_{(N-2)0N}$. $\cos^2 a_{N+1,1023\ldots N-1}$ depend on $\frac{1}{2}N(N-1)$ angles α_{102}, α_{103}, $\alpha_{10(N-2)}$, $\alpha_{10(N-2)}$, $\alpha_{10(N+1)}$, α_{203}, $\ldots$ $\alpha_{(N-2)0(N+1)}$. Taking into account (7) and (8), the inner product $\vec{u}(A_0, A_N)\cdot\vec{u}(A_0, A_{N+1})$ yields:

$$\alpha_{N0N+1} = \alpha_{N0N+1}(\alpha_{102}, \alpha_{103}, \alpha_{10N-2}, \alpha_{10N}, \alpha_{10(N+1)}, \ldots\alpha_{(N-2)0(N+1)}).$$

The angle $a_{i,1023\ldots(N-1)}$ is formed by $\vec{u}(A_0, A_i)$ and $\vec{u}(A_0, A_{i,1023\ldots(N-1)})$, where $A_{i,1023\ldots(N-1)}$ is the projection of A_i with respect to subspace of $A_1A_0A_2A_4\ldots A_{N-1}$ for $i = N, N+1$. The normal $\vec{n}_{1023\ldots(n-1)}$ is orthogonal to $A_1A_0A_2\ldots A_{N-1}$. Hence, we get the following: $\vec{n}_{0,123\ldots(N-1)} \cdot \vec{u}(A_0, A_i) = \sin a_{i,1023\ldots(N-1)}$ for $i = N, N+1$. The weighted floating condition (2) yields:

$$\sum_{i=1}^{N+1} B_i\vec{u}(A_0, A_i) = 0. \tag{9}$$

By taking the inner product of (9) with $\vec{n}_{0,123\ldots(N-1)}$, we derive the following:

$$\sum_{i=1}^{N+1} B_i\vec{u}(A_0, A_i) \cdot \vec{n}_{0,123\ldots(N-1)} = 0,$$

or

$$B_N \sin a_{N,1023\ldots(N-1)} = B_{N+1} \sin a_{N+1,1023\ldots(N-1)}. \tag{10}$$

By working similarly and by exchanging the indices $\{1, 2, 3, \ldots, N, N+1\}$ of the vertices A_i for $i = 1, \ldots, N+1$, we obtain:

$$B_{i_m} \sin a_{i_m,i_10i_2\ldots i_{N-1}} = B_{i_N} \sin a_{i_N,i_10i_2\ldots i_{N-1}} \tag{11}$$

for $m, N, i_m, i_N \in \{1, 2, \ldots, N+1\}$, $m \neq n$, $i_m \neq i_N$. By replacing (11) in (3), we obtain (6). Thus, the weights B_i depend on $\frac{1}{2}N(N+1) - 1$ given angles α_{i0j}, for $i, j = 1, 2, \ldots, N+1$, $i \neq j$ and $\alpha_{N0N+1} = \alpha_{N0N+1}(\alpha_{i0j})$. □

We set $\vec{a}_i \equiv \overrightarrow{A_0A_i}$. The corner $(A_0, A_1A_2\ldots A_n)$ with N edges is determined by $\vec{a}_1, \vec{a}_2, \ldots \vec{a}_N$. The N-dimensional polar sine of the corner $(A_0, A_1, A_2\ldots A_N)$ is defined in [7, p. 76]:

$$polsin_N(A_0, A_1, A_2\ldots A_N) = \frac{\left|[\vec{a}_1, \vec{a}_2, \ldots, \vec{a}_N]\right|}{|\vec{a}_1|\,|\vec{a}_2|\ldots|\vec{a}_N|},$$

where $[\vec{a}_1, \vec{a}_2, \ldots, \vec{a}_N]$ is the content of an N-dimensional parallelotope with sides $|\vec{a}_1|, |\vec{a}_2|, \ldots, |\vec{a}_N|$.

Theorem 3 (Weighted volume equalities in $\mathbb{R}^N$)

$$\frac{B_1}{a_1 Vol(A_0A_2A_3\ldots A_{N+1})} = \frac{B_2}{a_2 Vol(A_1A_0\ldots A_{N+1})} = \ldots \tag{12}$$
$$= \frac{B_{N+1}}{a_{N+1} Vol(A_1A_2\ldots A_0)} = \frac{\sum_{i=1}^{N+1} \frac{B_i}{a_i}}{Vol(A_1A_2\ldots A_{N+1})}.$$

Proof By multiplying both members of (10) with $\prod_{i=1}^{N+1} |\vec{a}_i| polsin_{N-1}(A_0, A_1A_2\ldots A_{N-1})$, we get the following:

$$B_N \prod_{i=1}^{N+1} |\vec{a}_i| polsin_{N-1}(A_0, A_1A_2 \ldots A_{N-1}) \sin a_{N,1023\ldots N-1}$$
$$= B_{N+1} \prod_{i=1}^{N+1} |\vec{a}_i| polsin_{N-1}(A_0, A_1A_2 \ldots A_{N-1}) \sin a_{N+1,1023\ldots N-1}. \quad (13)$$

The volumes of the N-dimensional simplexes $A_0A_1A_2\ldots A_{N-1}A_N$ and A_0A_1 $A_2\ldots A_{n-1}A_{n+1}$, are given by

$$N!Vol(A_0A_1A_2\ldots A_{N-1}A_N) = \prod_{i=1}^{N+1} |\vec{a}_i| polsin_N(A_0, A_1A_2\ldots A_N), \quad (14)$$
$$N!Vol(A_0A_1A_2\ldots A_{N-1}A_{N+1}) = \prod_{i=1}^{N+1} |\vec{a}_i| polsin_N(A_0, A_1A_2\ldots A_{N+1}). \quad (15)$$

By substituting (14), (15) in (13), we obtain:

$$\frac{B_{N+1}}{a_{N+1}Vol(A_0A_1\ldots A_{N-1}A_N)} = \frac{B_N}{a_N Vol(A_0A_1A_2\ldots A_{N-1}A_{N+1})}.$$

By exchanging the indices cyclically $i \to j$ for $i, j = 1, 2, \ldots, N, N+1$ and taking into account (11), we get (12). □

Definition 5 We call *dynamic plasticity* of a variable weighted Fermat tree (weighted network), whose endpoints correspond to a closed polytope in $\mathbb{R}^N$, which is formed by $(N+2)$ weighted line segments meeting at the weighted Fermat point A_0, the set of solutions of the $n+2$ variable weights with respect to the $(N+1)$-INVWF problem in $\mathbb{R}^N$, for a given constant value c, which correspond to a family of variable weighted networks that preserve the weighted Fermat point and the boundary of the closed polytope, such that the $N+1$ variable weights depend on a variable weight and the value of c.

We denote by $(B_i)_{12\ldots N+2}$ the weight which corresponds to the vertex that lies on the ray A_0A_i, for $i = 1, 2, \ldots, N+2$ and the weight $(B_j)_{i_1i_2\ldots i_{N+1}}$ corresponds to the vertex A_j that lies on the ray A_0A_j with respect to the N-simplex $A_{i_1}A_{i_2}\ldots A_{i_{N+1}}$, for $i_1, i_2, \ldots, i_{N+1} \in \{1, 2, \ldots N+2\}$ and $i_1 \neq i_2 \neq \ldots \neq i_{N+1}$.

Theorem 4 *The following equations indicate the dynamic plasticity of a weighted Fermat tree for $N+1$ weighted boundary closed polytopes with respect to the non-negative variable weights $(B_i)_{12\ldots N+2}$ in $\mathbb{R}^N$:*

$$(\frac{B_1}{B_{N+1}})_{12\ldots N+2} = (\frac{B_1}{B_{N+1}})_{12\ldots N+1}(1-(\frac{B_{N+2}}{B_{N+1}})_{12\ldots (N+2)}(\frac{B_{N+1}}{B_{N+2}})_{2\ldots N+2}), \quad (16)$$
$$(\frac{B_2}{B_{N+1}})_{12\ldots N+2} = (\frac{B_2}{B_{N+1}})_{12\ldots N+1}(1-(\frac{B_{N+2}}{B_{N+1}})_{12\ldots N+2}(\frac{B_{N+1}}{B_{N+2}})_{13\ldots N+2}), \quad (17)$$

...

$$(\frac{B_N}{B_{N+1}})_{12\ldots N+2}=(\frac{B_N}{B_{N+1}})_{12\ldots N+1}(1-(\frac{B_{N+2}}{B_{N+1}})_{12\ldots N+2}(\frac{B_{N+1}}{B_{N+2}})_{12\ldots (N-1)(N+1)(N+2)}). \quad (18)$$

Proof We consider the weighted floating case for $A_0 \notin \{A_1 A_2 \ldots A_{N+2}\}$:

$$\sum_{i=1}^{N+2} (B_i)_{12\ldots N+2} \vec{u}(A_0, A_i) = 0. \tag{19}$$

The normals $\vec{n}_{0,23\ldots N}$, $\vec{n}_{0,13\ldots N}$, and $\vec{n}_{0,12\ldots(N-1)}$ are orthogonal to the subspaces $A_2 A_3 \ldots A_N$, $A_1 A_3 \ldots A_N$, $A_1 A_2 \ldots A_N$, respectively. We set

$$sgn_{i,203\ldots N} = \begin{cases} +1, & \text{if } A_i \text{is above the subspace } A_2 A_0 \ldots A_N, \\ 0, & \text{if } A_i \text{ belongs to the subspace } A_2 A_0 \ldots A_N, \\ -1, & \text{if } A_i \text{ is below the subspace } A_2 A_0 \ldots A_N, \end{cases}$$

for $i = 1, N+1, N+2$. The inner product of (19) with $\vec{n}_{0,23\ldots N}$, $\vec{n}_{0,13\ldots N}$ and $\vec{n}_{0,12\ldots(N-1)}$ yields

$$\begin{aligned} &(B_1)_{12\ldots N+2} sgn_{1,203\ldots N} \sin a_{1,203\ldots N} \\ &+ (B_{N+1})_{12\ldots N+2} sgn_{N+1,203\ldots N} \sin a_{N+1,203\ldots N} \\ &+ (B_{N+2})_{12\ldots N+2} sgn_{N+2,203\ldots n} \sin a_{N+2,203\ldots N} = 0, \end{aligned} \tag{20}$$

$$\begin{aligned} &(B_2)_{12\ldots N+2} sgn_{2,103\ldots N} \sin a_{2,103\ldots N} \\ &+ (B_{N+1})_{12\ldots N+2} sgn_{N+1,103\ldots N} \sin a_{N+1,103\ldots N} \\ &+ (B_{N+2})_{12\ldots N+2} sgn_{N+2,103\ldots N} \sin a_{N+2,103\ldots N} = 0, \end{aligned} \tag{21}$$

$$\ldots$$

$$\begin{aligned} &(B_N)_{12\ldots N+2} sgn_{N,102\ldots N-1} \sin a_{N,102\ldots N-1} \\ &+ (B_{N+1})_{12\ldots N+2} sgn_{N+1,102\ldots N-1} \sin a_{N+1,102\ldots N-1} \\ &+ (B_{N+2})_{12\ldots N+2} sgn_{N+2,102\ldots N-1} \sin a_{N+2,102\ldots N-1} = 0. \end{aligned} \tag{22}$$

By substituting $B_{N+2} = 0$ in (20), (21), (22), we derive:

$$\left(\frac{B_1}{B_{N+1}}\right)_{12\ldots N+1} = -\frac{sgn_{N+1,203\ldots N} \sin a_{N+1,203\ldots N}}{sgn_{1,203\ldots N} \sin a_{1,203\ldots N}}, \tag{23}$$

$$\left(\frac{B_2}{B_{N+1}}\right)_{12\ldots N+1} = -\frac{sgn_{N+1,103\ldots N} \sin a_{N+1,103\ldots N}}{sgn_{2,103\ldots N} \sin a_{2,103\ldots N}}, \tag{24}$$

$$\left(\frac{B_N}{B_{N+1}}\right)_{12\ldots N+1} = -\frac{sgn_{N+1,102\ldots N-1} \sin a_{N+1,102\ldots N-1}}{sgn_{N,102\ldots N-1} \sin a_{N,102\ldots N-1}}. \tag{25}$$

By substituting $(B_1)_{12\ldots N+2} = 0$, $(B_2)_{12\ldots N+2} = 0$ and $(B_N)_{12\ldots N+2} = 0$ in (20), (21) and (22), respectively, we derive:

$$(\frac{B_{N+2}}{B_{N+1}})_{2...N+2} = -\frac{sgn_{N+1,203...N} \sin a_{N+1,203...N}}{sgn_{N+2,203...N} \sin a_{N+2,203...N}}, \quad (26)$$

$$(\frac{B_{N+2}}{B_{N+1}})_{13...N+2} = -\frac{sgn_{N+1,103...N} \sin a_{N+1,103...N}}{sgn_{N+2,103...N} \sin a_{n+2,103...N}}, \quad (27)$$

$$(\frac{B_{N+2}}{B_{N+1}})_{12...(N-1)(N+1)(N+2)} = -\frac{sgn_{N+1,102...N-1} \sin a_{N+1,102...N-1}}{sgn_{N+2,102...N-1} \sin a_{N+2,102...N-1}}. \quad (28)$$

By substituting (23), (24), (25), (26), (27) and (28) in (20), (21) and (22), we obtain (16), (17) and (18). □

A consequence of Theorem 4 is the following corollary, setting $\sum_{12...N+2} B \equiv \sum_{i=1}^{N+2} (B_i)_{12...N+2}$, $\sum_{i_1 i_2 ... i_{N+1}} B = (B_{i_1})_{i_1 i_2 ... i_{N+1}} + \ldots + (B_{i_{N+1}})_{i_1 i_2 ... i_{N+1}}$ for $i_1, \ldots i_{N+1} \in \{1, 2, \ldots N+2\}$.

Corollary 1 *If* $\sum_{12...N+2} B = \sum_{i_1 i_2 ... i_{N+1}} B$ *for all* $i_1, \ldots i_{N+1} \in \{1, 2, \ldots N+2\}$, *where* $\sum_{12...N+2} B := (B_{N+1})_{12...N+2}(1 + \sum_{i=1, i \neq N+1}^{N+2} (\frac{B_i}{B_{N+1}})_{1,2,...,N+2})$, *then* $(B_i)_{12...N+2} = a_i (B_{N+2})_{12...N+2} + b_i$, $i = 1, 2 \ldots, N+1$, *where*

$$(a_{N+1}, b_{N+1}) = \quad (29)$$

$$(\frac{(\frac{B_1}{B_{N+1}})_{12...N+1}(\frac{B_{N+1}}{B_{N+2}})_{2...N+2} + \ldots + (\frac{B_N}{B_{N+1}})_{12...N+1}(\frac{B_{N+1}}{B_{N+2}})_{12...(N-1)(N+1)N+2} - 1}{\sum_{i=1}^{N+1} (\frac{B_i}{B_{N+1}})_{12...N+1}},$$

$$(B_{N+1})_{12...N+1}),$$

$$(a_N, b_N) = (a_{N+1}(\frac{B_N}{B_{N+1}})_{12...N+1} - (\frac{B_N}{B_{N+1}})_{12...N+1}(\frac{B_{N+1}}{B_{N+2}})_{12...(N-1)(N+1)(N+2)},$$

$$(B_N)_{12...N+1}),$$

...

$$(a_1, b_1) = (a_{N+1}(\frac{B_1}{B_{N+1}})_{12...N+1} - (\frac{B_1}{B_{N+1}})_{12...N+1}(\frac{B_{N+1}}{B_{N+2}})_{23...(N+2)},$$

$$(B_1)_{12...N+1}).$$

Suppose that A_0 is an interior weighted Fermat point for the N-simplex with respect to the non-negative given weights $\{B_1(0), B_2(0), \ldots, B_N(0), B_{N+1}(0)\}$ in $\mathbb{R}^N$. Therefore, the topology of the branches $A_i A_0$ which meet at A_0 form a unique floating weighted Fermat tree. The unique solution of the N-INVWF problem for $A_1 A_2 \ldots A_{N+1}$ is responsible for the cancellation of the dynamic plasticity of simplexes. We assume that after time t an $N+2$ branch $A_0 A_{N+2}$ starts to grow from A_0 and the new branch $A_0 A_{N+2}$ is located inside the cone $C(ray(A_0 A_1), ray(A_0 A_3) \ldots, ray(A_0 A_{N+1}))$ and $A_0 A_2$ is located outside of $C(ray(A_0 A_1), ray(A_0 A_3) \ldots, ray(A_0 A_{N+1}))$. We assume that A_{N+1} is upper from the hyperplane formed by the first, second, ... and the $(N-1)$th ray and A_N, A_{N+2} are under this hyperplane.

Theorem 5 *An increase to the weight that corresponds to the* $(N+2)th$ *ray causes a decrease to the weight that corresponds to the* $(N+1)th$ *ray and a variation to the* *weight that corresponds to the* ith *ray depends on the difference* $(B_i)_{123...N(N+1)} - (B_i)_{123...N(N+2)}$, *for* $i = 1, 2, \ldots N$.

Proof First, we will show that $a_{N+1} < 0$. Taking into account (29), we obtain $a_{N+1} = -\frac{(B_{N+1})_{123...N+1}}{(B_{N+2})_{123...NN+2}} < 0$, because $(B_{N+1})_{123...N+1}$ and $(B_{N+2})_{123...NN+2} > 0$. Thus, we get the following:

$$(B_{N+1})_{123...N+2} = -\frac{(B_{N+1})_{123...N+1}}{(B_{N+2})_{123...NN+2}}(B_{N+2})_{123...N+2} + (B_{N+1})_{123...N+1}. \quad (30)$$

By inserting (30) into (20), (21), ... (22), we derive that

$$(B_i)_{123...N+2} = \frac{(B_i)_{123...N(N+2)} - (B_i)_{123...(N+1)}}{(B_{N+2})_{123...NN+2}}(B_{N+2})_{123...N+2} + (B_i)_{123...N+1}$$

for $i = 1, 2...N$. Therefore, the difference $\delta B_i = (B_i)_{123...N(N+2)} - (B_i)_{123...(N+1)}$ of the two positive weights, which corresponds to the ith branch A_0A_i of the boundary simplexes $A_1A_2 \ldots A_N A_{N+2}$ and $A_1A_2 \ldots A_N A_{N+1}$, yields the following two results:

(a) If $\delta B_i > 0$, the weight $(B_i)_{123...N+2}$ increases,
(b) If $\delta B_i < 0$, the weight $(B_i)_{123...N+2}$ decreases.

□

Proposition 2 *An increase to the weight that corresponds to the* $(N + 2)th$ *ray and a decrease to the weight that corresponds to the* ith *ray for* $i = 1, 3, \ldots, N$, *causes a decrease to the weight that corresponds to the* $(N + 1)th$ *ray and an increase to the weights that corresponds to the second ray.*

Proof By applying Theorem 5 and taking into account that $a_1, a_3, \ldots a_N < 0$, the length of the vectors $\vec{X}_i = (B_i)_{12...n+2}(\delta t)\vec{u}(A_0, A_i)$ are decreased after time δt. We proceed by arranging some vector terms of the weighted floating balancing condition for $\vec{X}_i$ (see Theorem 1):

$$\vec{X}_2 + \vec{X}_{n+1} = -(\sum_{i=1, i\neq 2}^{N} \vec{X}_i) - \vec{X}_{N+2}.$$

We observe that after time δt the length of the vector $\sum_{i=1, i\neq 2}^{N} \vec{X}_i$ is decreased. Thus, on the right hand side of the above equation a vector of reduced length is composed with a vector of increased length $\vec{X}_{N+2}$ and on the left hand side a vector of reduced length $\vec{X}_{N+1}$ is composed with $\vec{X}_2$ whose length is increased. Therefore, we obtain that $a_2 > 0$. □

Suppose that at time $t = 0$, an evolutionary multitree occurs such that $(N + 1)! \cdot k$ weighted Fermat trees correspond to k simplexes, which form a Fréchet N-multisimplex in $\mathbb{R}^N$ with $(N + 1)!$ permutation of the weights $\{B_1(0), B_2(0), \ldots, B_N(0), B_{N+1}(0)\}$ in $\mathbb{R}^N$ for boundary incongruent N-simplexes

$$(A_1)_k(A_2)_k \ldots (A_{N+1})_k, \quad k \leq \frac{\frac{1}{2}N(N+1)!}{(N+1)!}.$$

After time t, a $(N+2)th$ branch $(A_0)_k(A_{N+2})_k$ starts to grow from the weighted Fermat point $(A_0)_k$ and the new branch $(A_0)_k(A_{N+2})_k$ is located inside the cone $C(ray((A_0)_k(A_1)_k), ray((A_0)_k(A_3)_k)\ldots, ray((A_0)_k(A_{N+1})_k))$ and $(A_0)_k(A_2)_k$ is located outside $C(ray(((A_0)kA_1)_k), ray((A_0)_k(A_3)_k)\ldots, ray((A_0)_k(A_{N+1})_k))$.

Let $(A_{N+1})_k$ be upper from the hyperplane formed by the first, second, ... and the $(N-1)$th ray and $(A_N)_k$, $(A_{N+2)_k}$ are under this hyperplane.

Theorem 6 *An increase to the weight that corresponds to the $(N+2)th$ ray causes a decrease to the weight that corresponds to the $(N+1)th$ ray and a variation to the weight that corresponds to the ith ray depends on the difference $(B_i)_{123\ldots N(N+1)} - (B_i)_{123\ldots N(N+2)}$, such that the geometric structure of the WFFR multitree with respect to a boundary N-multisimplex, remains the same, for $i = 1, 2, \ldots N$, $1 \leq k \leq \frac{\frac{1}{2}N(N+1)!}{(N+1)!}$.*

Proof Using Theorem 5 starting from a WFFR multitree with respect to a boundary Fréchet N-multisimplex in $\mathbb{R}^N$, we get the plasticity of a WFFR multitree by adding the ray $(A_0)_k(A_{N+2})_k$ for $i = 1, 2, \ldots N$, $1 \leq k \leq \frac{\frac{1}{2}N(N+1)!}{(N+1)!}$. □

4 The Weighted Fermat-Steiner-Fréchet Multitree for a $\frac{1}{2}N(N+1)$-tuple of Positive Real Numbers Determining the Edge Lengths of Incongruent N-simplexes in $\mathbb{R}^N$

In this section, we deal with the solution (multitree) of the WFSFR problem (P(FSFFR)) for a given $\frac{1}{2}N(N+1)$ of positive real numbers determining incongruent N-simplexes in $\mathbb{R}^N$, by inserting $N(N-1)$ equality constraints derived by $N-1$ independent solutions for $N-1$ variable WF problems for the Fréchet N-multisimplex derived by incongruent boundary N-simplexes in $\mathbb{R}^N$, which correspond to the same $\frac{1}{2}N(N+1)$-tuple of positive real numbers (edge lengths) and $N-2$ equality constraints derived by two different expressions of each line segments connecting two consecutive weighted Fermat-Steiner points. By applying a Lagrange program, we can detect the WFSFR multitree for a given $\frac{1}{2}N(N+1)$-tuple of edge lengths determining incongruent N-simplexes (Fréchet N-multisimplex) in $\mathbb{R}^N$, by using the Dekster Wilker function. By seeking unweighted Fermat-Steiner-Fréchet multitrees with $N-1$ equally WFS points inside the Fréchet N-multisimplex, we can detect the most natural of $\frac{1}{2}N(N+1)$ consecutive natural numbers (Dekster-Wilker $\frac{1}{2}N(N+1)$-tuple) and it is achieved by seeking an upper bound for these equal weights, which yield a global WFS tree of minimum length for a boundary $N-simplex$ having the maximum volume among the derived incongruent N-simplexes in $\mathbb{R}^N$.

We describe the Dekster-Wilker Euclidean domain (see [5, 6]), which gives all incongruent N-simplexes in $\mathbb{R}^N$ derived by the same $\frac{1}{2}N(N+1)$-tuple of positive real numbers $\{a_{ij}\}$. Denote by $\ell = \max_{i,j} a_{ij}$, $s = \min_{i,j} a_{ij}$ and

$$\lambda_N(\ell) = \begin{cases} \ell\sqrt{1 - \frac{2(N+1)}{N(N+2)}} & for\ even\ N \geq 2, \\ \ell\sqrt{1 - \frac{2}{N+1}} & for\ odd\ N \geq 3. \end{cases}$$

Definition 6 (The Dekster-Wilker Euclidean domain [5, 6]) The Dekster-Wilker Euclidean domain $DW_{\mathbb{R}^N}(\ell, s)$ is a closed domain in $\mathbb{R}^2$ between the ray $s = \ell$ and the graph of a function $\lambda_N(\ell)$, $\ell \geq 0$, which is less than ℓ for $\ell \neq 0$.

Let $A_{0,1}, A_{0,2}, \ldots A_{0,N-1}$ be $N-1$ points inside the N-simplex $A_1A_2\ldots A_{N+1}$ in $\mathbb{R}^N$. Denote by $a_{(0,i),j}$ the length of the line segment $A_{0,i}A_j$, by $\beta_{(0,k),u}$ the Schläfli angle formed by normals of subpaces spanned by $\{A_{(0,k)}A_1A_2\ldots A_{u-2}\}$ and $\{A_1A_2\ldots A_{u-1}\}$ for $i = 1, 2, \ldots N$ $j = 1, 2, \ldots N+1$, $k = 1, 2, \ldots N-1$, $u = 4, 5, \ldots, N$. The WFSFR problem for a given $\frac{1}{2}N(N+1)$-tuple of edge lengths determining incongruent N-simplexes in $\mathbb{R}^N$ states that:

Problem 6 (The WFSFR problem in $\mathbb{R}^N$) Given a $2N$-tuple of weights

$$\{b_1, b_2, \ldots, b_{N+1}, b_{ST}, \ldots, b_{ST}\}$$

and a $\frac{1}{2}N(N+1)$-tuple of positive real numbers (edge lengths) $\{a_{ij}\}$, determining a Fréchet N-multisimplex $F(A_1A_2\ldots A_{N+1})$, find the position of $A_{0,1}$ and/or $A_{0,2}$ and/or $\ldots$ $A_{0,N-1}$ with given weights b_{ST} in $A_{0,1}$, b_{ST} in $A_{0,2}$, $\ldots$ b_{ST} in $A_{0,N-1}$ such that

$$f_0(T_S) = \sum\nolimits_{a_{i_k,j_k} \in T_S} b_k a_{i_k,j_k} \to \min.$$

Note that the volume of an $(N-1)$-simplex $A_1A_2\cdots A_N$ in $\mathbb{R}^{N-1}$ is given by the Caley-Menger determinant in terms of edge lengths ([23, (5.1), p. 125]):

$$\mathrm{Vol}(A_1A_2\cdots A_N)^2 = \frac{1}{(-1)^N 2^{N-1}((N-1)!)^2} \begin{vmatrix} 0 & 1 & 1 & \cdots & 1 & 1 \\ 1 & 0 & a_{12}^2 & \cdots & a_{1(N-1)}^2 & a_{1N}^2 \\ 1 & a_{21}^2 & 0 & \cdots & a_{2(N-1)}^2 & a_{2N}^2 \\ . & . & . & . & . & . \\ 1 & a_{N1}^2 & a_{N2}^2 & \cdots & a_{N(N-1)}^2 & 0 \end{vmatrix}.$$

Definition 7 (A non-degenerate WFS tree for $\{A_1A_2\ldots A_{N+1}\}$) A non-degenerate WFS tree T_S is a tree, which consists of some line segments between the $N-1$ weighted Fermat-Steiner points $A_{0,j}$, such that each $A_{0,j}$ has degree (connections) three and of $N+1$ line segments joining each boundary vertex A_i with some $A_{0,j}$.

We consider the following three types of weighted Fermat-Steiner points $A_{0,i}$:

- Type one, if it is connected with two boundary vertices and one WFS point.
- Type two, if it is connected with one boundary vertex and two WFS points.
- Type three, if it is connected with three WFS points.

Problem 7 (The WFSFR (P(FSFR)) problem in $\mathbb{R}^N$ with equality constraints)

$$f_0(\tilde{x}) = \sum_{a_{i_k,j_k} \in T_S} b_k a_{i_k,j_k},$$
$$f_{i,j}(\tilde{x}) = 0,\ i = 1, \ldots, (N-1), \quad j = 1, \ldots, N,$$
$$f_k(\tilde{x}) = 0, \quad k = 1, \ldots, N-2,$$

where $N-1$ is: # of neighboring WFS points $A_{0,i}$ type one, two or three,

$$f_0(\tilde{x}) = f_0(T_S) = \sum_{a_{i_k,j_k} \in T_S} b_k a_{i_k,j_k}, \tag{31}$$

$$f_{1,1}(\tilde{x}) = \frac{w_{(0,1),1}}{a_{(0,1),1} \operatorname{Vol}(A_{0,1} \ldots A_{N+1})} \tag{32}$$
$$- \frac{1 - \sum_{i=1}^{N} w_{(0,1),i}}{a_{(0,1),N+1}(a_{(0,1),1}, a_{(0,1),2}, a_{(0,1),3}, \beta_{(0,1),4}, \ldots, \beta_{(0,1),N}) \operatorname{Vol}(A_{0,1} A_1 \ldots A_N)},$$

...

$$f_{1,N}(\tilde{x}) \tag{33}$$
$$= \frac{w_{(0,1),N}}{a_{(0,1),N}(a_{(0,1),1}, a_{(0,1),2}, a_{(0,1),3}, \beta_{(0,1),4}, \ldots, \beta_{(0,1),N}) \operatorname{Vol}(A_1 \ldots A_{0,1} A_{N+1})}$$
$$- \frac{1 - \sum_{i=1}^{N} w_{(0,1),i}}{a_{(0,1),N+1}(a_{(0,1),1}, a_{(0,1),2}, a_{(0,1),3}, \beta_{(0,1),4}, \ldots, \beta_{(0,1),N}) \operatorname{Vol}(A_{0,1} A_1 \ldots A_N)},$$

...

$$f_{N-1,1}(\tilde{x}) = \frac{w_{(0,N-1),1}}{a_{(0,N-1),1} \operatorname{Vol}(A_{0,N-1} \ldots A_{N+1})} - \frac{1 - \sum_{i=1}^{N} w_{(0,N-1),i}}{a_{(0,N-1),N+1}) \operatorname{Vol}(A_{0,N-1} A_1 \ldots A_N)}, \tag{34}$$

where

$$a_{(0,N-1),N+1}$$
$$= a_{(0,N-1),N+1}(a_{(0,N-1),1}, a_{(0,N-1),2}, a_{(0,N-1),3}, \beta_{(0,N-1),4}, \ldots, \beta_{(0,N-1),N})$$

...

$$f_{N-1,N}(\tilde{x}) = \frac{w_{(0,N-1),N}}{a_{(0,N-1),N}) \operatorname{Vol}(A_1 \ldots A_{0,N-1} A_{N+1})} - \frac{1 - \sum_{i=1}^{N} w_{(0,N-1),i}}{a_{(0,N-1),N+1} \operatorname{Vol}(A_{0,N-1} A_1 \ldots A_N)}, \tag{35}$$

where

$$a_{(0,N-1),N} =$$

$$= a_{(0,N-1),N}(a_{(0,N-1),1}, a_{(0,N-1),2}, a_{(0,N-1),3}, \beta_{(0,N-1),4}, \ldots, \beta_{(0,N-1),N})$$

$$f_k(\tilde{x}) = a_{(0,k),(0,k+1)}(\tilde{x};\ \triangle A_i A_{0,k+1} A_j) - a_{(0,k),(0,k+1)}(\tilde{x};\ \triangle A_{k+2} A_{0,k} A_l), \quad (36)$$

or

$$f_k(\tilde{x}) = a_{(0,k),(0,k+1)}(\tilde{x};\ \triangle A_i A_{0,k+1} A_j) - a_{(0,k),(0,k+1)}(\tilde{x};\ \triangle A_{k+2} A_{0,k} A_l),$$

or

$$f_k(\tilde{x}) = a_{(0,k),(0,k+1)}(\tilde{x};\ \triangle A_i A_{0,k+1} A_j) - a_{(0,k),(0,k+1)}(\tilde{x};\ \triangle A_{k+2} A_{0,k} A_{k+l}),$$

or

$$f_k(\tilde{x}) = a_{(0,k),(0,k+1)}(\tilde{x};\ \triangle A_k A_{0,k+1} A_j) - a_{(0,k),(0,k+1)}(\tilde{x};\ \triangle A_{k+2} A_{0,k} A_{k+l}) \quad (37)$$

for $k = 1, 2, \ldots N - 2$.

We need the following lemma, which is derived as a special case of the generalized cosine law in $\mathbb{R}^3$ ([28, Formula 27, p. 997]), by using an equality condition for the dihedral angles $\alpha_{g_i} = \alpha$ formed by $\triangle A_j A_k A_i$ and $\triangle A_j A_k A_l$.

Lemma 6 (Generalized Cosine law in $\mathbb{R}^2$) *The line segment a_{i0} depends on a_{j0}, a_{k0} in $\mathbb{R}^2$:*

$$a_{i0}^2 = a_{k0}^2 + a_{ki}^2 - 2a_{ki}\Big(\sqrt{a_{k0}^2 - h_{0,jk}^2}\cos(\alpha_{jki}) + h_{0,jk}\sin(\alpha_{jki})\Big),$$

or

$$a_{i0}^2 = a_{j0}^2 + a_{ji}^2 - 2a_{ji}\Big(\sqrt{a_{j0}^2 - h_{0,jk}^2}\cos(\alpha_{kji}) + h_{0,jk}\sin(\alpha_{kji})\Big),$$

where $h_{0,jk} = \frac{a_{j0}a_{k0}}{a_{jk}}\sqrt{1 - \left(\frac{a_{j0}^2 + a_{k0}^2 - a_{jk}^2}{2a_{j0}a_{k0}}\right)^2}$ *for* $i, j, k = 3, 4, \ldots, N+1, i \neq j \neq k$.

We take into account that:

- (31) is the objective function of the WFS problem for $A_1 A_2 \ldots A_{N+1}$ in $\mathbb{R}^N$ having $N - 1$ WFS points $A_{0,i}$ such that the following is valid.

I. Type one weighted Fermat-Steiner point $A_{0,i}$

$$\alpha_{j(0,i)k} = \arccos\left(\frac{b_{ST}^2 - b_i^2 - b_j^2}{2b_i b_j}\right),$$

$$\alpha_{j(0,i)(0,i+1)} = \arccos\left(\frac{b_k^2 - b_j^2 - b_{ST}^2}{2b_j b_{ST}}\right),$$

$$\alpha_{k(0,i)(0,i+1} = \arccos\left(\frac{b_j^2 - b_k^2 - b_{ST}^2}{2b_k b_{ST}}\right),$$

$A_{0,i}$ is the WFS point with respect to the boundary triangle $\triangle A_j A_k A_{0,i+1}$.

II. Type two WFS point $A_{0,i}$

$$\alpha_{(0,i+1)(0,i)j} = \arccos\left(\frac{-b_j}{2b_{ST}}\right),$$

$$\alpha_{(0,i+1)(0,i)(0,i+2)} = \arccos\left(\frac{b_j^2 - 2b_{ST}^2}{2b_{ST}^2}\right),$$

$$\alpha_{(0,i+1)(0,i)j} = \arccos\left(\frac{-b_j}{2b_{ST}}\right),$$

$A_{0,2}$ is the WFS point with respect to the boundary triangle $\triangle A_{0,i+1} A_{0,i+2} A_j$.

III. Type three WFS point $A_{0,i}$

$$\alpha_{(0,i+1)(0,i)(0,i+2)} = \alpha_{(0,i+1)(0,i)(0,i+3)} = 120^\circ,$$

$A_{0,i}$ is the unweighted FS point with respect to the boundary triangle $\triangle A_{0,i+1} A_{0,i+2} A_{0,i+3}$.

• (32)–(33) deal with the solution of the WF problem for $A_1 \ldots A_{N+1}$ with weights

$$w_{(0,1),1},\ \ldots,\ w_{(0,1),N},\quad w_{(0,1),N+1} = 1 - \sum_{i=1}^{N} w_{(0,1),i},$$

which is determined by applying Theorem 3 for the unique WF point $A_{0,1}$.

• (34)–(35) deal with the solution of the WF problem for $A_1 \ldots A_{N+1}$ with weights

$$w_{(0,N-1),1},\ w_{(0,N-1),2},\ \ldots,\ w_{(0,N-1),N},\ w_{(0,N-1),N+1} = 1 - \sum_{i=1}^{N+1} w_{(0,N-1),i},$$

which is determined by applying Theorem 3 and the unique WF point $A_{0,N-1}$.

• (36)–(37) is a derivation of possible expressions of $a_{(0,i),(0,i+1)}$ with respect to the boundary $\triangle A_i A_{0,k+1} A_j$, $\triangle A_{k+2} A_{0,k} A_l)$, $\triangle A_k A_{0,k+1} A_j$, $\triangle A_{k+2} A_{0,k} A_{k+l}$ by applying the generalized cosine law in $\mathbb{R}^2$ given in Lemma 6.

• The Schläfli angle $\beta_{(0,k),u}$ was discovered in [22]. The distance function $a_{(0,i),j}$ is

$$a_{(0,i),j} = a_{(0,i),j}(a_{(0,i),1}, a_{(0,i),2}, a_{(0,i),3}, \beta_{(0,i),4}, \beta_{(0,i),5}, \ldots, \beta_{(0,i),j})$$

for $i = 1, 2, \ldots, N-1$, $j = N, N+1$. In $\mathbb{R}^3$, we derived that $a_{(0,i),4}$, depends on $a_{(0,i),1}$, $a_{(0,i),2}$, $a_{(0,i),3}$, because $\beta_{(0,i),4}$ can be expressed explicitly as a function with respect to $a_{(0,i),1}$, $a_{(0,i),2}$, $a_{(0,i),3}$. In $\mathbb{R}^4$, we derived that $a_{(0,i),4}$, $a_{(0,i),5}$, depend on $a_{(0,i),1}$, $a_{(0,i),2}$, $a_{(0,i),3}$, $\beta_{(0,i),4}$, because we derived an implicit function with respect to $a_{(0,i),1}$, $a_{(0,i),2}$, $a_{(0,i),3}$, $\beta_{(0,i),4}$ and $\beta_{(0,i),4}$ that cannot be solved explicitly with respect to $a_{(0,i),1}$, $a_{(0,i),2}$, $a_{(0,i),3}$. Thus, Schläfli angles $\beta_{(0,i),k}$ are embodied in the computation of the distances $a_{(0,i),N}$, $a_{(0,i),N+1}$, for $N \geq 4$ and $k = 4, 5, \ldots, N$.

Theorem 7 (Lagrange multiplier rule for WFSFR multitree in $\mathbb{R}^N$) *If an admissible point $\tilde{x}_i$ yields a WFSFR multitree for $1 \leq i \leq \frac{\frac{1}{2}N(N+1)!}{(N+1)!}$, which corresponds to a Fréchet N-multisimplex derived by a $\frac{1}{2}N(N+1)$-tuple of edge lengths determining up to $\frac{\frac{1}{2}N(N+1)!}{(N+1)!}$ incongruent N-simplexes constructed by the Dekster-Wilker domain $DW_{\mathbb{R}^N}(\ell, s)$, then there are $\lambda_{0i}, \lambda_{1i}, \ldots \lambda_{(N^2-1)i}$ such that the following is valid:*

$$\frac{\partial \mathcal{L}_i(\tilde{x}_i, \tilde{\lambda}_i)}{\partial x_{ji}} = 0, \quad j = 1, 2, \ldots, 2N(N-1), \tag{38}$$

where

$$\tilde{x}_i = \{a_{(0,1),1}, a_{(0,1),2}, a_{(0,1),3}, \beta_{(0,1),4}, \ldots, \beta_{(0,1),N}, \ldots, a_{(0,N-1),1}, a_{(0,N-1),2},$$

$$a_{(0,N-1),3}, \beta_{(0,N-1),4}, \ldots, \beta_{(0,N-11),N},$$

$$w_{(0,1),1}, \ldots, w_{(0,1),N}, w_{(0,2),1}, \ldots, w_{(0,2),N}, \ldots,$$

$$w_{(0,N-1),1}, \ldots, w_{(0,N-1),N}\} \quad (\tilde{\lambda}_i = \{\lambda_0, \lambda_1, \ldots, \lambda_{N^2-1}\}).$$

Proof By setting $f_{p,q} \equiv f_k$, we consider that $\frac{\partial (f_k)_i}{\partial x_{ji}}$ are continuous in each parallelepiped Π_i, for $1 \leq i \leq \frac{\frac{1}{2}N(N+1)!}{(N+1)!}$, $k = 0, 1, 2, \ldots, N^2 - 1, j = 1, 2, \ldots, 2N(N-1)$ and by applying Lagrange multiplier rule, we obtain the Lagrangian vector $\tilde{\lambda}_i = \{\lambda_{0i}, \lambda_{1i}\lambda_{2i}, \ldots, \lambda_{(N^2-1)i}\}$ such that (38) holds. □

Denote by

$$a(N) = \begin{cases} \frac{(\frac{1}{2}N(N+1)-1)\sqrt{1-\frac{2(N+1)}{N(N+2)}}}{1-\sqrt{1-\frac{2(N+1)}{N(N+2)}}} & for\ even\ N \geq 2, \\ \frac{(\frac{1}{2}N(N+1)-1)\sqrt{1-\frac{2}{N+1}}}{1-\sqrt{1-\frac{2}{N+1}}} & for\ odd\ N \geq 3. \end{cases}$$

The function $a(N)$ is obtained by the function $\ell(N)$ of Dekster-Wilker (see [5, p. 352]) using the inequality

$$\frac{a(N)}{a(N) + \frac{1}{2}N(N+1) - 1} \geq \ell(N).$$

Theorem 8 *The most natural $\frac{1}{2}N(N+1)$-tuple of numbers from $\frac{1}{2}N(N+1)$ consecutive natural numbers $\{a + \frac{1}{2}N(N+1) - 1, \ldots, a+1, a\}$ for $a \geq a(N)$ is a $\frac{1}{2}N(N+1)$-tuple of edge lengths having the maximum volume (maximum*

$\frac{1}{2}N(N+1)$-tuple) among the $\frac{\frac{1}{2}N(N+1)!}{(N+1)!}$ incongruent N-simplexes, which corresponds to a Fermat-Steiner tree of minimum total weighted length (global minimum solution) such that the upper bound for the weight B_{ST} is determined by the rest Fermat-Steiner minimal trees having larger or equal weighted minimal total length.

Proof By applying Theorem 7 for the Dekster-Wilker $\frac{1}{2}N(N+1)$-tuple of edge lengths $\{a+\frac{1}{2}N(N+1)-1,\ldots,a+1,a\}$ for $a \geq a(N)$ forming $\frac{\frac{1}{2}N(N+1)!}{(N+1)!}$ incongruent N-simplexes in $\mathbb{R}^N$, we obtain a class of Fermat-Steiner trees for $b_i = 1$, for $i = 1, 2, \ldots, N+1$ and $b_{ST} = 1$, which yields a Fermat-Steiner-Fréchet multitree in $\mathbb{R}^N$. By selecting the proper $\frac{1}{2}N(N+1)$-tuple of edge lengths, which yields the maximum volume from all $\frac{\frac{1}{2}N(N+1)!}{(N+1)!}$, incongruent N-simplexes in $\mathbb{R}^N$, we consider a variable weighted Fermat-Steiner tree having three equally weighted Fermat Steiner points with weight B_{ST} and the same boundary weights $b_i = 1$. By perturbing the length weighted Fermat-Steiner tree structure for the N-simplex $(A_1A_2\ldots A_{N+1})_{\max}$ having the maximum volume, we can derive an upper bound for the variable weight B_{ST}, which is calculated by comparing the perturbed length tree structure compared with the other Fermat-Steiner tree structures, that belong to the unweighted Fermat-Steiner-Fréchet multitree in $\mathbb{R}^N$. Hence, this particular arrangement of the $\frac{1}{2}N(N+1)$ consecutive natural numbers for $a \geq a(N)$ with the upper bound $(B_{ST})_s$ yields the most natural weighted $\frac{1}{2}N(N+1)$-tuple of natural numbers referring to weighted Fermat-Steiner trees with the minimum mass transfer. □

5 Intermediate Weighted Fermat-Steiner-Fréchet Multitrees for a $\frac{1}{2}N(N+1)$-tuple of Positive Real Numbers that Determine the Edge Lengths of Incongruent N-simplexes in $\mathbb{R}^N$

In this section, we deal with the solution (multitree) of the intermediate weighted Fermat-Steiner-Fréchet problem (P(I.FSFR)) for a given $\frac{1}{2}N(N+1)$ of positive real numbers determining incongruent N-simplexes in $\mathbb{R}^N$, by inserting $N(l+s)$ equality constraints derived by $l+s < N-1$ independent solutions for $l+s$ variable WF problems for the Fréchet N-multisimplex derived by incongruent boundary N-simplexes in $\mathbb{R}^N$, which correspond to the same $\frac{1}{2}N(N+1)$-tuple of positive real numbers (edge lengths) and $l+s-1$ equality constraints derived by two different expressions of each line segments connecting two consecutive WFS points. By applying a Lagrange program, we can detect I.WFSFR multitrees for a given $\frac{1}{2}N(N+1)$-tuple of edge lengths determining incongruent N-simplexes (Fréchet N-multisimplex) in $\mathbb{R}^N$, by using the Dekster-Wilker function.

First, we give the definitions of an intermediate weighted Fermat-Steiner (I.WFS) tree and an I.WFS multitree for a given $\frac{1}{2}N(N+1)$-tuple of positive real numbers determining the edge lengths of incongruent N-simplexes in $\mathbb{R}^N$.

Definition 8 (A non-degenerate I.WFS tree for $\{A_1 A_2 ... A_{N+1}\}$) A non-degenerate I.WFS tree T_{IS} is a tree, which consists of some line segments between the WFS points $A_{0,j}$, and WF points $P_{0,j}$ whose # $< N-1$, such that each $A_{0,j}$ has degree three, each $P_{0,j}$ has degree more than three and of $N+1$ line segments joining each boundary vertex A_i with some $A_{0,j}$, or $P_{0,j}$.

Definition 9 An I.WFS multitree MT_{IS} is a union of I.WFS trees, which correspond to incongruent N-simplexes derived by the same Dekster-Wilker $\frac{1}{2}N(N+1)$-tuple of positive real numbers determining edge lengths.

We consider an I.WFS multitree in $\mathbb{R}^N$ having incongruent boundary simplexes, with s a fixed # of WFS points and l a fixed # of WF points in $\mathbb{R}^N$, such that $s+l < N-1$.

Problem 8 (The I.WFS (P(I.FSFR)) problem in $\mathbb{R}^N$ with equality constraints)

$$f_0(\tilde{x}) = \sum\nolimits_{a_{i_k,j_k} \in T_S} b_k a_{i_k,j_k},$$
$$f_{i,j}(\tilde{x}) = 0, \quad i = 1, \ldots, l+s), \quad j = 1, \ldots, N,$$
$$f_k(\tilde{x}) = 0, \quad k = 1, \ldots, l+s-1,$$

where s is # of WFS points $A_{0,i}$ type one, two or three and l is # of WF points $A_{0,i}$ of degree (connections) more than three,

$$f_0(\tilde{x}) = f_0(T_S) = \sum\nolimits_{a_{i_k,j_k} \in T_S} b_k a_{i_k,j_k}, \tag{39}$$

$$f_{1,1}(\tilde{x}) = \frac{w_{(0,1),1}}{a_{(0,1),1} \operatorname{Vol}(A_{0,1} \ldots A_{N+1})} \tag{40}$$
$$- \frac{1 - \sum_{i=1}^{N} w_{(0,1),i}}{a_{(0,1),N+1}(a_{(0,1),1}, a_{(0,1),2}, a_{(0,1),3}, \beta_{(0,1),4}, \ldots, \beta_{(0,1),N}) \operatorname{Vol}(A_{0,1} A_1 \ldots A_N)},$$

$\ldots$

$$f_{1,N}(\tilde{x}) =$$
$$\frac{w_{(0,1),N}}{a_{(0,1),N}(a_{(0,1),1}, a_{(0,1),2}, a_{(0,1),3}, \beta_{(0,1),4}, \ldots, \beta_{(0,1),N}) \operatorname{Vol}(A_1 \ldots A_{0,1} A_{N+1})}$$
$$- \frac{1 - \sum_{i=1}^{N} w_{(0,1),i}}{a_{(0,1),N+1}(a_{(0,1),1}, a_{(0,1),2}, a_{(0,1),3}, \beta_{(0,1),4}, \ldots, \beta_{(0,1),N}) \operatorname{Vol}(A_{0,1} \ldots A_N)}, \tag{41}$$

$\ldots$

$$f_{l+s,1}(\tilde{x}) = \frac{w_{(0,l+s),1}}{a_{(0,l+s),1} \operatorname{Vol}(A_{0,l+s} \ldots A_{N+1})} -$$
$$- \frac{1 - \sum_{i=1}^{N} w_{(0,l+s),i}}{a_{(0,l+s),N+1}) \operatorname{Vol}(A_{0,l+s} A_1 \ldots A_N)}, \tag{42}$$

where

$$a_{(0,l+s),N+1} =$$

$$= a_{(0,l+s),N+1}(a_{(0,l+s),1}, a_{(0,l+s),2}, a_{(0,l+s),3}, \beta_{(0,l+s),4}, \ldots, \beta_{(0,l+s),N}),$$

...

$$f_{l+s,N}(\tilde{x}) = \frac{w_{(0,l+s),N}}{a_{(0,l+s),N}) \operatorname{Vol}(A_1 \ldots A_{0,l+s} A_{N+1})} - \\ - \frac{1 - \sum_{i=1}^{N} w_{(0,l+s),i}}{a_{(0,l+s),N+1} \operatorname{Vol}(A_{0,l+s} A_1 \ldots A_N)}, \tag{43}$$

where

$$a_{(0,l+s),N} = a_{(0,l+s),N}(a_{(0,l+s),1}, a_{(0,l+s),2}, a_{(0,l+s),3}, \beta_{(0,l+s),4}, \ldots, \beta_{(0,l+s),N}),$$

$$f_k(\tilde{x}) = a_{(0,k),(0,k+1)}(\tilde{x};\ \text{boundary simplex 1}) \\ -a_{(0,k),(0,k+1)}(\tilde{x};\ \text{boundary simplex 2}) \tag{44}$$

for $k = 1, 2, \ldots l + s - 1$.

We take into account that:

• (39) is the objective function of the I.WFS problem for $A_1 A_2 \ldots A_{N+1}$ in $\mathbb{R}^N$ having $N - 1$ WFS and WF points $A_{0,i}$ such that:

There are l weighted Fermat-Steiner points $A_{0,i}$ of degree three with respect to a boundary triangle and s weighted Fermat points $A_{0,i}$ of degree more than three with respect to a boundary r-simplex in $\mathbb{R}^N$, for $r < N$, which can be derived by the local minimality criterion of Lemma 1.

• (40)–(41) deal with the solution of the WF problem for $A_1 \ldots A_{N+1}$ with variable weights

$$w_{(0,1),1}, \ldots, w_{(0,1),N} \text{ and } w_{(0,1),N+1} = 1 - \sum_{i=1}^{N} w_{(0,1),i},$$

which is determined by applying Theorem 3 for the unique WF point $A_{0,1}$. ...

• (42)–(43) deal with the solution of the WF problem for $A_1 \ldots A_{N+1}$ with variable weights

$$w_{(0,l+s),1}, w_{(0,l+s),2}, \ldots, w_{(0,l+s),N} \text{ and } w_{(0,l+s),N+1} = 1 - \sum_{i=1}^{N+1} w_{(0,l+s),i},$$

which is determined by applying Theorem 3 and the unique WF point $A_{0,l+s}$.

• (44) is a derivation of possible expressions of $a_{(0,i),(0,i+1)}$ with respect to boundary r-simplexes for $2 \le r < N$ by applying the generalized cosine law in $\mathbb{R}^2$ given in Lemma 6 combined with Theorem 3 for $A_{i_1} \ldots A_{i_{r+1}}$, with $r + 1$ fixed given weights taken from the set $\{b_1, b_2, \ldots, b_{N+1}, b_{ST}\}$.

The following Lagrangian program detects an I.WFSFR multitree in $\mathbb{R}^N$.

Theorem 9 (Lagrange multiplier rule for I.WFSFR multitree in $\mathbb{R}^N$) *If the admissible point $\tilde{x}_i$ yields an I.WFSFR multitree for $1 \leq i \leq \frac{\frac{1}{2}N(N+1)!}{(N+1)!}$, which correspond to a Fréchet N-multisimplex derived by a $\frac{1}{2}N(N+1)$-tuple of edge lengths determining upto $\frac{\frac{1}{2}N(N+1)!}{(N+1)!}$ incongruent N-simplexes constructed by the Dekster-Wilker domain $DW_{\mathbb{R}^N}(\ell, s)$, then there are numbers $\lambda_{0i}, \lambda_{1i}, \ldots \lambda_{((l+s-1)(N+1)+1)i)}$ such that:*

$$\frac{\partial \mathcal{L}_i(\tilde{x}_i, \tilde{\lambda}_i)}{\partial x_{ji}} = 0 \tag{45}$$

for $j = 1, 2, \ldots, 2N(l + s - 1)$,

$$\tilde{x}_i = \{a_{(0,1),1}, a_{(0,1),2}, a_{(0,1),3}, \beta_{(0,1),4}, \ldots, \beta_{(0,1),N}, \ldots, a_{(0,l+s),1}, a_{(0,l+s),2},$$

$$a_{(0,l+s),3}, \beta_{(0,l+s),4}, \ldots, \beta_{(0,l+s),N},$$

$$w_{(0,1),1}, \ldots, w_{(0,1),N}, w_{(0,2),1}, \ldots, w_{(0,2),N}, \ldots,$$

$$w_{(0,l+s),1}, \ldots, w_{(0,l+s),N}\},$$

$\tilde{\lambda}_i = \{\lambda_{0i}, \lambda_{1i}, \ldots, \lambda_{((l+s-1)(N+1)+1i)}\}$.

Proof By setting $f(p, q) \equiv f_k$ we consider that $\frac{\partial (f_k)_i}{(x_{ji})}$ are continuous in each parallelepiped Π_i, for $1 \leq i \leq \frac{\frac{1}{2}N(N+1)!}{(N+1)!}$, $k = 0, 1, 2, \ldots, N^2 - 1$, $j = 1, 2, \ldots, 2N(l + s - 1)$ and by applying Lagrange multiplier rule, we obtain the Lagrangian vector $\tilde{\lambda}_i = \{\lambda_{0i}, \lambda_{1i}\lambda_{2i}, \ldots, \lambda_{((l+s-1)(N+1)+1i)}$, such that (45) holds. □

Example 1 Let $\{a_{ij}\}$ be a given 15-th tuple of positive real numbers determining incongruent 5-simplexes using Dekster-Wilker conditions in $\mathbb{R}^5$ and $A_1 A_2 \ldots A_6$ be a member of the Fréchet 5-multisimplex in $\mathbb{R}^5$. We consider a fixed tree topology that contains two WFS points $A_{0,1}$, $A_{0,2}$ of degree three and one weighted Fermat point of degree four.

The I.WFS tree for $A_1 A_2 A_3 A_4 A_5 A_6$ in $\mathbb{R}^5$ consists of the line segments

$$\{A_{0,1}A_1, A_{0,1}A_2, A_{0,1}A_{0,2}, A_{0,2}A_3, A_{0,2}A_{0,3}, A_{0,3}A_4, A_{0,3}A_5, A_{0,3}A_6\}.$$

By applying the Lagrangian program of Theorem 9, we can detect I.WFSFR multitrees for incongruent 5-simplexes in $\mathbb{R}^5$ derived by a given Dekster-Wilker 15-tuple of edge lengths.

6 "Mutation" of an Intermediate Weighted Fermat-Steiner-Fréchet Multitree for m Closed Polytopes in $\mathbb{R}^N$

Here, we introduce the "mutation" of an I.WFSFR multitree for m boundary polytopes, by applying the plasticity solutions of the m-INVWF problem for $A_1A_2\ldots A_{m+1}$ in $\mathbb{R}^N$, enriched by a two way mass transportation network. The dynamic plasticity solutions of the m-INVWF problem in $\mathbb{R}^N$ combined with conditions derived by optimal mass transport and storage gives the "mutation of an I.WFFR multitree for boundary m closed polytopes in $\mathbb{R}^N$. We start by deriving the geometric and dynamic plasticity for m closed polytopes in $\mathbb{R}^N$ for $m \geq N+1$.

Definition 10 We call *geometric plasticity* of a WF tree whose endpoints correspond to a closed polytope in $\mathbb{R}^N$, which is formed by $m+1$ variable line segments meeting at the WF point A_0, for $m+1$ given values of the weights, the set of solutions of the $m+1$ variable lengths of the line segments, which correspond to a family of weighted networks that preserve the WF point A_0.

We assume that we select B_i that correspond to each vertex A_i, such that the weighted floating inequalities of Theorem 1 hold:

$$\left\| \sum_{j=1, j\neq i}^{m+1} B_j \vec{u}(A_j, A_i) \right\| > B_i,$$

for $i, j = 1, 2, \ldots, m+1$. Thus, the WF point A_0 of an m closed polytope $A_1A_2\ldots A_m$ in $\mathbb{R}^N$ is the unique intersection point of $m+1$ line segments A_0A_m.

Theorem 10 (Geometric plasticity of $m$closed polytopes in $\mathbb{R}^n$) *If we select a point A_i' on the ray defined by A_0A_i with corresponding weight B_i such that*

$$\left\| \sum_{j=1, j\neq i}^{m+1} B_j \vec{u}(A_j', A_i') \right\| > B_i,$$

for $i, j = 1, 2, \ldots, m+1$, then the corresponding weighted Fermat point $A_0' \equiv A_0$.

Proof The weighted floating inequalities

$$\left\| \sum_{j=1, j\neq i}^{m+1} B_j \vec{u}(A_j, A_i) \right\| > B_i, \quad \left\| \sum_{j=1, j\neq i}^{m+1} B_j \vec{u}(A_j', A_i') \right\| > B_i,$$

yield, respectively,

$$\sum_{i=1}^{m+1} B_i \vec{u}(A_0, A_i) = \vec{0}, \quad \sum_{i=1}^{m+1} B_i \vec{u}(A_0, A_i') = \vec{0}.$$

Thus, we derive that $A_0' \equiv A_0$. □

We assume that $m+1$ line segments A_0A_i intersect at A_0 in $\mathbb{R}^N$, for $m \geq N+1$. We note that:

(1) for $n = 2$, m angles α_{i0j} determine the dynamic plasticity equations of $A_1A_2\dots A_{m+1}$ in $\mathbb{R}^2$.
(2) for $n = 3$, $m + (m-1)$ angles α_{i0j} determine the dynamic plasticity equations of $A_1A_2\dots A_{m+1}$ in $\mathbb{R}^3$.
(3) for $n > 3$, $\sum_{i=1}^{N-1}(m+1-i)$ angles α_{i0j} determine the dynamic plasticity equations of $A_1A_2\dots A_{m+1}$ in $\mathbb{R}^N$.

We set $\sum_{12\dots m+1} B \equiv \sum_{i=1}^{m+1}(B_i)_{12\dots m+1}$, $\sum_{i_1i_2\dots i_{N+1}} B = (B_{i_1})_{i_1i_2\dots i_{N+1}} + \dots + (B_{i_{N+1}})_{i_1i_2\dots i_{N+1}}$ for $i_1,\dots i_{N+1} \in \{1,2,\dots m+1\}$.

Lemma 7 *If* $\sum_{12\dots m+1} B = \sum_{i_1i_2\dots i_{n+1}} B$ *for every* $i_1,\dots i_{n+1} \in \{1,2,\dots m+1\}$, *where* $\sum_{12\dots m+1} B := (B_{N+1})_{12\dots m+1}(1 + \sum_{i=1,i\neq N+1}^{m}(\frac{B_i}{B_{N+1}})_{1,2,\dots m+1})$, *then*

$$(B_i)_{12\dots m+1} = \sum_{j=N+2}^{m} a_{i,j}(B_j)_{12\dots m+1} + b_i, \quad i = 1,2\dots,N+1,$$

where

$$(a_{N+1,j},\, b_{N+1}) =$$
$$\left(\frac{(\frac{B_1}{B_{N+1}})_{12\dots N+1}(\frac{B_{N+1}}{B_j})_{2\dots j} + \dots + (\frac{B_N}{B_{N+1}})_{12\dots N+1}(\frac{B_{N+1}}{B_j})_{12\dots(N-1)(N+1)j} - 1}{\sum_{i=1}^{N+1}(\frac{B_i}{B_{N+1}})_{12\dots N+1}},\right.$$
$$\left.(B_{N+1})_{12\dots N+1}\right),$$

$$(a_{N,j},\, b_N) = (a_{N+1,j}(\frac{B_N}{B_{N+1}})_{12\dots N+1} - (\frac{B_N}{B_{N+1}})_{12\dots N+1}(\frac{B_{N+1}}{B_j})_{12\dots(N-1)(N+1)j},$$
$$(B_N)_{12\dots N+1}),$$

...

$$(a_{1,j},\, b_1) = (a_{N+1,j}(\frac{B_1}{B_{N+1}})_{12\dots N+1} - (\frac{B_1}{B_{N+1}})_{12\dots N+1}(\frac{B_{N+1}}{B_j})_{23\dots j},$$
$$(B_1)_{12\dots N+1}), \quad j = n+2,\dots m+1.$$

By assuming mass flow continuity and by applying the geometric and dynamic plasticity of m polytopes in $\mathbb{R}^N$, (Theorem 10, Lemma 7) the corresponding WF tree solutions yield some mass transportation networks, in which the weights correspond to an instantaneous collection of images of masses, which satisfy some specific conditions.

Definition 11 We call (m, k) "mutation" of an I.WFS tree with respect to boundary am polytope the plasticity solutions of the m-INVWF problem for $A_1A_2\dots A_{m+1}$ in $\mathbb{R}^N$, enriched by a two way mass transportation network, such that k masses (weights) are transferred in k directions from $A_i \to A_0$, for $i = 1,\dots k$ (inflow) creating a storage at A_0 and $N-k$ masses are transferred in $m-k$ directions from $A_0 \to A_j$ (outflow), for $j=k+1,\dots m+1$ and reversely $m-k$ masses (new weights) are transferred back to A_0 along the same $m-k$ directions creating new storage at A_0, and k masses (new weights) are transferred back from A_0 to A_i, $i=1,2,\dots k$.

We denote by B_i a mass flow which is transferred from A_i to A_0 for $i=1,2,\dots k$ by B_0 a residual weight which remains at A_0 and by $B_{k+1},\dots B_{m+1}$ a mass flow which is transferred from A_0 to $A_{k+1},\dots,A_{m+1}$. We denote by $\tilde{B}_i$ a mass flow which is transferred from A_0 to A_i for $i=1,2,\dots k$ by $\tilde{B}_0$ a residual weight which remains at A_0 and by $\tilde{B_{k+1}},\dots,B_{m+1}$ a mass flow which is transferred from $A_{k+1},\dots A_{m+1}$ to A_0. Thus, we derive the following:

$$\sum_{i=1}^{k} B_i = \sum_{k+1}^{m+1} B_i + B_0, \tag{46}$$

$$\sum_{i=1}^{k} \tilde{B}_i + \tilde{B}_0 = \sum_{i=k+1}^{m+1} \tilde{B}_i. \tag{47}$$

By adding (46) and (47) and by letting $\bar{B}_0 = B_0 - \tilde{B}_0$ we get the following:

$$\sum_{i=1}^{k} \bar{B}_i = \sum_{i=k+1}^{m+1} \bar{B}_i + \bar{B}_0 \tag{48}$$

such that

$$\sum_{i=1}^{m+1} \bar{B}_i = c, \tag{49}$$

where c is a positive real number. Thus, we obtain the following theorem as a consequence of Lemma 7 under the condition for the weights $\bar{B}_{12\dots m+1}$ taken from (48), (49), which deal with the (m,k) "mutation" of m closed polytopes in $\mathbb{R}^N$.

Theorem 11 ((m,k) "mutation" of an intermediate weighted Fermat-Steiner tree for boundary m closed polytopes in $\mathbb{R}^N$) *If* $\sum_{12\dots m+1} \bar{B} = \sum_{i_1 i_2\dots i_{n+1}} \bar{B}$, *for every* $i_1,\dots i_{n+1} \in \{1,2,\dots m+1\}$, *where* $\sum_{12\dots m+1} \bar{B} := (\bar{B}_{n+1})_{12\dots m+1}(1+\sum_{i=1,i\neq n+1}^{m+1} (\frac{\bar{B}_i}{\bar{B}_{N+1}})_{1,2,\dots m+1})$, *then*

$$(\bar{B}_i)_{12\dots m+1} = \sum_{j=n+2}^{m+1} a_{i,j}(\bar{B}_j)_{12\dots m+1} + b_i, \quad i=1,2\dots,N+1,$$

where

$$(a_{N+1,j}, b_{N+1}) =$$

$$\left(\frac{(\frac{\bar{B}_1}{\bar{B}_{N+1}})_{12\dots N+1}(\frac{\bar{B}_{N+1}}{\bar{B}_j})_{2\dots j} + \dots + (\frac{\bar{B}_N}{\bar{B}_{N+1}})_{12\dots N+1}(\frac{\bar{B}_{N+1}}{\bar{B}_j})_{12\dots(N-1)(N+1)j} - 1}{\sum_{i=1}^{n+1}(\frac{\bar{B}_i}{\bar{B}_{N+1}})_{12\dots N+1}},\right.$$

$$(\bar{B}_{N+1})_{12\dots N+1}),$$

$$(a_{N,j}, b_N) = (a_{N+1,j}(\frac{\bar{B}_N}{\bar{B}_{N+1}})_{12\ldots N+1} - (\frac{\bar{B}_N}{\bar{B}_{N+1}})_{12\ldots N+1}(\frac{\bar{B}_{N+1}}{\bar{B}_j})_{12\ldots(N-1)(N+1)j},$$

$$(\bar{B}_N)_{12\ldots N+1}),$$

...

$$(a_{1,j}, b_1) = (a_{N+1,j}(\frac{\bar{B}_1}{\bar{B}_{N+1}})_{12\ldots N+1} - (\frac{\bar{B}_1}{\bar{B}_{N+1}})_{12\ldots N+1}(\frac{\bar{B}_{N+1}}{\bar{B}_j})_{23\ldots j},$$

$$(\bar{B}_1)_{12\ldots N+1}),$$

for $j = n+2, \ldots m+1$, *under the conditions for the weights:*

$$\sum_{i=1}^{k} (\bar{B}_i)_{12\ldots m+1} = \sum_{i=k+1}^{m+1} (\bar{B}_i)_{12\ldots m+1} + (\bar{B}_0)_{12\ldots m+1},$$

$$\sum_{i=1}^{m+1} (\bar{B}_i)_{12\ldots m+1} = c.$$

Suppose that at time $t = 0$, an I.WFSFR multitree occurs with respect to a given $\frac{1}{2}N(N+1)$-tuple of positive real numbers determining the edge lengths of incongruent N-simplexes generates by the Dekster-Wilker domain in $\mathbb{R}^N$. Then $N+2, \ldots, m+1$ rays start to grow from the weighted Fermat-point $(A_0)_l$, which creates a two-way mass transport, in order to obtain an (m, k) mutation of an I.WFSFR multitree having one WF point of degree $m+1$.

Theorem 12 ((m,k) "Mutation" of an I.WFSFR multitree in $\mathbb{R}^N$) *The equations of (m,k) "mutation" of an I.WFSFR multitree in* $\mathbb{R}^N$ *are derived by adding* $(A_0)_l(A_{N+2})_l$, $(A_0)_l(A_{N+3})_l, \ldots, (A_0)_l(A_{m+1})_l$ *rays to a given WFFR multitree with respect to a boundary Fréchet N-multisimplex, such that k masses (weights) are transferred in k directions from* $(A_i)_l \to (A_0)_l$, *for* $i = 1, \ldots k$ *(inflow) creating a storage at* $(A_0)_l$ *and* $N-k$ *masses are transferred in* $m-k$ *directions from* $(A_0)_l \to (A_j)_l$ *(outflow), for* $j = k+1, \ldots m+1$ *and reversely* $m-k$ *masses (new weights) are transferred back to* $(A_0)_l$ *along the same* $m-k$ *directions creating new storage at* $(A_0)_l$, *and k masses (new weights) are transferred back from* $(A_0)_l$ *to* $(A_i)_l$, $i = 1, 2, \ldots k$, $1 < l \leq \frac{\frac{1}{2}N(N+1)!}{(N+1)!}$.

Proof By adding $(A_0)_l(A_{N+2})_l, (A_0)_l(A_{N+3})_l, \ldots, (A_0)_l(A_{m+1})_l$ rays to a given WFFR multitree with respect to a boundary Fréchet N-multisimplex in $\mathbb{R}^N$ and by applying Theorem 11, we derive the (m, k) plasticity equations of an I.FFR multitree for m boundary closed polytopes in $\mathbb{R}^N$. □

7 Constructive Tree Weights for the Vertices of a Fréchet N-multisimplex in $\mathbb{R}^N$

In this section, we obtain an ϵ-approximation of the value of the weight B_{N+1}, which corresponds to the vertex A_{N+1} of an N-simplex in $\mathbb{R}^N$ circumscribed in a $(N-1)$-dimensional sphere S^{N-1} of radius r, and center O, by applying the N-INWF problem in $\mathbb{R}^N$. For $\epsilon \to 0$, the limiting floating weighted Fermat tree solution coincides with the absorbing weighted Fermat tree solution of Theorem 1. An application of this method is an approximation for the weights of a Fréchet N-multisimplex via an ϵ approximation of each corner of incongruent N-simplexes with a multiweighted FFR multitree in $\mathbb{R}^N$.

Let $A_1 A_2 \dots A_{N+1}$ be an N-simplex circumscribed in $S^{N-1}(0,r)$. We consider a point $A_0 \in [A_{N+1}, O]$, and we denote by $\epsilon = |A_{N+1}A_0|$.

Theorem 13 *The weights $B_1(\epsilon), \dots, B_{N+1}(\epsilon)$ are uniquely determined by*

$$B_i(\epsilon)=\frac{C}{1+\left|\frac{\sin\alpha_{i,0k_1k_2\dots k_{N-1}}}{\sin\alpha_{k_N,0k_1k_2\dots k_{N-1}}}\right|+\left|\frac{\sin\alpha_{i,0k_1k_2\dots k_{N-2}k_N}}{\sin\alpha_{k_{N-1},0k_1k_2\dots k_{N-2}k_N}}\right|+\dots+\left|\frac{\sin\alpha_{i,0k_2\dots k_{N-1}k_N}}{\sin\alpha_{k_1,0k_2\dots k_{N-1}k_N}}\right|} \tag{50}$$

for $i, k_1, k_2, \dots, k_N = 1, 2, \dots, N+1$ and $k_1 \neq k_2 \neq \dots \neq k_N$, and the weighted Fermat tree $\{A_1A_0, A_2A_{0,}, \dots A_{N+1}A_0$ is an ϵ approximation of the absorbing weighted Fermat tree $\{A_1A_{N+1}, A_2A_{N+1}, \dots A_NA_{N+1}\}$ such that $A_0 \equiv A_{N+1}$, with a weighted error estimate for B_{N+1} :

$$\left|\sqrt{\sum_{i=1}^{N} B_i^2(\epsilon) + 2\sum_{i,j=1,i<j}^{N} B_i(\epsilon)B_j(\epsilon)\cos\alpha_{i(N+1)j}} - B_{N+1}(\epsilon)\right|. \tag{51}$$

Proof By applying the cosine law and the sine law in $\triangle A_O A_i A_j$ and $\triangle O A_i A_0$, we derive that $\frac{N(N+1}{2} - 1$ angles satisfy $\alpha_{i0j} = \alpha_{i0j}(\epsilon)$. By substituting these angles in 6, we obtain a unique solution of $B_i(\epsilon)$ given by (50). By replacing $B_i(\epsilon)$ for $i = 1, 2, \dots, N$ in the weighted absorbing condition of Theorem 1, we get an error estimate for B_{N+1}, see (51). □

8 Bessel Plasticity and "Mutation" of an Intermediate Weighted Fermat-Steiner-Fréchet Multitree for m Polytopes in $\mathbb{R}^N$

In this section, we will describe the evolutionary structure of an I.WFSFR multitree for m boundary polytopes in $\mathbb{R}^N$ with random weights following a Bessel motion. We start by constructing a Bessel motion in the sense of McKean ([12, 19]). We consider the m-dimensional Brownian motion with sample paths $\mathbf{b}(t)$, $(m \geq 2, t \geq 0)$ and generator $\mathcal{G} = \frac{1}{2}(\sum_{i=1}^{m} \frac{\partial^2}{\partial \mathbf{b}_i^2})$. The radial part $\mathbf{r}(t) = |\mathbf{b}(t)|$ $(t \geq 0)$ is the Bessel motion with generator $\mathcal{G}^+ = \frac{1}{2}(\frac{d^2}{d\mathbf{r}^2}) + \frac{N-1}{\mathbf{r}}\frac{d}{d\mathbf{r}}$. The probability of the event B $P.(B)$ as a function of the starting point $\mathbf{a} = \mathbf{b}(0)$ of the Brownian path if $t_1 < t_2$ is given by:

$$P.([\mathbf{r}(t_2) \leq l | \mathbf{r}(s) : s \leq t_1]) = \int_{|\mathbf{b}-\mathbf{b}(t_1)|<l} (2\pi(t_2 - t_1))^{-m/2} e^{-\frac{|\mathbf{b}-\mathbf{b}(t_1)|^2}{2(t_2-t_1)}} d\mathbf{b}_1 \ldots d\mathbf{b}_m.$$

The Bessel motion $[\mathbf{r}(t), P.]$ is a Markov process, which depends upon $\mathbf{r}(t_1)$.

Lemma 8 (Non-negative solutions of Bessel Motion, see p. 318–319 in [19]) *The solution of the singular integral equation*

$$\mathbf{r}(t) = \mathbf{b}(t) + \frac{N-1}{2}\int_1^t \mathbf{r}^{-1} ds, \quad t \geq 0,$$

is a Bessel motion starting at $\mathbf{r}(0) = \mathbf{b}(0)$, *which takes non-negative real values by neglecting a class of Brownian paths* $\mathbf{b}$ *of Wiener measure 0.*

By taking into account the solution of the N-INVWF problem for polytopes in $\mathbb{R}^N$, we derive the equations of Bessel plasticity of a weighted Fermat-tree for a boundary polytope in $\mathbb{R}^N$. The Bessel plasticity of a weighted Fermat tree characterizes the combinatorial plasticity (random weights) of non-random polytopes in $\mathbb{R}^N$.

Theorem 14 *The following equations point out the Bessel plasticity of an I.WFS tree with one WF point for a boundary* $N+1$ *weighted closed polytope with respect to the non-negative random weights* $(B_i)_{12\ldots N+2}$, *which depend on the Bessel motion* $\mathbf{r}_{N+2}(t) \equiv (B_{N+2})_{12\ldots N+2}$ *in* $\mathbb{R}^N$:

$$(\frac{B_1}{B_{N+1}})_{12\ldots N+2} = (\frac{B_1}{B_{N+1}})_{12\ldots N+1}(1 - (\frac{\mathbf{r}_{N+2}(t)}{B_{N+1}})_{12\ldots(N+2)}(\frac{B_{N+1}}{B_{N+2}})_{2\ldots N+2}), \quad (52)$$

$$(\frac{B_2}{B_{N+1}})_{12\ldots N+2} = (\frac{B_2}{B_{N+1}})_{12\ldots N+1}(1 - (\frac{\mathbf{r}_{N+2}(t)}{B_{N+1}})_{12\ldots n+2}(\frac{B_{N+1}}{B_{N+2}})_{13\ldots N+2}), \quad (53)$$

...

$$(\frac{B_N}{B_{N+1}})_{12\ldots N+2} = (\frac{B_N}{B_{N+1}})_{12\ldots N+1}(1 - (\frac{\mathbf{r}_{N+2}(t)}{B_{N+1}})_{12\ldots N+2}(\frac{B_{N+1}}{B_{N+2}})_{12\ldots(N-1)(N+1)(N+2)}) \quad (54)$$

such that $\sum_{i=1}^{n+2}(B_i)_{12\ldots N+2} = c$.

Proof By inserting $\mathbf{r}_{N+2}(t) \equiv (B_{N+2})_{12\ldots N+2}$ into (16), (17) and (18), we obtain (52), (53) and (54), respectively. □

Suppose that at time $t = 0$, an I.WFSFR multitree occurs with respect to a given $\frac{1}{2}N(N+1)$-tuple of positive real numbers determining the edge lengths of incongruent N-simplexes generates by the Dekster-Wilker domain in $\mathbb{R}^N$. Then an $(N+2)$th ray start to grow from the weighted Fermat-point $(A_0)_l$.

If the weight $(B_{N+2})_l$ follows a Bessel motion, we derive the Bessel plasticity of a weighted multitree for a boundary $N+1$ closed polytope in $\mathbb{R}^N$.

Theorem 15 (Bessel plasticity of an I.FSFR multitree for an $N+1$ boundary closed polytope in $\mathbb{R}^N$) *The Bessel plasticity equations of an I.FSFR multitree for an $N+1$ boundary closed polytope in $\mathbb{R}^N$ are given by the Bessel plasticity equations of a WF tree, such the random weight that corresponds to the $(N+2)$th ray follows a Bessel motion.*

Proof By applying Theorem 14 for an I.Fermat-Steiner-Fréchet multitree for an $N+1$ boundary closed polytope in $\mathbb{R}^N$, which is derived by adding a ray $(A_0 A_{N+2})$ with a random weight following a Bessel motion to I.FSFR multitree for an N boundary Fréchet multisimplex in $\mathbb{R}^N$, we get the Bessel plasticity equations of the generated I.FSFR multitree in $\mathbb{R}^N$. □

Remark 1 The Bessel plasticity of the I.FSFR multitree for an $N+1$ boundary closed polytope in $\mathbb{R}^N$ may cause a distortion of the length structure of the initial I.FSFR multitree for an N boundary Fréchet multisimplex in $\mathbb{R}^N$.

References

1. Blumenthal L.M.: A budget of curiosa metrica, Amer. Math. Monthly, f66 : 6, (1959), 453–460
2. Boltyanski V., Martini H., Soltan V.: Geometric methods and optimization problems, Kluwer, Dordrecht/Boston/London (1999)
3. Cieslik D.: Steiner minimal trees. Nonconvex optimization and its applications, 23. Kluwer Academic Publishers, Dordrecht, 1998
4. Courant R. Robbins H.: What is mathematics? An elementary approach to ideas and methods. New York, NY: Oxford University Press (1996)
5. Dekster B.V., Wilker J.B.: Edge lengths guaranteed to form a simplex. Arch. Math. 49 (1987), 351–366
6. Dekster B.V., Wilker J.B.: Large spherical simplexes. J. Geom. 42, No. 1-2 (1991), 59–92
7. Eriksson F.: The law of sines for tetrahedra and n-simplices, Geom. Dedicata. 7 (1978), 71–80
8. Fréchet M.:, Sur la définition axiomatique d'une classe d'espaces vectoriels distancies applicables vectoriellement sur l'espace de Hilbert. Ann. Math. 36 (1935), 705–718
9. Herzog F.: Completely tetrahedral sextuples, Amer. Math. Monthly, 66 : 6, (1959), 460–464
10. Gilbert E.N., Pollak H.O.: Steiner minimal trees, SIAM Journal on Applied Mathematics. 16 (1968), 1–29
11. Gueron S., Tessler R.: The Fermat-Steiner problem, Amer. Math. Monthly 109 (2002) 443–451
12. Itô K., McKean H.P.: Diffusion processes and their sample paths, Repr. of the 1974 ed. Classics in Mathematics, Springer-Verlag, Berlin, (1974)
13. Ivanov A.O., Tuzhilin A.A.: Geometry of minimal networks and the one-dimensional Plateau problem. Russ. Math. Surv. 47 (1992), No. 2, 59–131
14. Ivanov A.O., Tuzhilin A.A.: Minimal networks. The Steiner problem and its generalizations. CRC Press, Boca Raton, FL, 1994
15. Ivanov A.O., Tuzhilin A.A.: Weighted minimal binary trees. Russian Math. Surveys 50 (1995), no. 3, 623–624
16. Ivanov A.O., Tuzhilin A.A.: Linear networks and convex polytopes. J. Math. Sci., New York 104 (2001), No. 4, 1283–1288.

17. Ivanov A.O., Tuzhilin A.A.: Branching solutions to one-dimensional variational problems. Singapore: World Scientific 2001
18. Jarník V., Kössler M.: Sur les graphes minima, contenant n points donnes. Čas. Mat. Fys. 63, 223–235
19. McKean H.P.: The Bessel motion and a singular integral equation, Mem. Coll. Sci., Univ. Kyoto, Ser. A 33 (1960), 317–322
20. Melzak Z. A.: On the problem of Steiner, Canad. Math. Bull. 4 (1961), 143–148
21. Rubinstein J. H., Thomas D. A. and Weng J.: Minimum networks for four Points in Space, Geom. Dedicata. 93 (2002), 57–70
22. Schläfli L.: Collected mathematical works. Vol. I. (Gesammelte mathematische Abhandlungen. Bd. I. Herausgegeben vom Steiner-Schläfli-Komitee der Schweizerischen Naturforschenden Gesellschaft.) German Basel: Birkhäuser Verlag, (1949)
23. Sommerville D. M. Y.: An introduction to the geometry of n-dimensions. New York: Dover Publications, (1958)
24. Sturm R.: Ueber den punkt kleinster entfernungsumme von gegebenen punkten. J. Reine Angew. Math. XCVII (1884), 49–62
25. Uspensky J. V.: Theory of equations. New York, McGraw-Hill, (1948)
26. Zachos A. and Zouzoulas G.: The weighted Fermat-Torricelli problem for tetrahedra and an "inverse" problem, J. Math. Anal. Appl. 353 1, (2009), 114–120
27. Zachos A.N.: The weighted Fermat-Torricelli-Menger problem for a given sextuple of edge lengths determining tetrahedra, J.Math. Chem. 54 7, (2016), 1447–1460
28. Zachos A.N.: The plasticity of some mass transportation networks in the three-dimensional Euclidean space, J. Convex Anal., 27 (3), (2020), 989–1002
29. Zachos A.N.: An evolutionary design of weighted minimum networks for four points in the three-dimensional Euclidean space, Analysis (Munchen) 41:2, (2021), 79–112

MIX
Papier aus verantwortungsvollen Quellen
Paper from responsible sources
FSC® C105338

If you have any concerns about our products,
you can contact us on
ProductSafety@springernature.com

In case Publisher is established outside the EU,
the EU authorized representative is:
Springer Nature Customer Service Center GmbH
Europaplatz 3, 69115 Heidelberg, Germany

Printed by Libri Plureos GmbH
in Hamburg, Germany